*Advances in
Engine Technology*

WILEY – EC AERONAUTICS RESEARCH SERIES

Advances in Onboard Systems Technology
Edited by **A. del Core**

Advances in Techniques for Engine Applications
Edited by **R. Dunker**

Advances in Engine Technology
Edited by **R. Dunker**

Advances in Optronics and Avionics Technologies
Edited by **M. Garcia**

Advances in Acoustics Technology
Edited by **J. M. Martin Hernandez**

EUROPEAN COMMISSION
Directorate-General XII
Science, Research and Development

Advances in Engine Technology

Edited by

R. Dunker
European Commission, DGXII, Brussels

European Commission Aeronautics Research Series

JOHN WILEY & SONS
Chichester • New York • Brisbane • Toronto • Singapore

Publication EUR 15344 EN of the
European Commission,
Directorate-General XII for Science Research and Development,
Brussels

© ECSC-EEC-EAEC, Brussels-Luxembourg, 1993.

LEGAL NOTICE
Neither the European Commission nor any person acting on
behalf of the Commission is responsible for the use
which might be made of the following information.

Published in 1995 by John Wiley & Sons Ltd,
    Baffins Lane, Chichester,
    West Sussex PO19 1UD, England

Telephone: *National* (01243) 779777
    International (+44) 1243 779777

All rights reserved.

No part of this book may be reproduced by any means,
or transmitted, or translated into a machine language
without the written permission of the publisher.

Other Wiley Editorial Offices

John Wiley & Sons, Inc., 605 Third Avenue,
New York, NY 10158-0012, USA

Jacaranda Wiley Ltd, 33 Park Road, Milton,
Queensland 4064, Australia

John Wiley & Sons (Canada) Ltd, 22 Worcester Road,
Rexdale, Ontario M9W 1L1, Canada

John Wiley & Sons (SEA) Pte Ltd, 37 Jalan Pemimpin #05-04,
Block B, Union Industrial Building, Singapore 2057

British Library Cataloguing in Publication Data

A catalogue record for this book is available from the British Library

ISBN 0 471 95102 1

Typeset in 10/12pt Palatino by Keytec Typesetting Ltd, Bridport, Dorset, UK
Printed and bound in Great Britain by Bookcraft (Bath) Ltd

Contents

Foreword	vii
1 Tip Clearance Effects in Advanced Axial Flow Compressors	**1**
F. Corlais, P. Ivey and K. Papailiou	
0 Abstract	2
1 Introduction	3
2 Research objectives	3
3 Research activities	6
4 Research results	41
5 Conclusions	95
6 References	100
2 Low Emission Combustor Technology (LOWNOX I)	**105**
F. Joos and G. Pellischek	
0 Abstract	106
1 Introduction	110
2 Research objectives	111
3 Research activities	115
4 Research results	125
5 Conclusions	143
6 References	149
3 Investigation of the Wake Mixing Process behind Transonic Turbine Inlet Guide Vanes with Trailing Edge Coolant Flow Ejection	**153**
C. H. Sieverding et al.	
List of symbols	154
0 Abstract	156
1 Introduction	158
2 Aerodynamic design of the BRITE turbine inlet guide vane	159
3 Mechanical design of blades and cascades	175
4 Description of test facilities and measuring systems	184
5 Straight cascade tests	195
6 Annular cascade tests	235

7 Viscous flow calculations	284
8 Comparison of experimental and numerical results	320
9 Conclusions	338
10 References	339
Appendices	342
Index	361

Foreword

Over the last thirty years the European aeronautical industry has achieved a respected and internationally successful position. Aeronautical products made in Europe are able to secure large market shares, and in some instances have even become dominant. Nevertheless, the level of global competition continues to be set by the United States of America, which is traditionally committed to preeminence in this field.

Despite the importance of the European dimension, it is only in the comparatively recent past that the European Community has started to play a significant role in the technological challenge in aeronautics.

In 1988, the major European aircraft manufacturers presented the EUROMART study (European Cooperative Measures for Aeronautical Research and Technology) to the European Commission. The study identified areas of research which were considered to be critical to the future competitiveness of the industry in world markets. Separate reports, conveying views on the content of a European research activity in aeronautics, were submitted by representative groups of the European Aero-Engine Manufacturers and the European Aerospace Equipment Manufacturers and Systems Suppliers.

Following these extensive consultations with industry, the Commission launched its first dedicated aeronautical research programme in March 1989. This initiative became known as the aeronautics 'pilot phase' and was in fact Area 5: Specific Activities Relating to Aeronautics, of the BRITE/EURAM-Programme, itself a Specific Programme within the Community Second Framework Programme for Research and Technology Development (1989–92).

The action is being continued under the 3rd EC Framework Programme (1991–1994) and will be pursued under the 4th EC Framework Programme (1994–1998).

The pilot phase comprised 28 projects, covering four technology areas—aerodynamics, acoustics, airborne systems and equipment, and propulsion systems.

The pilot phase comprised 28 projects, covering four technology areas—aerodynamics, acoustics, airborne systems and equipment, and propulsion systems.

Despite a relatively small budget, 35 MECU over 2 years, this exploratory action achieved considerable success in stimulating wide-ranging cooperation between all types of actors. These included airframe, engine and equipment manufacturers, small and medium-sized enterprises, research centres and universities from the Member States of the Community and also from EFTA Countries.

This EC-Aeronautics Research Series provides the opportunity to present the results of those research projects which have reached completion and to illustrate the achievements of this pilot action.

This volume includes three aero-engine projects covered under the propulsion systems work area of the pilot phase.

The aim of the 'tip-clearance' project was to develop validated computational codes capable of evaluating tip clearance effects in multi-stage axial flow compressors. Three aero-engine manufacturers and two university institutes combined their efforts to enhance the understanding of tip-clearance phenomena, using experiments and theoretical modelling, in order to improve the design methodologies of the rear stages of high-pressure core compressors for civil applications.

Within the LOWNOX project, the challenge was to respond to current and prospective concerns regarding the atmospheric impact of aircraft emissions, particularly of oxides of nitrogen (NO_x) at high altitude. This project, in which eight European aero-engine manufacturers are jointly engaged, supported by universities and aeronautical research establishments, in total 20 partners, aims to develop low or even ultra-low emission combustor technology. The research work in the pilot-phase project was aimed at fundamental combustion investigations including measurement techniques, the assessment of the pollutant emissions to be expected by future aero engines—around the turn of the century—and providing guidelines for future emission standards. The most promising reduction methods have been evaluated and selected.

The 'turbine-wake mixing' project coordinated by a research institute, combined the efforts of a second research establishment, a university and three aero-engine companies at a fundamental level. The aim of the project was to study the flow-mixing process behind a transonic turbine inlet guide vane with coolant flow ejection either from the pressure side or through the trailing edge. In addition to a better physical understanding of the mixing process through the full integration of experimental and theoretical tools, the project was able to provide valuable benchmark test cases for the validation of two-dimensional and three-dimensional Navier–Stokes codes.

Thanks are due to all partners of the three consortia, particularly to the scientists, engineers and managers who contributed to the success of these aeronautical projects, reflecting the high level of collaboration attained under a common interest of advancing the European Community technology base in this crucial sector.

Thanks are especially due to the authors of the three technical project reports.

The notable technical successes achieved in these projects illustrate the added value of collaboration at European level.

R. Dunker
Brussels
December 1993

1 Tip clearance effects in advanced axial flow compressors

F. Corlais*, P. Ivey† and K. Papailiou‡

This report, for the period April 1990 to October 1992, covers the activities carried out under the BRITE/EURAM Area 5: "Aeronautics" Research Contract No. AERO-CT90-0021 (Project: AERO-P1036) between the Commission of the European Communities and the following:

*Société Nationale d'Etude et de Construction de Moteurs d'Aviation, Snecma (coordinator)
Rolls Royce Plc, RR
Société Turboméca, TM
†Cranfield Institute of Technologies, CIT
‡National Technical University of Athens, NTUA/LTT
Contact: Mr. F. Corlais, Snecma, Centre Villaroche, Dept. YKMG, F-77550 Moissy Cramayel, France (Tel: +33/1/6059 8516, Fax: +33/1/6059 7843)

Advances in Engine Technology. Edited by R. Dunker
© 1995 John Wiley & Sons Ltd.

0 Abstract

Three aero-engine manufacturers (Snecma, Rolls-Royce, Turbomeca) and two universities (Cranfield, NTUA) have merged their efforts, capacities and knowledge to increase the understanding of tip-clearance phenomena and therefore improve the design of the rear stage of high-pressure civil core compressors. To meet such an objective two approaches have been chosen: experiments and theoretical modelling.

Firstly, a four-stage low-speed research compressor (LSRC) has been comprehensively instrumented with conventional pressure taps and probes. Laser two focus (L2F) has been used as well to get detailed pressure and velocity maps of a tip-clearance vortex developing in a multi-stage environment. Two sizes of tip gap have been completed and analysed: comprehensive and accurate sets of data extremely close to both hub and casing have been successfully obtained, and interesting features of the flow are revealed.

The demonstration of the three-dimensional anemometer has been unsuccessful in the time available after problems in the areas of probe alignment and signal processing. Successful use of this instrument is only just beginning on bench tests, but its employment in a turbomachine is recognized as having high relative risk. Therefore an alternative has been identified from a laser manufacturer to obtain detailed three-dimensional measurements for validations of three-dimensional theoretical approach.

To conclude on experimental studies, a simple experimental high-speed cascade rig (HSCR) has been partially prepared to understand the basic tip-clearance leakage behaviour; the inlet scroll which delivers the upstream yaw angle has been designed and manufactured, a rotating hub arrangement which simulates the relative motion between casing and rotating blade rows has been studied and designed (detailed drawings have been achieved). Finally a draft blade has been designed and will be finalised with respect to the inlet scroll performances.

Also, a database has been built after a literature study. As anticipated, only a few experimental cases have been found interesting since many parameters are lacking, which underlines the need to do the above experiments. This database not only provides a new means of communication between partners but it has been used to validate the theoretical model.

Secondly, a simple two-dimensional model based on a throughflow approach and a solid-body rotation of the tip-leakage vortex has been achieved and tested against a selection of the database experimental test cases. The model does not predict the tip leakage underturning to an acceptable level of accuracy and overestimates the spanwise extent of the leakage vortex effects: a great deal of effort is required to improve its accuracy and reduce its CPU time consumption for industrial purposes.

A three-dimensional Navier–Stokes pressure-corrected solver has been modified to account for actual tip clearance geometry and therefore improve its numerical prediction accuracy. Four geometries have been tested: their results helped to visualize the complex flow patterns in the clearance aera; some of them are compared to those of the first-level model. Work has also be done to reduce the CPU time consumption which is now of the order of 1 ms/node/iteration. Improvement in turbulence modelling as well as in grid orthogonality are mandatory.

This project can be seen as the pilot phase of the following AC3A part A IMT Area 3 aeronautics programme numbered 2031.

1 Introduction

The competitiveness of the world aeronautics industry forces engine manufacturers to design machines with ever-increasing performance and reliability and decreasing production and maintenance cost. For the compression system of the engine, this trend implies obtaining higher pressure ratios and efficiencies with a smaller number of stages, while ensuring trouble-free operation. A particularly promising path has been identified to achieve such progress, and this forms the subject of this project.

The objectives of this programme are to understand and master the complex three-dimensional viscous flows in the blade clearance region, which have a strong impact on both overall performance and surge margin.

It is well known that increasing the tip clearance of the compressor rotor or stator has an appreciable effect on losses. A figure often quoted is that of a 2% drop in efficiency, when the tip clearance is increased by 1% of the blade height. It is also well established that the stability limit of the compressor is influenced by tip-clearance effects and furthermore the casing treatments which probably modify tip-clearance flows may be used to displace the stability line. Finally, experiments have demonstrated that changes in the geometry of the blade tip and use of 'end bends' can also have a substantial effect upon performance.

By developing reliable tip-clearance models, this programme has yielded improved compressor aerodynamic design software, capable of more accurate performance predictions. The following qualitative benefits can be expected: improved compressor performance (higher pressure rise per stage, higher efficiency) obtained more quickly (design 'right first time').

Improved compressor performance translates into

- *fuel consumption reductions*: in practical terms, 0.75% improvement in average compressor efficiency might be expected from a successful conclusion to this project;
- *lighter, cheaper, more compact engines* (fewer stages, shorter compressors);
- *large reductions in development costs*: the ability to design 'right first time' will lead to a consequent reduction in development programme timescales.

2 Research objectives

The objectives of the proposed research are to develop validated theoretical software capable of evaluating tip-clearance effects in multi-stage axial-flow compressors [1]. Some of the experiments necessary for the validation have been also carried out.

The resulting software is compatible with the current and future industrial prediction codes, used to analyse and design advanced multi-stage axial flow compressors.

The present research has led to new capabilities:

- Improve the current advanced industrial design tool's (quasi-three-dimensional) capabilities in predicting tip clearance effects;

- Improve the future advanced industrial design solver's (fully three-dimensional) capabilities in predicting tip clearance effects;

- Increase understanding of tip-clearance effects for situations which are beyond reach with present-day knowledge;

- Generate experimental data using existing advanced instrumentation at Cranfield in order to support present and future theoretical development. Of these data, some will be acquired from a new facility conceived for this project, which will also be available for future use;

- Develop new experimental capabilities for studying complete three-dimensional viscous flows, at NTUA. Use them to generate experimental results during the second phase of the programme;

- Increase understanding of complex three-dimensional viscous flow situations and their modelling;

- Increase international collaboration.

It is important to note that this research by its nature provides benefits not only in the basic elucidation of specific axial compressor clearance phenomena, but also more generally in that a broader understanding of three-dimensional viscous flows obtained through the experimental and theoretical work proposed will assist the comprehension of a broad range of aero-thermodynamic applications.

To meet the objectives outlined above the following programme has been completed. It could be divided into two main parts, a theoretical one comprising two levels of modelling and an experimental one which is divided in two ways to overcome some of the presently insurmountable difficulties in performing detailed measurements in a multi-stage environment at an acceptable level of cost:

- *Task 1*: to constitute a small database using both existing open literature experimental results and those produced during the project work. This is required to produce material for modelling validation not only during the project but at any time in the future;

- *Task 2*: to define the important parameters that determine the environment in which tip-clearance flows develop in multi-stage axial-flow compressors.

The industrial partners have then specified the range of values which are compatible with their current experience in order to ensure that the experiments conducted in the present work are relevant and meaningful;

- *Task 3*: to model the tip-clearance effects on radial variations of circumferentially averaged flows. Special attention is given to ensuring that this first-level model is compatible with the existing circumferentially averaged through-flow industrial codes. Whilst the second level is inherently capable of a much more precise description of the flow field, the view is taken here that both levels will be used at various stages of the design procedure in the years to come, almost irrespective of the state of development of the second-level tools. The reason for this lies in the need, at least in the early stages of a compressor design evaluation, to utilize computationally efficient prediction softwares over a wide range of possible operating points;

- *Task 4*: to get experimental data in a multi-stage environment utilizing the existing large-scale low-speed four-stage compressor test rig of Cranfield. Interblade pneumatic measurements and two-dimensional laser anemometry have been taken at stages one and three; rotor and stator surface static pressure have been measured as well. Those experimental investigations have been made for two tip-clearance sizes. These detailed and well-controlled experimental data are required for the validation of the two-level models developed during the project;

- *Task 5*: to validate the first-level model in the industrial partner's existing through-flow codes and then assess the validity of the model;

- *Task 6*: to develop the first stage of the second-level model which is expected to solve three-dimensional Navier–Stokes equations in a multi-stage environment. For this and on the basis of an existing three-dimensional Navier–Stokes solver a simple eddy viscosity model has been employed to model the flow turbulence, although the code has been prepared to accept a two equation model. The code has been applied to the experimental cases selected in Task 1, and the calculated results compared to the measured data. In this way the model has been modified so that it predicts the essential features of tip-clearance flow;

- *Task 7*: to demonstrate a three-dimensional laser anemometry capability suitable to measure the complex three-dimensional flow phenomena due to the shear layer complexity and the small scale of the structures emergent in the tip gap area. The demonstration has been undertaken on the final rig build. This measurement technique will provide test data which offer the ability to both validate level 2 models and gain an insight into the true flow field of the tip-clearance region. It may be essential to possess this form of information to further enhance both the level 1 and 2 models;

- *Task 8*: to design an appropriate cascade facility that is needed to get detailed three-dimensional experimental data. These detailed and well-controlled experimental data are required both for the validation of the level 2 model developed during the project and to reveal basic mechanisms of tip leakage flows. Design, technical studies and partial construction of the annular cascade rig (scroll manufacturing) have been achieved during this phase of the project;

- *Task 9*: to determine the industrial and scientific benefits achieved during the project, highlight the areas for future research and make recommendations for a follow-on programme. Results of those thoughts are fully expanded in Section 5.

3 Research activities

3.1 Task 1: database

The role of NTUA in Task 1 was to establish a data base of test cases of axial compressors, containing

(a) full geometrical data for each test case;

(b) flow field data, coming from measurements or predictions.

The test cases were selected so that information about tip-clearance effects, in particular, was present and could be included.

The data for each compressor were taken either from the open literature (for the cases that such data existed) or were provided by individual partners. The final database contains both the test cases elaborated by NTUA and the ones prepared by the other partners, according to the format established by NTUA.

The main goals set for constitution of the database were the following:

- Include all geometrical data necessary for performing first- and second-level flow calculations. As a starting point, the format usually employed by throughflow calculation codes was taken into account. In this way a minimal effort is involved for transformation of data from the database to the particular format employed by the code of each partner. Subsequently, data needed for more detailed flow calculations are included;

- Include available experimental data, for the purpose of validating predictions, while data for predictions can also be included, for reference purposes;

- Establish such a format that exchange of information is possible with minimum documentation.

The database test cases are summarized in Table 1.1 in Section 4.1, and an example is shown in Figure 14 in Section 4.1.

3.1.1 Format of the database

NTUA produced an initial format for the database, which was then finalized, after interaction with the rest of the partners, in order to ensure compatibility with the information needed by their codes. The final format is common for everybody, and a pre- and post-processor might be essential in order to transform data to individual codes format.

The format was standardized in such a way that common symbols for all quantities are used. For this purpose a common list of symbols was established, after a proposition by NTUA and exchange of views with the partners. This list was then circulated among the partners.

Then a format of the files constituting the database was established by organizing the data in separate files, according to their nature. In particular, the database was organized in the form of a directory containing subdirectories:

(a) *One subdirectory containing information about the format of files.* It is named Filefrmt and is included in order to have a self-sufficient package. Each one of the files it contains gives information about the format of the corresponding data file, in subdirectories with data. The files have all the name 'name' and extension as explained below;

(b) *Subdirectories containing the actual data for each compressor test case.* Each one of them contains files with the data for one compressor test case. All such data files are in ASCII format. The name of each file is a generic one, identifying the test case (for example the generic name 'PWLSR' indicates the test case of the Pratt & Wittney Low Speed Rotor).

The extensions of the ASCII files indicate the kind of information which is included in each one of them:

- name.ID   File with identification data
- name.BIF   File with brief information
- name.GDT   File with detailed data for flow calculation
- name.DDT   File with aerodynamic data distributions
- name.ADT   File with additional data, if needed for additional calculations

Appropriate keys are included in the data files, for identifying the origin and kind of data (for example, data from measurement or prediction, etc). The

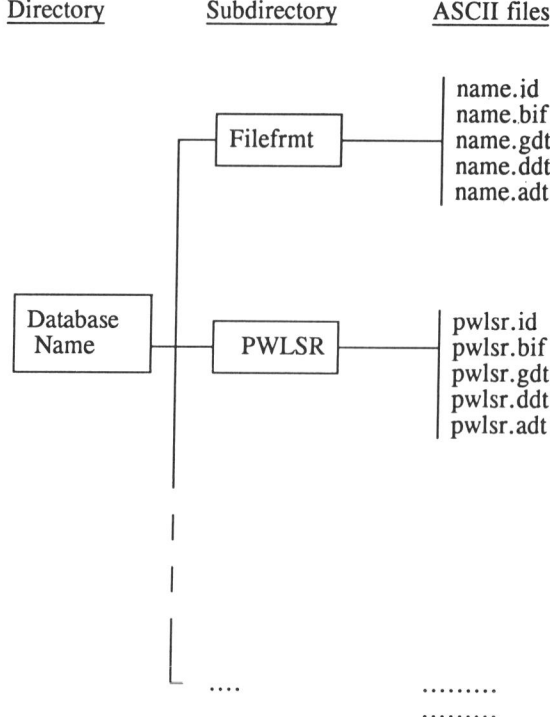

Figure 1.1 Database organization.

structure of files containing aerodynamic data is such that updating with additional data from new experiments or calculations is possible.

A schematic of the organization of the database is given in Figure 1.1.

3.1.2 Documentation

During the project more detailed information concerning the structure and the content of the database has been produced and is included in related reports. The meeting reports compiled by NTUA are given below, for the reader interested in further information about the database: 3-NTUA-1, 4-NTUA-1, 5-NTUA-1, 6-NTUA-1.

3.2 Task 2: parameters influencing tip-clearance flows

A text describing the parameters expected to be significant was provided by NTUA and finalized after an exchange of views with partners. Representative ranges of values for multi-stage high speed compressors were provided by the industrial partners. A combined list was then established (see Table 1.2, in Section 4.2).

3.3 Task 3: first-level modelling

3.3.1 Preliminaries

In this task the development of a theoretical tool for the prediction of the tip-clearance flow behaviour was targeted. The radial variations of the circumferentially averaged flow quantities were to be calculated in a manner compatible with the existing meridional through-flow calculation codes of the industrial partners.

The proposed tip-clearance model is based on a multiple zone modelling of the different end-wall effects. In this respect, the tip-clearance flow is introduced as a third zone in the already existing two-zone model, which is used for the calculation of the secondary flow effects. The superposition principle is thus adopted and the proposed tip-clearance method calculates the difference between the real flow field and that constituted when the external and secondary flow fields are combined together.

The study of the tip-clearance flow effects can be divided into two subproblems:

(a) the flow inside the tip clearance, and

(b) the induced velocity and pressure fields inside the blade passage and downstream of the blade row.

The pressure difference between the pressure and suction sides of the blade at the tip level drives the flow through the tip clearance. An almost potential flow field characterizes the entrance of the gap, leading to the formation of a *'vena contracta'*. The mixing region, following the *'vena contracta'*, is responsible for the losses occurring inside the tip gap.

The leakage jet flow at the gap exit interacts with the secondary flow, having a direction from the pressure to the suction side of the passage, rolls up and forms the leakage vortex. The additional mass, coming out of the tip clearance, enters the vortex, increasing its radius and strength.

In order to develop theoretical tools for the prediction of the above phenomena, the work of previous researchers was reviewed. Rains' [29], Lakshminarayana's [30] and Ohayon's [31] investigations were used as a starting point. However, a different approach was adopted and additional work has been done, leading to relatively simple models, which can provide a good description of the complicated tip leakage flow.

In contrast to other research, in the present work a complete theoretical calculation procedure was constituted, which, when integrated with a meridional calculation method, can provide the variations of the flow quantities inside and downstream of the tip clearance.

The basic models as well as the implementation of these models into a calculation procedure will be presented in the following sections. However, detailed description of the present work can be found in previous reports [32]–[38].

3.3.2 Basic models

Gap flow model Rains [29] advanced the assumption that the flow through the tip gap of a compressor blade may be simulated on the basis of two-dimensional considerations. He acknowledged the predominant role of the pressure difference across the gap against the pressure gradient along the blade chord on both sides of the blade and he assumed that the gap exit flow results from this pressure difference in the direction of the normal to the tip blade camber line.

According to Rains' two-dimensional model, the flow across the blade is characterized by the formation of a *vena contracta* at the gap entrance, followed by a mixing region, with the longitudinal momentum component being conserved through the gap.

Wadia and Booth [39], Booth *et al.* [40], as well as Moore and Tilton [41], validated Rains' basic model and improved it, providing good results for turbine test cases. Very good agreement between the Rains model and experiment was obtained also by Storer [42] for a compressor linear cascade.

The simple model adopted here and schematically presented in Figure 1.2 is very similar to the ones mentioned above and uses Rains' basic assumptions in order to calculate both the mass flow rate through the gap and the correspond-

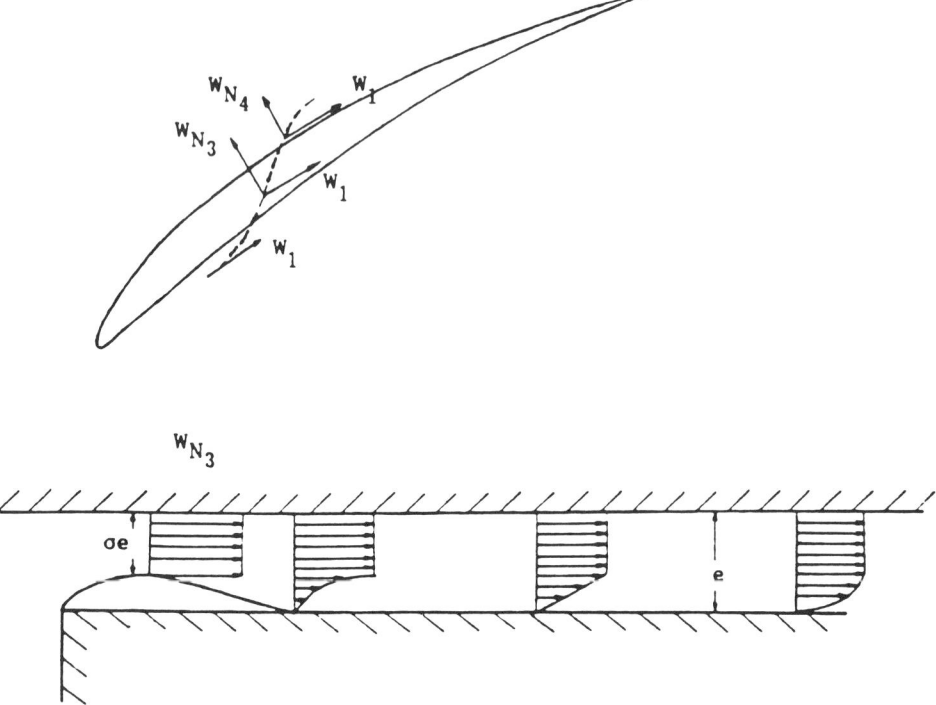

Figure 1.2 Proposed modelling of the flow inside the tip gap.

ing total pressure losses. Following the formation of the *vena contracta*, a loss-producing region appears, characterized by one of the simple profiles present in Figure 1.2. The profile shape at different gap positions is assumed to depend upon the value of the non-dimensional (upon the gap height) length. The gap losses are calculated by mixing out the existing at the gap exit velocity profile.

This simple model, as will be seen below, gives with good accuracy the mean exit jet velocity (and the corresponding mass flow rate of the tip clearance jet), which is critical for the prediction of the features of the tip-clearance vortex. At the same time, it provides an acceptable estimate of the gap losses.

According to this model, the gap exit jet flow is driven by the value of the static pressure at the gap exit in respect to the total pressure at the entrance of the flow to the gap, or, if conservation of the longitudinal momentum is assumed, by the static pressure difference existing between the pressure and the suction sides. A potential flow field is assumed to prevail near the entrance to the gap, leading to the formation of the *vena contracta*. This no-loss region is followed by a mixing region, which is responsible for the non-uniform jet velocity profiles at the gap exit. The presence of a *vena contracta* has been identified by the experimental investigations of Sjolander and Amrud [43] and Yaras *et al.* [44] for turbine blades among others. Storer's [42] experiments on a compressor cascade suggest that the length scale of reattachment inside the clearance gap is primarily a function of the tip gap height. Similar results were obtained by Moore *et al.* [45] by performing laminar and turbulent flow calculations for an idealized two-dimensional tip gap geometry. These results, compared with Graham's (1985) measurements for different gap heights, suggest that the tip-gap height to blade-thickness ratio could be a useful criterion for the reconstruction of the jet profile at the tip-clearance gas exit. The adoption of different simple profiles, based on the blade-thickness to gap-height ratio, is used in the present model to determine the mass flow rate through the gap. As said above, the losses occurring inside the gap are equalled to the mixed out losses, which may be computed by applying momentum considerations to each of the profile shapes adopted. It is thus assumed that all the losses are taking place inside the mixing region after the *vena contracta*.

The potential solution of a flow at rest entering a slot gives the value for the contraction ratio σ (Fig. 1.3), where

$$\sigma = \pi/(\pi + 2) = 0.611 \tag{1.1}$$

An analysis of Milne-Thomson [46] for symmetric slots, where the centreline of the slot represented the end wall, was used in order to derive equation (1.1). The experimental results of Moore and Tilton [41] and their analysis provide support for the existence of a *vena contracta* at the entrance region of the tip gap with the value of the contraction ratio mentioned above.

Applying the Bernoulli theorem from position 1 to position 3 (Figure 1.2) and considering that the momentum of the flow through the gap is conserved in the direction of the mean camber line one gets

$$P_1 = P_3 + 0.5\rho W_{N_3}^2 \tag{1.2}$$

where N denotes the normal to the camberline direction.
Applying the continuity equation from station 3 to station 4 one gets

$$W_{N_3}\sigma e = \int_{O_4}^{e} W_{N_4}\,dy \tag{1.3}$$

For the various gap exit profile shapes considered, different expressions result from the above equation. Considering the 'parabolic' profile (see Figure 1.2)

$$\frac{W_{N_3}}{W_{N_{4max}}} = 1 + \frac{2}{3}\left(\frac{1}{\sigma} - 1\right) \tag{1.4}$$

Applying the momentum equation in the normal to the mean camber line direction from position 3 to position 4 (Figure 3) one gets

$$(P_3 - P_4)\,e = -\rho W_{N_3}^2(\sigma e) + \rho \int_{O_4}^{e} W_{N_4}^2\,dy \tag{1.5}$$

For the 'parabolic' profile considered, this expression yields

$$P_3 - P_4 = -\rho\sigma W_{N_3}^2 + W_{N_{4max}}^2 \rho\sigma\left[1 + \frac{8}{15}\left(\frac{1}{\sigma} - 1\right)\right] \tag{1.6}$$

Substituting the continuity into the momentum equation developed above, and taking into account Bernoulli's equation as well as the conservation of the longitudinal component of momentum, we can develop a relation between the jet velocity at the gap exit and the pressure difference between suction and pressure sides at the tip region level. For the case of the 'parabolic' profile this relation reads

$$\frac{W_{N_{4max}}}{W_1} = 0.92\sqrt{-Cp_{s_4}} \tag{1.7}$$

where

$$Cp_{s_4} = \frac{P_4 - P_1}{\frac{1}{2}\rho W_1^2} \tag{1.8}$$

A similar relation may be developed, which provides the mean value of the normal component of jet velocity at the clearance gap exit, which in its general form reads

$$\frac{\overline{W}_{N_4}}{W_1} = C_D\sqrt{-Cp_{s_4}} \tag{1.9}$$

Research activities

C_D, the discharge coefficient, takes the following values, according to the adopted gap jet profiles:

$$\begin{array}{ll} \text{uniform} & 0.84 \\ \text{parabolic} & 0.80 \\ \text{triangular} & 0.76 \\ \text{inverse parabolic} & 0.72 \end{array} \quad (1.9a)$$

The pressure difference used for the calculation of the mass flow rate through the gap is the one modified by the presence of the leakage vortex. It is further modified near the leading edge region of the blade, as proposed by Yaras et al. [44], using a linear variation of the pressure difference for the first 30% of the chord.

In order to calculate the losses inside the gap it is assumed that the loss production is occurring at the mixing region downstream from the 'vena contracta'. Then, one gets for the losses

$$\Delta \dot{E} = \left[\frac{1}{2}\Delta s \int_{O_4}^{e} W_{N_4} W_4^2 dy - \frac{1}{2}\Delta s \sigma e [W_{N_3}^3 + W_{S_3}^2 W_{N_3}]\right] \quad (1.10)$$
$$+ \left[\int_{O_4}^{e} \frac{P_4}{\rho} W_{N_4} \Delta s\, dy - \frac{P_3}{\rho}\Delta s \sigma e W_{N_3}\right]$$

where Δs is a finite distance along the mean camber line of the blade. The work due to the stresses in the shear layers near the wall was ignored in the above equation. For the considered 'parabolic' profile, equation (1.10) reads

$$\Delta \dot{E} = \left[\frac{1}{2}\Delta s e [W_{N_{4max}}^3 \left(\frac{16}{35}(1-\sigma)+\sigma\right) + W_{N_{4max}} W_{S_4}^2 (\tfrac{2}{3}(1-\sigma)+\sigma)]\right.$$
$$\left. -\frac{1}{2}\Delta s \sigma e [W_{N_3}^3 + W_{S_3}^2 W_{N_3}]\right] \quad (1.11)$$
$$+\left[\frac{P_4}{\rho}\Delta s e W_{N_{4max}}(\tfrac{2}{3}(1-\sigma)+\sigma) - \frac{P_3}{\rho}\Delta s \sigma e W_{N_3}\right]$$

Taking into account the moment of momentum equation and the conservation of momentum in the direction of the blade camber line, finally equation (1.11) yields

$$\Delta \dot{E} = \tfrac{1}{2}\Delta s e W_1^3 B(-Cp_{S_4})^{3/2} \quad (1.12)$$

with B taking the following values:

$$\begin{array}{ll} \text{uniform} & -0.24 \\ \text{parabolic} & -0.19 \\ \text{triangular} & -0.15 \\ \text{inverse parabolic} & -0.11 \end{array}$$

The mixed losses inside the tip clearance can be computed using equation (1.12) with constant B taking the above values, according to the shape of the profile at the gap exit.

Simple vortex model The calculation of the induced kinematic field as well as the estimation of the losses downstream of the gap exit is based mainly on the correct computation of the strength and position of the leakage vortex. The formation and the evolution of this vortex presents many similarities with that of the finite wing case and previous researchers insisted on using prediction methods tailored to this last case.

Lifting line theory, originally developed to predict lift forces on finite wings, was introduced by Lakshminarayana [47] and others in the turbomachinery field, in order to predict the effects of the leakage vortex in the kinematic field and the amount of loss occurring inside the blade passage. The fact that the tip clearance vortex strength is not equal to the blade bound vorticity led Lakshminarayana and Horlock [48] to introduce a factor K, which expressed the percentage of the bound vorticity, which was retained inside the tip gap, with the other part shed into the tip vortex in the case of tip clearance. An experimental relation was proposed by Lakshminarayana and Horlock [48], relating K with the clearance-to-chord ratio and values of K down to 0.4 were attributed to it. A similar approach was adopted by Lewis and Yeung [49], who developed their relation for calculating the retained lift and thus the vortex strength, which was found to be different from that of Lakshminarayana. Additional experiments exist which confirm that the circulation of the leakage vortex is a fraction of the blade's bound circulation, but do not actually prove or disprove the 'retained lift' theory, which stipulates that the shed out vorticity is produced by part of the vortex lines of the blade, the other part going towards the end wall and producing the existing pressure difference along the gap.

On the other hand various researchers, such as Yamamoto [50], Inoue and Kurumaru [51], Sjolander and Amrud [43] and others demonstrated that the strength of the leakage vortex seems to increase with increasing clearance gap. This increase in the amount of vorticity of the leakage vortex is followed by a rising of the vortex diameter. In addition, the data of Inoue and Kurumaru [51] demonstrate that the shed circulation depends also on the relative wall speed. In the case of a compressor this relative wall motion increases the shed vorticity, while in the case of a turbine the relative wall motion decreases the amount of vorticity shed inside the blade passage. This fact was confirmed by Yaras et al. [52] in their experiment on a turbine cascade, where the relative wall motion was simulated with a moving belt. The above experimental evidence seems to suggest that the key factor for the determination of the amount of shed vorticity in the blade passage is the mass flow rate through the tip clearance gap. It increases with an increase of the gap and is enhanced through relative wall motion in the compressor case, while the opposite is true for the turbine case. This working hypothesis has been adopted in the present work and a model was conceived, which is based on it.

The flow field near the end wall is characterized by the interaction of two opposing flows. The secondary flow, having a direction from the pressure to the suction side of the passage, and the leakage jet flow, coming out of the gap at an angle to the main flow. As a result of this interaction very near the end wall, the leakage flow rolls down away from the end wall, along a separation line, and forms the leakage vortex. This vortex interacts in the outer flow with the tip-side passage vortex, rotating contrary to the leakage vortex. At the same time, further downstream, the additional mass from the tip clearance jet, entering the vortex, increases its radius and its strength. The complexity of this flow situation suggests that some drastic assumptions should be done in order to derive a simple model.

In order to constitute this model, experimental evidence was used. In fact, flow visualization suggests that the mass leaving the gap exit enters in its quasi-totality inside the leakage vortex. The performed measurements (Lakshminarayana [30]), Inoue and Kurumaru [51], Yaras and Sjolanden, [53] etc.) indicate that the vortex structure is close to a solid-body rotation. These features are retained in the adopted model. In order to follow in a schematic way the flow situation depicted by flow visualization and available experimental evidence, it was assumed that the mass flow coming out of the gap is wrapping up around the existing solid-body rotation vortex, increasing its radius and moment of momentum in the direction of the vortex axis. The most elementary form of this model is schematically presented in Figure 1.3, where, at the same time, an elementary control volume between two successive computational stations is defined. In order to complete the picture, it is added that it is assumed that the vortex mass flow rate enters and exits the control volume uniformly and practically in the direction of the vortex axis, while the static pressure is considered uniform in these two surfaces. From inlet to exit inside the control volume, the flow quantities are assumed to vary linearly. It can be easily seen that basic features of the flow, such as the ones mentioned above, are retained by this simple model. On the other hand, as will be seen below, this model presents an alternative mechanism for specifying the

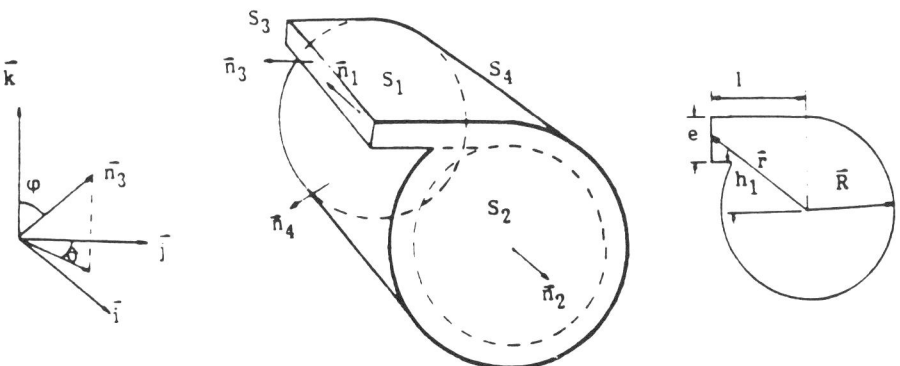

Figure 1.3 Tip-clearance vortex model and the corresponding control volume for the application of the moment of momentum equation.

strength of the vortex, which now depends strongly upon the mass flow rate coming out of the clearance gap.

Considering the assumptions outlined above, we apply the moment of momentum theorem to the control volume R of Figure 1.4, for steady-state conditions and ignoring the gravity effects. Considering the component of the moment of momentum equation in the direction of the axis of the rigid body and neglecting the contribution of the shear stress tensor we finally have

$$\omega_1 \int_0^{2\pi}\int_0^{R_1} \rho r^3 V_{1_i} \, dr \, d\phi - \omega_2 \int_0^{2\pi}\int_0^{R_2} \rho r^3 V_{2_i} \, dr \, d\phi + \int_0^{\Delta s}\int_{h_1}^{h_1+e} \rho V_{3_n} V_{3_j} \, dh \, ds$$

$$= \int_0^{\Delta s}\int_{R_1}^{R_1+e} (l\cos\phi + h\sin\phi\cos\theta) P \, dh \, ds \quad (1.13)$$

where V_{3_n} is the component in the direction of n_3.

The mass balance for the control volume R reads

$$\left| \int_{(S_1)} \rho \, ds_1 \, n_1 \cdot V_1 \right| + \left| \int_{(S_3)} \rho \, ds_3 \, n_3 \cdot V_3 \right| = \left| \int_{(S_2)} \rho \, ds_2 \, n_2 \cdot V_2 \right| \quad (1.14)$$

If we assume that the density is constant in the considered volume and uniform distributions of the velocity magnitude are assumed on the surfaces (S_1) and (S_2), the following equations are derived from equations (1.13) and (1.14) respectively:

$$\rho\left[\omega_1 V_{1_i} \pi \frac{R_1^4}{2} - \omega_2 V_{2_i} \frac{\pi R_2^4}{2} - V_{3_n} V_{3_j} \Delta s \frac{(h_1 + h_2)}{2} e\right]$$

$$= P_3 \int_0^{\Delta} s\left[\int_{h_1}^{h_1+e}(l\cos\phi + h\sin\phi\cos\theta)\,dh\right]ds \quad (1.15)$$

$$\pi R_2^2 V_{2_i} = \pi R_1^2 V_{1_i} + V_{3_n}\Delta se \quad (1.16)$$

The circulation associated with a solid body rotation of rotational speed α is at a radius R

$$\Gamma = \oint_c dr \, V = 2\pi\omega R^2 \quad (1.17)$$

Combination of equations (1.15), (1.16), (1.17) yields the expression of the vortex strength $\Gamma 2$ at station 2, in respect to existing conditions at station 1:

$$\Gamma_2 = \frac{1}{V_{2_i} R_2^2}[\Gamma_1 V_{1_i} R_1^2 + 2V_{3_n} V_{3_j} \Delta_s (h_1 + h_2)\,e]$$

$$- \frac{4}{V_{2_i} R_2^2}\frac{P_3}{\rho} \int_0^{\Delta_s}\int_{h_1}^{h_1+e}(l\cos\phi + h\sin\phi\cos\theta)\,dh\,ds \quad (1.18)$$

While equation (1.18) may be used for the calculation of the strength of the vortex, the mass balance equation (1.16) provides a basis for the calculation of the vortex core radius.

Revised vortex model The rolling up of the tip clearance jet and the formation of the leakage vortex is followed by a diffusion of vorticity, leading to an increase of the vortex radius and a reduction of the maximum value of vorticity downstream from the trailing edge. Calculations performed on the basis of the simple model described above, demonstrated reasonable agreement with experiment. However, discrepancies were present, especially downstream of the blade region. For this reason an effort was made in order to incorporate the diffusion process mentioned above into the model, rendering it somewhat more sophisticated. At the same time an accurate prediction of the total pressure losses occurring inside the blade passage due to the leakage vortex presence was attempted. The corresponding theoretical reasoning is described in this section.

As was described in the previous section, the calculation of the leakage vortex radius is involved in the computation of the shed vorticity. The leakage flow enters the vortex increasing its radius, while the diffusion process itself adds an additional increase. For this reason a diffusion model must be introduced in the calculation procedure.

During the evolution of the leakage vortex, a reduction of static pressure is produced inside it. This modification of the pressure field can be easily distinguished in the measured static pressure difference distributions at the tip region of the blade suction side. The diffusion of the leakage vortex produces a rise in the static pressure along the centerline of the vortex and the pressure deficit is reduced downstream of the trailing edge (Yaras and Sjolanden [53], Inoue and Kurumaru [51]).

Yaras *et al.* [52], [53] used Lamb's [54] formulation for the diffusion of a line vortex, for the calculation of the induced kinematic field of the leakage vortex, with very promising results. Storer [42] used Newman's [55] formulation for the same problem, to calculate the blockage caused by the presence of the leakage vortex. His measurements of the axial velocity and total pressure loss distributions, downstream from the trailing edge, suggested that the adoption of the simple model of a line vortex can adequately describe the phenomenon.

The models developed by Lamb [54] and Newman [55] for the diffusion of a line vortex are used in the present work to simulate the diffusion of the leakage vortex and predict the pressure disturbance as well as the total pressure loss distribution due to the tip-clearance presence. The corresponding formulation, described in the appendix to this section is involved also in the computation of the shed vorticity, as the vortex radius increases due to the diffusion process. Consequently, the calculation of the strength of the circulation is influenced (see equation (33) of the complete tip clearance model).

Complete tip-clearance model The already described model of the tip-clearance gap provides us with the mean jet velocity magnitude and direction at the gap exit, as well as the total pressure loss occurring inside the gap. For this calculation, the static pressure difference between pressure and suction sides at the tip level is needed, as this is modified by the tip-clearance vortex presence. As the jet velocity is the key element for the vorticity shed from the gap inside the leakage vortex and its computation must be done accurately, the complete calculation must necessarily be iterative.

The second model provides the necessary elements for the leakage vortex formation, its strength, its evolution characteristics and total pressure losses, but it doesn't provide its position in space. In the present calculation procedure Owen's [56] model for the calculation of the eddy viscosity is used.

For the calculation of the position of the leakage vortex a similar approach to that of Chen *et al.* [57] was utilized, who employed a slender body approximation of the leakage vortex, decomposing the velocity field into independent throughflow and crossflow parts, in order to provide a generalized description of the clearance vortex evolution. In Chen's *et al.* [57] work the three-dimensional steady flow is treated as a two-dimensional unsteady one, using a vortex method to reconstruct the vortex shape in planes normal to the blade camber.

In the present work the formation of the leakage vortex and the calculation of the induced kinematic field is treated also in successive planes, adopting the decomposition assumption of Chen *et al.* [57]. In order to achieve better cooperation between a meridional flow calculation procedure, which will be described later, and the present model, the successive calculation planes are parallel and normal to the axis of the machine. In this way, each plane corresponds to a specific axial position and the peripherally mean values can be calculated easily.

The computation of the shed vorticity is performed between two successive planes. A linear variation of the leakage mass flow rate is assumed between the two planes. The resulting vortex strength corresponds to a vorticity distribution in the second of the two successive planes. Lamb's [54] model of the diffusion of a line vortex is used for the positioning of the vorticity at the current calculation plane, which is expressed as

$$\omega(r, R) = \frac{\Gamma}{\pi R^2} \exp\left(-\frac{r^2}{R^2}\right) \qquad (1.19)$$

where R is the radius of the leakage vortex, calculated using the mass balance equation and taking into account the diffusion process. However, the axis of the leakage vortex is not perpendicular to the calculation planes, so that the angle of the vortex trace must be taken into account. In that way the component of vorticity normal to the calculation planes is used and the axisymmetric distribution of equation (1.19) is transformed into an elliptic one, through projecting on the normal to the axis calculation planes.

The vortex trace and the peripheral position of the vortex centre at the current calculation plane are computed using the self-induced velocity field in the previous plane. The time step resulting from the flow velocity at the vortex centre, considered in the absence of tip-clearance effects, and the distance between the calculation planes are used in order to compute the peripheral displacement of the vortex centre. The radial position is estimated empirically assuming that the vortex core edge reaches up to a height between the endwall and the blade tip. It is interesting to note that the computation of the vortex position based on the induced velocity field, as described above, provides both the peripheral and the radial position of its centre. However, comparison between theory and experiment revealed an important disagreement for the radial position of the leakage vortex. This value alone in the complete computational procedure is specified empirically.

The computation of the induced velocity field necessitates the solution of a Poisson equation of the form

$$\Delta \Psi = -\omega \qquad (1.20)$$

For calculation planes inside the blade passage a zero normal to the wall velocity is used as boundary condition, for all the boundary nodes except those at the tip-clearance gap exit, where the leakage mass flow rate determines the boundary values of the stream function.

The zero normal to the wall velocity condition is used at planes downstream of the trailing edge, only for the hub and tip boundaries. At the other two boundaries, along which the flow is periodic, the stream function distribution is calculated using the induced velocities from the leakage vortex, taking into account its mirror vortex and the two corresponding periodic vortices (Figure 1.4). The periodic vortex arrangement is used also in the calculation of the vorticity distribution at planes downstream of the trailing edge, as corresponding calculation results presented in Figure 1.5 demonstrate. A stretched grid is used with most of the nodes at the leakage vortex region, in order to minimize the total number of nodes without losing accuracy. Typical grids and corresponding induced velocity profiles are presented in Figures 1.6 and 1.7(a,b) respectively for the case presented in Figure 1.5.

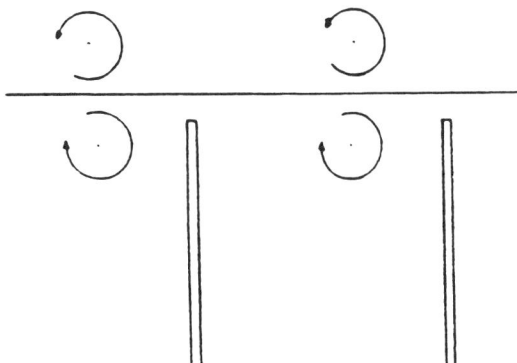

Figure 1.4 Schematic representation of the periodic and mirror vortices, for the planes downstream from the trailing edge.

Figure 1.5 Vorticity distribution inside the calculation domain, for the plane downstream from the trailing edge.

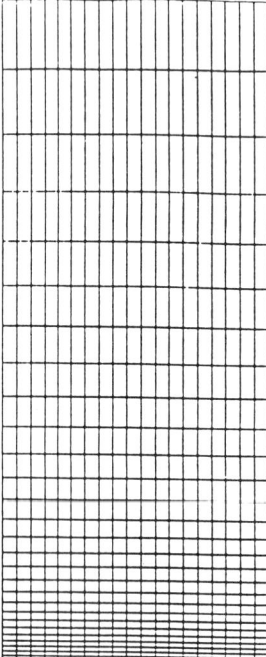

Figure 1.6 Typical mesh used for the computation of the induced kinematic field.

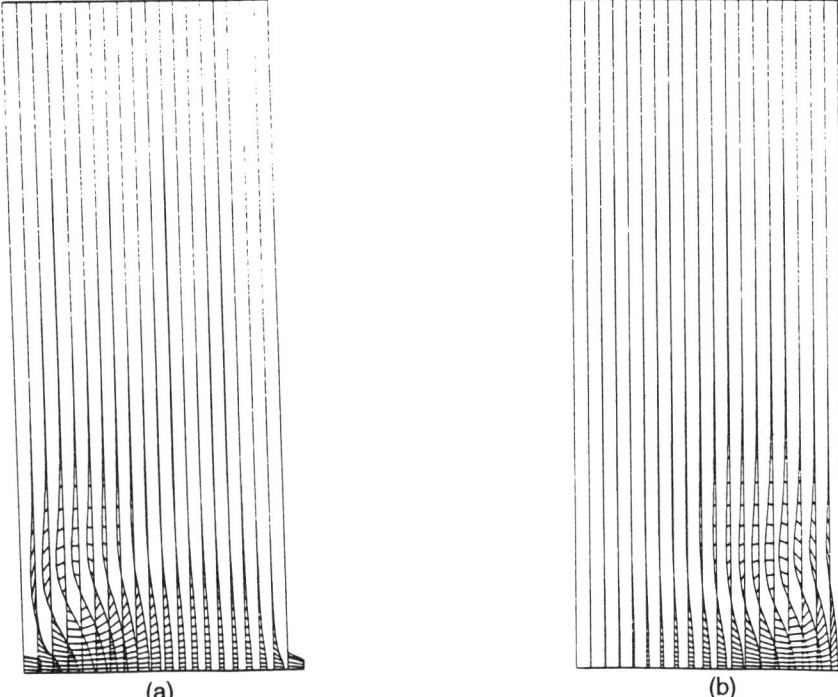

Figure 1.7 Induced velocity flow field inside (a) and downstream (b) of the blade row.

3.3.3 Incorporation of the models into a calculation procedure

The complete calculation procedure will be described in this section, with the remark that the tip-clearance flow will be introduced as a modification to the basic flow, existing in the absence of tip-clearance effects. The tip clearance flow is introduced, consequently, as a third zone in the already existing two-zone model, which is used for the calculation of the secondary flow effects.

The flow field without tip-clearance effects is established through the meridional flow calculation procedure, including secondary flow effects. Peripherally mean flow quantities are calculated and, for each streamtube considered, it is possible to compute, through momentum considerations, the forces acting on the blade surface (per unit surface), defined in the general case as

$$F_u = \frac{N}{2\pi}\left[(P_p - P_s) + (\tau_{uu_s} - \tau_{uu_p}) - R\left(\tau_{mu_s}\frac{\partial \theta_s}{\partial m} - \tau_{mu_p}\frac{\partial \theta_p}{\partial m}\right)\right. \\ \left. - R\left(\tau_{nu_s}\frac{\partial \theta_s}{\partial n} - \tau_{nu_p}\frac{\partial \theta_p}{\partial n}\right)\right] \quad (1.21)$$

$$F_m = \frac{N}{2\pi}\left[(\tau_{mu_s} - \tau_{mu_p}) + R\left(P_s\frac{\partial\theta_s}{\partial m} - P_p\frac{\partial\theta_p}{\partial m}\right) - R(\tau_{mm_s}\frac{\partial\theta_s}{\partial m} - \tau_{mm_p}\frac{\partial\theta_p}{\partial m})\right. \quad (1.22)$$

$$\left. - R\left(\tau_{mn_s}\frac{\partial\theta_s}{\partial n} - \tau_{mn_p}\frac{\partial\theta_p}{\partial n}\right)\right]$$

$$F_n = \frac{N}{2\pi}\left[(\tau_{nu_s} - \tau_{nu_p}) + R\left(P_s\frac{\partial\theta_s}{\partial n} - P_p\frac{\partial\theta_p}{\partial n}\right) - R\left(\tau_{mn_s}\frac{\partial\theta_s}{\partial m} - \tau_{mn_p}\frac{\partial\theta_p}{\partial m}\right)\right. \quad (1.23)$$

$$\left. - R\left(\tau_{nn_s}\frac{\partial\theta_s}{\partial n} - \tau_{nn_p}\frac{\partial\theta_p}{\partial n}\right)\right]$$

These forces can be computed using the meridional flow calculation results. When the forces F_u, F_m, F_n are known, then the pressure differences $(P_p - P_s)$ can be computed from equations (1.21), (1.22) and (1.23), if it is assumed that the shear stresses at the blade surfaces can be neglected.

This pressure difference provided by the meridional flow calculation procedure does not include the contribution of the leakage vortex. This last contribution may be computed integrating equation (1.31) of the Appendix:

$$\frac{\partial P}{\partial r} = \frac{\rho\Gamma^2}{4\pi^2 r^3}\left[1 - 2\exp\left(-\frac{r^2}{R^2}\right) + \exp\left(-\frac{2r^2}{R^2}\right)\right] \quad (1.24)$$

using the undisturbed flow value for the static pressure at the boundary. The vortex strength is needed for this computation, so that an iterative procedure must be set up in order to provide the various flow quantities, which need for their calculation the modified pressure difference at the blade tip. In the present work the complete calculation procedure which was set up conserved during iterations the same undisturbed flow, calculated once and for all during its initialization. The main reason for this was that the strong gradients of the flow quantities, resulting from the tip-clearance calculation procedure, induced instabilities to the computational procedure itself. Consequently, convergence was accepted when the assumed and computed pressure difference between pressure and suction sides were found to be the same. This calculation procedure is described in more detail immediately below.

At the considered station the blade thickness is known and a first guess for the static pressure difference may be made from the results of the secondary flow calculation procedure. Then according to the local blade thickness to tip gap height ratio, a simple exit velocity profile is adopted and the mean jet velocity and the total pressure losses occurring inside the gap are calculated. The conservation of the longitudinal component of momentum through the gap provides also the angle of the jet velocity. The peripherally mean value of the longitudinal component of velocity at the tip clearance height, provided by the secondary flow calculation, is used for this purpose. A linear variation of the jet velocity component, between the previous and the considered station is assumed for the calculation of the mass flow rate through the corresponding part of the gap. This flow rate is used to compute the increase in the vortex core radius between the two successive stations, assuming that the complete leakage flow enters the leakage vortex. An additional increase of the core radius, due to the diffusion process, is computed using equation (1.33) of the Appendix:

$$\Delta R^2 = 4v\frac{\Delta z}{W} \qquad (1.25)$$

where Δz is measured along the vortex axis, and W is the component of the free-stream velocity at the same direction.

The vortex model is then applied providing the amount of vorticity at the considered axial position. The vortex trace at the previous calculated position is used in order to determine the direction of the vortex axis. The calculation of the induced kinematic field is then carried out, with the positioning of the vortex core performed as described in previous sections.

The total pressure loss profile is then estimated, using the formulation described in the Appendix. More particularly, equation (1.41) is used to provide the total pressure difference between the cases with and without tip-clearance effects:

$$\Delta P_t = \Delta P + \tfrac{1}{2}\rho[2Ww - w^2 - v^2 - u^2] \qquad (1.26)$$

The static pressure at the suction side is modified due to the leakage vortex presence, according to the discussion above involving equation (1.24), and with the new static pressure difference the gap flow characteristics are computed again.

3.3.4 Appendix on line vortex diffusion model

The basic model for which Lamb [54] gave a mathematical solution may be discussed briefly as follows. We consider a uniform flow parallel to the axis $r = 0$ (with velocity equal to W). At time $t = 0$, the vorticity is zero everywhere, except along the axis $r = 0$, where there exists a line vortex of strength Γ. As the time passes vorticity is diffused radially and, if the problem is considered in successive planes normal to the vortex axis, a core is formed. Those planes lie at distances z equal to $z = Wt$, where t is the time needed to reach the corresponding plane if we are moving with speed equal to W. Then the distribution of vorticity reads

$$\omega(r,t) = \frac{\Gamma}{4\pi vt}\exp\left(-\frac{r^2}{4vt}\right) \qquad (1.27)$$

where v is the kinematic viscosity, r is the radial distance from the vortex axis and t is the time measured from the beginning of the diffusion process.

The peripheral velocity distribution is then given by

$$v(r,t) = \frac{\Gamma}{2\pi r}\left(1 - \exp\left(-\frac{r^2}{4vt}\right)\right) \tag{1.28}$$

For $r < 2(vt)^{0.5}$ the motion is a rigid body rotation with angular velocity $\Gamma/8\pi vt$, while for $r > 2(vt)^{0.5}$ the motion is irrotational, similar to that of a line vortex. This leads to a vortex core radius of

$$R = 2(vt)^{0.5} \tag{1.29}$$

We consider the problem not as a three-dimensional steady one but as a two-dimensional unsteady one, assuming that the time steps of the diffusion process correspond to distances Δz between successive planes normal to the vortex axis, fulfilling the relation

$$\Delta z = \Delta t \cdot W \tag{1.30}$$

where W is the velocity of the mainstream along the vortex axis.

From equation (1.29) we have, for two successive time steps,

$$R_2^2 - R_1^2 = \Delta R^2 = 4v(t_2 - t_1) \Rightarrow \Delta R^2 = 4v\Delta t \tag{1.31}$$

or, from equation (1.30),

$$\Delta R^2 = 4v\frac{\Delta z}{W} \tag{1.32}$$

The turbulent nature of the tip vortex flow can be incorporated by replacing viscosity v with $v + v_e$, where v_e is the appropriate eddy viscosity value. Yaras and Sjolander [53] use Owen's [56] model for v_e, which reads

$$\frac{v_e}{v} = \Lambda^2 \sqrt{\frac{\Gamma}{v}} \tag{1.33}$$

with Λ taking values between 0.7 and 1.2. The value of 0.95 has consistently been used throughout the present work.

Equation (1.32) provides a way of estimating the enlargement of the leakage vortex core due to the diffusion of its vorticity.

A similar solution to the problem of a line vortex diffused radially was proposed by Newman [55]. Choosing cylindrical polar coordinates r, θ, z, with Oz the axis of the line vortex, he used simplified forms of the three momentum equations plus the continuity equation in order to provide the distributions of the three velocity components and pressure at successive planes normal to the vortex axis.

Denoting the radial velocity by u, the rotational velocity by v and the longitudinal by w' and putting $w' = W - w$, where W is the axial velocity in the undisturbed fluid, the following relation was developed:

Research activities

$$v = \frac{\Gamma}{2\pi r}\left[1 - \exp\left(-\frac{Wr^2}{4vz}\right)\right] \quad (1.34)$$

Considering equations (1.29) and (1.30) we have

$$v = \frac{\Gamma}{2\pi r}\left[1 - \exp\left(-\frac{r^2}{R^2}\right)\right] \quad (1.35)$$

which is identical to equation (1.28). Newman's analysis is interesting, because expressions for the other velocity components, as well as for the static pressure, are developed. The two other velocity components may be finally expressed as

$$w = \frac{A}{z}\exp\left(-\frac{r^2}{R^2}\right) \quad (1.36)$$

$$u = -\frac{A}{2z^2}\exp\left(-\frac{r^2}{R^2}\right) \quad (1.37)$$

where A is the constant of integration.

The expression for the static pressure reads

$$\frac{\partial P}{\partial r} = \frac{\rho \Gamma^2}{4\pi^2 r^3}\left[1 - 2\exp\left(-\frac{r^2}{R^2}\right) + \exp\left(-\frac{2r^2}{R^2}\right)\right] \quad (1.38)$$

The numerical integration of the above equation provides the distribution of pressure inside the vortex using as boundary condition the value of the undisturbed free stream static pressure.

If $P_{t\infty}$ and P_t are the total pressure in the free stream and inside the vortex respectively we have

$$P_{t\infty} - P_t = P_\infty - P + \tfrac{1}{2}\rho[W^2 - (w'^2 + v^2 + u^2)] \quad (1.39)$$

Substituting w' in the above equation the total pressure loss is expressed as

$$\Delta P_t = \Delta P + \tfrac{1}{2}\rho[2Ww - w^2 - v^2 - u^2] \quad (1.40)$$

This provides the distribution of total pressure losses inside the vortex core which is diffused radially, using equations (1.35)–(1.38) in order to calculate the values of the various quantities appearing on the right-hand side.

3.4 Task 4: experiments on LSRC

The original proposal for this investigation was constrained by both time and the finance available. An appropriate proposal subset was selected; the subset detail pertinent to Cranfield is to 'complete a first series of experiments on the multistage compressor and demonstration of the 3D anemometer'. Success within this subset has been mixed.

For two values of tip clearance/height ratio, overall characteristics, blade surface pressures, interstage area traverses and streamline curvature analysis are performed. The three-dimensional structure of other flow in the tip clearance region is inferred from laser anemometry measurements in the blade-to-blade region.

3.4.1 Large-scale low-speed research compressor construction

The general arrangement of the four-stage LSRC is shown schematically in Figure 1.8. The compressor is driven by a thyristor controlled 150 HP d.c. motor through a 5.71:1 epicyclic reduction gearbox, and is coupled to the gearbox output by a strain gauge torque meter. The motor, gearbox and torque meter constitute the drive section which is housed inside the compressor exhaust ducting. A motorized conical valve to control the mass flow through the compressor is located at the exhaust exit.

The so-called 'working section' consists of a compressor, inlet flare and bullet, and a manguard which also acts as an air filter. The compressor itself consists of concentric casing rings of 1.21 m inner diameter which support 72 IGVs, 4 identical rows of stator vanes (72 per row) and 72 OGVs. The rotating drum is also a composite of rings, of 1.03 m outer diameter, which support four identical rows of rotor blades (79 per row). Consequently four identical stages are established having (for this investigation) a stage loading ($\Delta H/U^2$ of 0.45) and a flow coefficient (V_a/U of 0.7).

Both set of rings are located axially by tie bolts. However, each stator row can be traversed in a circumferential direction. This provides relative movement between the measurement plane and the stator vane, which permits area traverse measurements. For this investigation the first and third stages are to be studied. The rotating drum is driven by the drive section via end discs, which also serve as planes for dynamic balancing of the compressor.

3.4.2 Instrumentation

Historically, detailed interstage measurements of total and static pressure have been obtained from this compressor using miniature 23° wedge probes. These probes are relatively stable and are well understood away from solid surfaces, see Smout [74]. For an investigation of tip-clearance flows measurements adjacent to the casing are required and Cranfield undertook an additional set of measurements using a miniature RR design of a Cobra Probe, see Figure 1.9. These probes were manufactured and calibrated at Cranfield. Cobra probe results form the core of the streamline curvature analysis discussed later.

The compressor is comprehensively instrumented for the measurement of static pressure. Obtaining detailed and accurate characteristics, both overall

Research activities

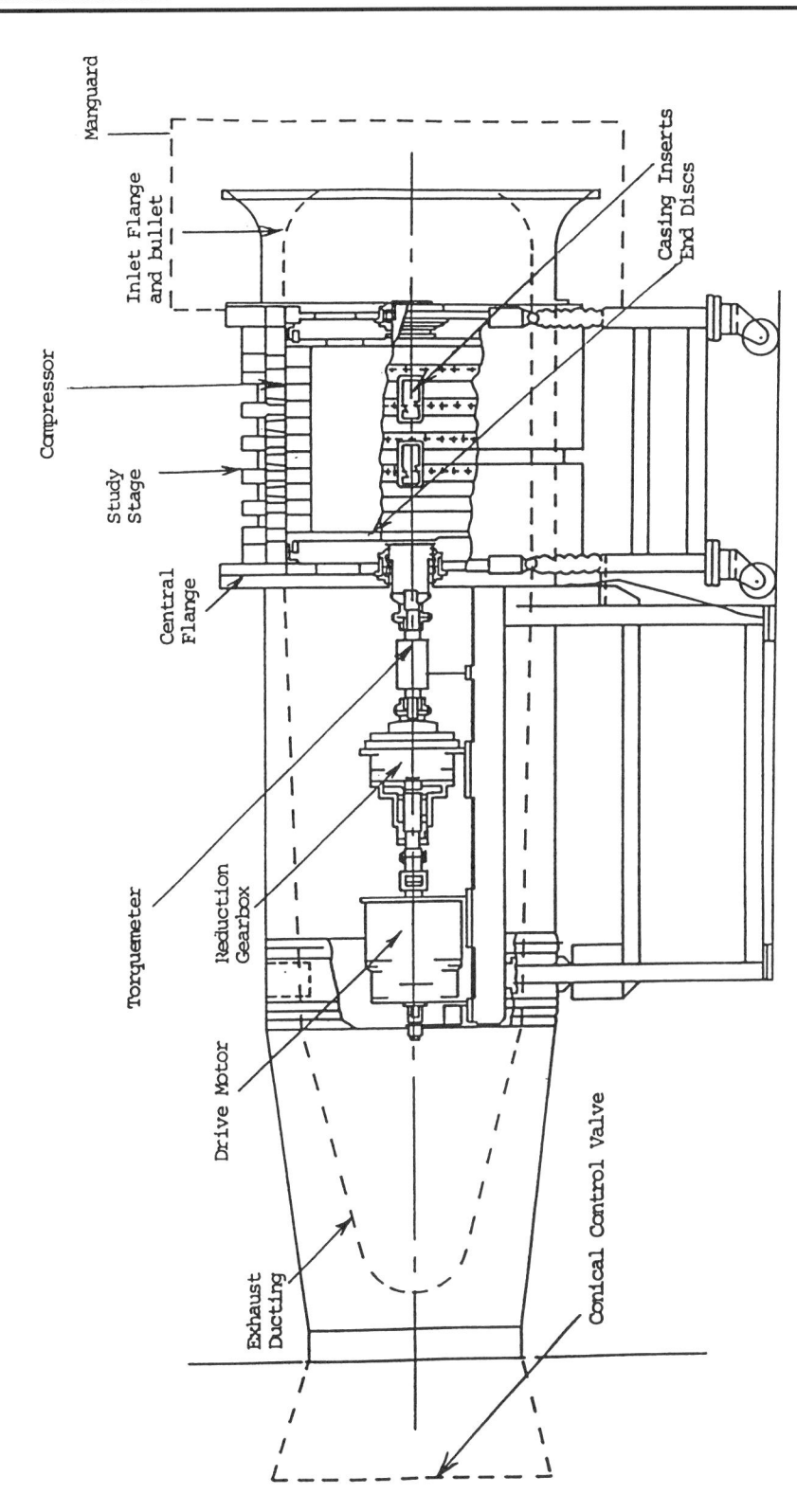

Figure 1.8 General arrangement of the low-speed research compressor at the Cranfield Institute of Technology.

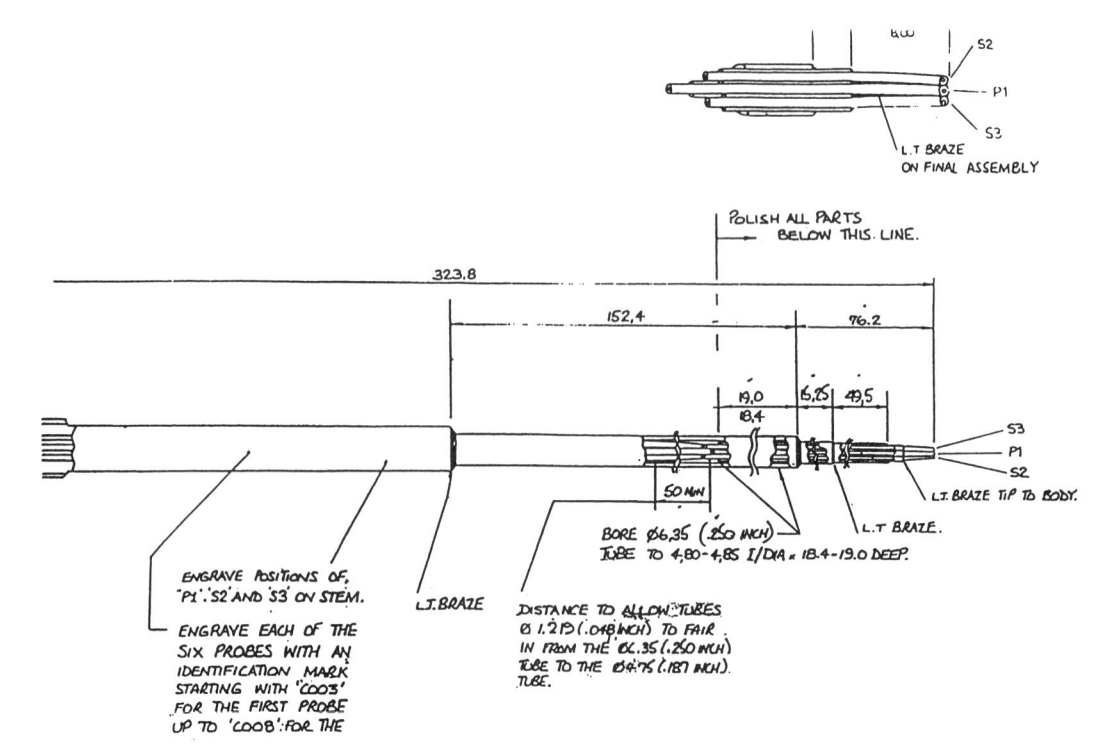

Figure 1.9 General arrangement of the Rolls-Royce Design Cobra Probes adapted for the LSRC.

and stage, is routine. Measurement of the chordwise distribution of static pressure, for both rotors and stators, is also available.

This BRITE/EURAM investigation has permitted an examination of the blade-to-blade region using L2F anemometry. The laser anemometry (L2F) equipment comprises a Lexel Model 95 2 watt argon ion laser, transmission and receiving optics, a Malvern 10 ns correlator and a data store. The measuring volume in the flow comprises two equal intensity beams with a spacing between them of 450 μm and an 'effective' length of 300 μm. These beams were orientated to measure the axial and tangential components of velocity. In the presence of a large radial component, measurements are either subject to bias or are not attainable.

Initial attempts to 'locally' seed the flow with silicon oil were unsuccessful; however, measurements were obtained using oil seeded with the flow at the inlet manguard. Optical access to the flow was obtained using windows inserted into the stationary casing, see Figure 1.10. Within the rotor blade-to-blade region, eight circumferentially discrete measurements were obtained using an electronic strobe unit which was triggered by an optical pick-up in a passing rotor-blade.

Finally, comprehensive measurements have been performed with two tip-clearance sizes: corresponding results are shown in Section 4.4.

Research activities

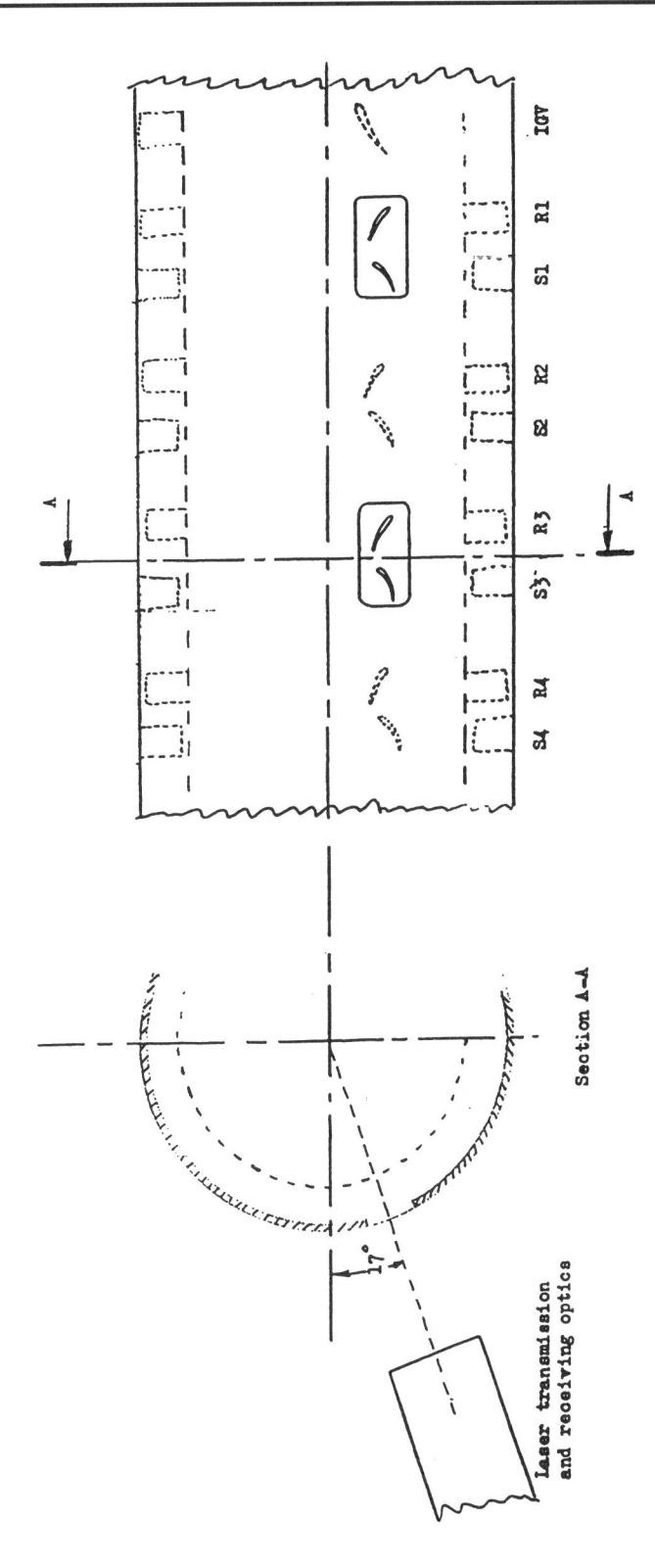

Figure 1.10 Window positioning in the Cranfield compressor.

3.5 Task 5: implementation of the first-level mode

3.5.1 Preliminaries

Industrial partners have provided informations of their current first-level methods to NTUA in order to anticipate on the pre-treatment software. This action had been performed as part of Task 3 of this project and was achieved during the first five months of the programme.

3.5.2 Reception of the code

Several versions of the first-level model have been sent to the industrial partners. The first draft version was provided during the seventh meeting, whereas the third was provided during the last meeting in Paris by the end of June; it upgrades the modelling of the outlet air angle and gives a radial pressure-loss distribution which is important with respect to throughflow analysis. Some modifications have also been made to reduce the CPU time consumption.

3.5.3 Implementation

A specific software has been achieved in industrial computational environments in such a way that each new version of the model could be easily implemented in the process. The flow data required for each computation are directly read in the industrial proprietary throughflow output datafile. The remaining parameters have been extracted from the code to allow an easier parametric study of the behaviour of the code. Then an output file has been implemented to allow use of customary post-treatment. The following results have been obtained.

3.5.4 Parametric study

Besides the geometrical and flow data, some parameters which either pilot the computational grid or are integrated in the code are collected in an input file. Since small changes of their values can have an appreciable influence on the radial profiles it has been found useful to make a parametric study of them. The aim of the project is to get a first-level model which will be used to analyse and design advanced multistage compressors; thus grid parameters have to be fairly well controlled.

The influence of the following parameters have been computed using the first version of the first-level model :

    ie   number of mesh points inside the tip clearance;
    rdx DX/DY ratio (DX, DY are the mesh steps);

ia   number of subspaces between successive stages of input data;
ip   number of centroids used for minimizing calculation effort;
ERR   maximum error of the Poisson equation calculation;
ns   number of sources used at the tip clearance gap exit region;
rr1 radial displacement of vortex core centre before the trailing edge.

This study has been conducted on Kyushu database test case at 2% of tip-clearance gap (gap/chord ratio).
To conclude this parametric study the following conditions for the main parameters which influence the tip-clearance model have been stated (further details can be found in the 8-SN-1 meeting report):

ie   > 4 or 5
rdx > 4
ia   > 20
ip   no effect
ERR   small effects
ns   no effect

It should be pointed out that '*ip*' and '*ns*' do not seem to influence either the jet velocity or the disturbed static pressure along the blade chord; therefore NTUA values are used. Regarding the maximum error of the Poisson equation, only small effects are noticeable; consequently, the maximum value of 10^{-4} has been chosen to reduce the CPU time.

3.5.5 Computation with database test cases

While the first-level model has been implemented as a routine in the industrial software, tests against database cases have been performed by each partner, and they are shown in Section 4.5.

3.6 Task 6: first stage of second-level code

3.6.1 Introduction

In the context of this task, LTT/NTUA undertook the modification and improvement of an existing three-dimensional Navier–Stokes code (named TURBO3D), originally developed in Cranfield. The improved code (referred to as the second-level tool) is intended to be used for validation of the first-level tool, developed in Task 3 of the project. Additionally, the code would be used in order to help understand the physics of the flow, as described by experiments. Furthermore, the code was to be used as a test-bed for different physical and numerical models, which may help the development and improvement of similar tools owned by the industrial partners.
Computational methods are now a useful tool for the designer, while

complicated and once expensive tools, such as three-dimensional Navier–Stokes solvers, become affordable for industrial use due to the increase in computer power. The quality of the numerical predictions for three-dimensional flows in turbomachines and in particular for cases with tip clearance depends strongly on how accurately the blade tip region is modelled. A literature survey reveals that the use of a thin blade, sometimes referred to as 'pinching' of the blade tip, was widely used by various researchers since it allows the use of standard H-type grids. However, it results in a poor description of the blade tip geometry and the use of skewed meshes in the tip region. Kunz *et al.* [59] showed that the accurate modelling of the blade-tip geometry, achieved through the use of an embedded H-type grid, allows better modelling of the tip-gap flow as well as of the vortex trajectory. Moore and Moore [60] and Hah and Reid [61] have used similar grids. Combining grids of a different kind, like the patched O and H type grids of Choi and Knight [62] or the overlapping C and H grids of Watanabe *et al.* [63] offer an alternative way of solving the problem.

In this task, particular attention was paid to the accurate modelling of the tip geometry in an attempt to maintain the true flat tip surface and its sharp edges. Limitations in computer resources forced NTUA to maintain, for the moment, the H-type grid topology and propose and assess a solution method that does not increase considerably the computing cost. The technique of pinched-tip geometries was already incorporated in the TURBO3D code provided by Cranfield, and consequently a large amount of work was undertaken in order to implement the new grid topology. The way this was done required a complete restructuring of the code.

It is also worth noting that the complete restructuring of the code allowed for a better adaption of the 'new' code on the mini-supercomputer of LTT/NTUA. The original code was programmed for working on a typical workstation with mild capabilities in computer memory. So, auxiliary calculations were repeated, instead of storing them, leading to an economy in computer memory at the expense of computing time. Working on the vector-parallel computer (Alliant FX-80) of LTT/NTUA, the 64 Mb of central memory allowed for a considerable economy in computing time, spent in repeated auxiliary calculations. Thus, in a first phase in the present task, our attempts were oriented towards the adaptation of the code structuring in our vector-parallel computer, using universal parallelization and vectorization techniques. Even if this subtask was not really foreseen, it proved to be very useful, since it led to affordable computational times. The actual version of the code requires approximately 0.8 ms/node/iteration on a four processor cluster of Alliant FX-80, which falls within the well-accepted computational times presented by other computers. It is finally to be said that our computations were performed with a number of grid points of the order of 104 to 105. Our computer capabilities permit calculations with up to 150 000 nodes approximately.

Turbulence modelling is another topic that mainly influences the quality of computational results. The use of a $k-\varepsilon$ model along with wall functions techniques to bridge the gap between the solid wall and the first adjacent node was maintained, since, apart from its well-known limitations, it resulted in a

Research activities

certain economy in the number of nodal points required. With present experience, an extension to the low Reynolds k–ε model seems mandatory. This, along with the introduction of algebraic–Reynolds stress models which is foreseen for the continuation of this project, will give the correct modelling of flow phenomena like separation, curvature and rotation effects.

Some of the improvements done in the code were necessary in order to overcome problems encountered when trying to model the flow in the four geometries chosen for testing and validating the code. The four geometries are listed below, as well as the various cases examined:

(1) Linear Compressor Cascade of Storer, for 0%, 2% and 4% of chord-tip clearances;

(2) Linear Compressor Cascade of Flot for 0% and 1.1% of chord-tip clearances;

(3) Rotating Compressor of Inoue, for 0.5 mm, 3.0 mm and 5.0 mm clearances;

(4) Rotating Compressor of Cranfield for 1.8% of span clearance.

The code produced results which showed good agreement with experimental measurement and helped visualize complex flow patterns in the clearance region. Any discrepancies between prediction and experiments are discussed in the report. Improvements need to be made to the turbulence model as well as to the orthogonality of the grids used. In some of these cases, results were provided for support of Task 3.

Here, a brief description of the topics covered under Task 6 will be presented. The reader is asked to refer to the corresponding progress reports for the complete coverage of each topic.

3.6.2 Brief description of the method

The steady turbulent flow is governed by the time-averaged equations of mass and momentum. In a rotating Cartesian coordinate system these equations are as follows :

continuity

$$\frac{\partial}{\partial x_j}(\rho u_j) = 0 \tag{1.41}$$

and momentum

$$\frac{\partial}{\partial x_j}(\rho u_j u_i) = \frac{\partial}{\partial x_j}\mu\left[\frac{\partial u_i}{\partial x_j} + \frac{\partial u_j}{\partial x_i} - \tfrac{2}{3}\delta_{ij}\frac{\partial u_k}{\partial x_k}\right] - \frac{\partial p}{\partial x_i} + F_i \tag{1.42}$$

where δ_{ij} is the Kronecker delta and F_i is the body force containing the Coriolis and centripetal forces,

$$\frac{\partial}{\partial x_j}(\rho u_j u_i) = \frac{\partial}{\partial x_j}\mu\left[\frac{\partial u_i}{\partial x_j} + \frac{\partial u_j}{\partial x_i} - \tfrac{2}{3}\delta_{ij}\frac{\partial u_k}{\partial x_k}\right] - \frac{\partial p}{\partial x_i} + F_i \qquad (1.43)$$

$$\mathbf{F}_i = -2\rho\mathbf{\Omega} \times \mathbf{u} - \rho\mathbf{\Omega} \times (\mathbf{\Omega} \times \mathbf{r}) \qquad (1.44)$$

For compressible flow, the energy and perfect gas relations are used to determine density and rothalpy, defined as

$$I = c_p T + \tfrac{1}{2}\mathbf{u}\cdot\mathbf{u} - \tfrac{1}{2}(\mathbf{\Omega}\times\mathbf{r})\cdot(\mathbf{\Omega}\times\mathbf{r}) \qquad (1.45)$$

The Reynolds stresses $\rho \overline{u'_i u'_j}$ which arise from the decomposition of the flow field into its mean and fluctuating parts are evaluated from

$$-\rho\overline{u_i u_j} = \mu_T\left[\frac{\partial u_i}{\partial x_j} + \frac{\partial u_j}{\partial x_i}\right] - \tfrac{1}{2}\rho\delta_{ij}\overline{u_k u_k} \qquad (1.46)$$

where $\mu_T = C_D \rho(k^2/\varepsilon)$ according to the 'two-equation' k–ε model of turbulence, where k, ε satisfy the following equations

$$\frac{\partial}{\partial x_j}(\rho u_j k) = \frac{\partial}{\partial x_j}\left[\frac{\mu_T}{\sigma_k}\frac{\partial k}{\partial x_j}\right] + G - \rho\varepsilon + G_c \qquad (1.47)$$

$$\frac{\partial}{\partial x_j}(\rho u_j \varepsilon) = \frac{\partial}{\partial x_j}\left[\frac{\mu_T}{\sigma_\varepsilon}\frac{\partial \varepsilon}{\partial x_j}\right](C_1 G - C_2\rho\varepsilon + G_c)\frac{\varepsilon}{k} \qquad (1.48)$$

where the generation term

$$G = \mu_T\left[\frac{\partial u_i}{\partial x_j} + \frac{\partial u_j}{\partial x_i}\right]\frac{\partial u_i}{\partial x_j}, \qquad G_c = 9\mathbf{\Omega}\tau_\Omega - \frac{9w\tau_\theta}{2R} \qquad (1.49)$$

where τ_Ω and τ_θ are the local shear stress components normal to the Coriolis acceleration and streamline R is the local streamline radius of curvature.

Wall functions are used to bridge the gap between the solid walls and the adjacent grid line. Zero Neumann type conditions are used for the calculation of the turbulent kinetic energy (k) while the turbulent energy dissipation (ε) at the nodes adjacent to the blade walls nodes is determined through the relationship

$$\varepsilon = \frac{[\kappa C_D^{1/2}]^{3/2}}{k\delta s} \qquad (1.50)$$

The above-mentioned equations for mass, momentum and turbulence scalars can be expressed in the form of a general transport equation for a dependent variable Φ :

$$\frac{\partial(\rho u_j \Phi)}{\partial x_j} = \frac{\partial}{\partial x_j}\left[\Gamma \frac{\partial \Phi}{\partial x_j}\right] + S(x,y,z) \qquad (1.51)$$

where Γ is the diffusion coefficient and S the source term.

By using the grid transformation $\xi = (x,y,z)$, $\eta = \eta(x,y,z)$, $\zeta = \zeta(x,y,z)$, the above transport equation takes the form to

$$\frac{\partial(\rho U_j \Phi)}{\partial \xi_j} = \frac{\partial}{\partial \xi_j}\left[J\Gamma g^{ik}\frac{\partial \Phi}{\partial \xi_k}\right] + J \cdot S(\xi,\eta,\xi) \qquad (1.52)$$

where J is the Jacobian of the transformation and $U_j = (U,V,W)$ are the contravariant components of velocity

$$U^i = J\frac{\partial \xi_i}{\partial x_j}u_j \qquad (1.53)$$

g^{ij} are the components of the metric tensor

$$g^{ij} = \frac{\partial \xi_i}{\partial x_k}\frac{\partial \xi_j}{\partial x_k} \qquad (1.54)$$

The governing equations are integrated over control volumes and are evaluated using the finite difference approximations of Patankar and Spalding. Pressure is established through corrections which arise from the satisfaction of local and global continuity, according to the SIMPLE algorithm (Patankar and Spalding [64]). An advantage of the present method is that a non-staggered grid is used for storing all variables. Pressure oscillations that would have resulted from this arrangement are eliminated using the pressure correction scheme of Rhie and Chow [65].

3.6.3 Modifications—Improvements

Implementation of a square-edge blade tip In the original code the blade was pinched at the tip. In this way the number of nodes in the blade-to-blade direction was constant from the hub to the shroud. However, the blade geometry in the tip region was inaccurate and the grid was highly skewed. The accurate representation of the blade geometry in the tip region implies the modelling of a square-edged blade. This was achieved by filling the clearance gap between the blade tip and the shroud with an H-type grid. Instead of tackling this extra domain using a multi-domain approach, the single domain approach of the original code was maintained by 'blending' the clearance grid with the main grid that covers the whole calculation domain. The resulting grid is shown in Figure 1.11.

In the resulting calculation domain, the dimensions of transversal sections that lie upstream and downstream of the blades are different to those lying

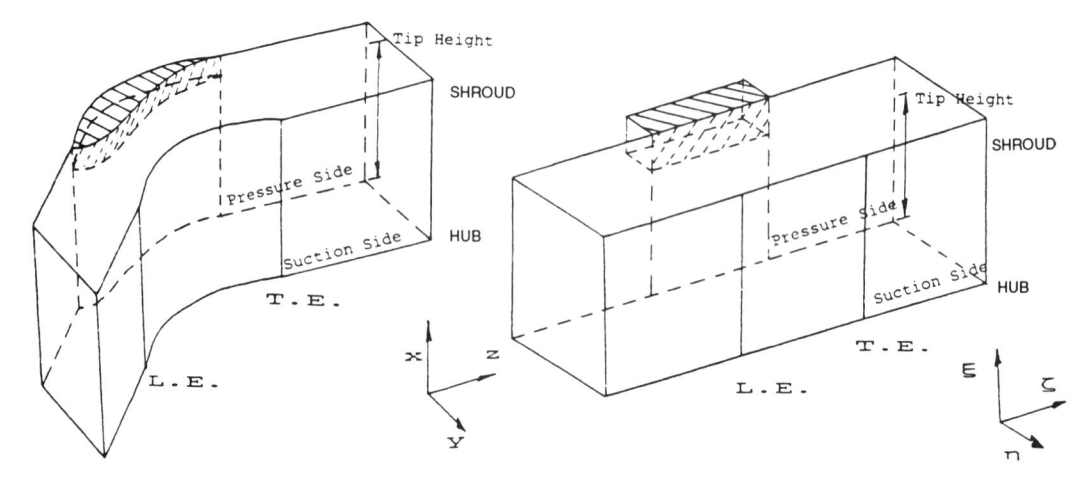

Figure 1.11 Actual and transformed computational grid including the clearance grid (hatched region).

between the leading and trailing edges of the blade. This problem was tackled by including the clearance grid in the string used by the code in storing flow variables for the calculation nodes. The determination of this string for the particular complicated grid is cumbersome and one cannot afford to perform it every time this string is required. This string is thus stored and is a function of the i,j,k indices of each node. In order to specify fully the topology of the grid, the nodes that lie east, west, front and back of each node need to be determined, an operation which is not straightforward with the extra clearance grid. These indices are thus determined once and are stored as functions of the node's string index mentioned above. All flow variables and geometrical data are stored as functions of the aforementioned string. This string contains only the nodes of the main and the clearance grids, avoiding all other nodes that would fill a parallelepiped that would contain the whole of the extended grid. In this way the increase in memory requirement, due to the additional clearance grid, is minimal.

An important factor in obtaining a solution for a particular test case, and indeed an accurate one, is the calculation grid used. This was constructed in a different way for each of the four geometries examined, depending on the way each blade was specified. Once the blade's profile is defined, a two-dimensional H-type linear grid is constructed around it. This, in the case of a linear cascade, is stacked in the spanwise direction producing the three-dimensional grid. For three-dimensional twisted rotor blades, different two-dimensional linear grids are constructed for various radii and are wrapped on cylindrical surfaces in order to produce the three-dimensional grid. The clearance grid is a simple linear H-type grid defined by notional extensions of the blade's pressure and suction sides. Grid density is naturally high close to solid walls. The size of the clearance grid in the through-flow and radial directions is the same as that of the main grid, but in the pressure to suction-side direction it can be specified independently.

Accurate leading and trailing edge modelling In tackling the different blade profiles, problems arose when the blade geometry was characterized by a thick leading or trailing edge. This was due to the fact that the original code approximated the leading/trailing edges by a notional triangle; this was equivalent to omitting the geometrical leading/trailing edge and bridging the space between the last/first periodic point and the first/last blade node through a particular finite volume. In this case, the problem of calculating the metrics at the leading/trailing edge is circumvented, but information is lost concerning the local flow patterns which influence in a certain degree the loading of the blade.

The choice of this notional triangle as well as its orientation influences the quality of the results obtained even in the case of thin leading edges. This drawback was overcome by retaining the real leading and trailing edges and at the same time circumventing the ambiguity of the leading/trailing edge metrics through appropriate one-sided differencing within the finite-volume integration.

For the sake of simplicity, the analysis will be presented in two dimensions, the extension to three being straightforward. The discussion will be divided into two parts, one related to the satisfaction of the mass conservation equation and the other to the analysis of non-orthogonal cross-terms in the diffusion part of the transport equations.

Mass conservation analysis The hatched region in Figure 1.12 shows the finite-volume which is adjacent to the blade at its leading edge. The cell is drawn in the transformed domain and its side is 1.5 times its base, which coincides with the solid boundary. In this way, the finite volumes of the non-boundary nodes completely cover the computational domain.

The continuity equation, when discretized over the hatched finite volume, reads

$$[(J\rho W^1)_e - (J\rho W^2)_w]\Delta\zeta + [(J\rho W^3)_f - (J\rho W^3)_b]\Delta\xi = 0 \qquad (1.55)$$

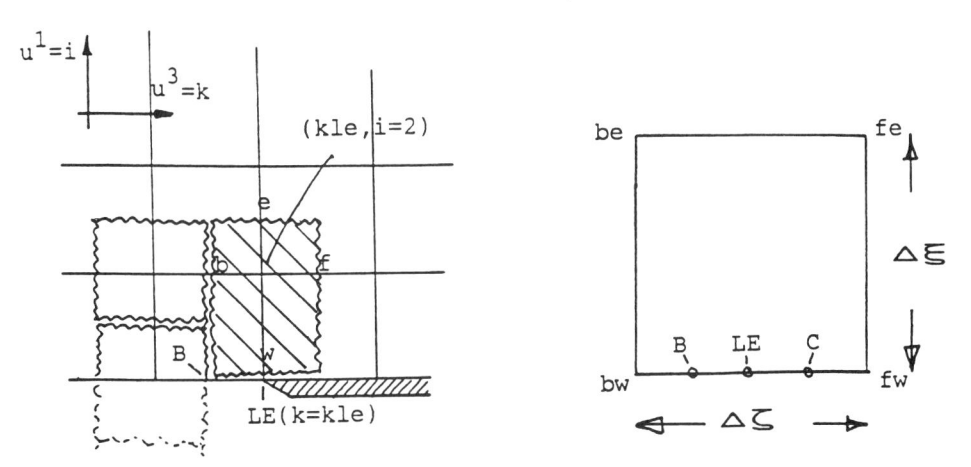

Figure 1.12 Treatment of finite volume adjacent to the leading edge.

where e,w,f,b stand for the east, west, forward and backward face of the cell, as indicated in Figure 1.13. For any internal cell, this equation uses contravariant velocity components at the mid-nodes found by an interpolation scheme which is compatible with the accuracy of the whole discretization; there, of course, $\Delta\xi = \Delta\eta = 1$. In the particular cell of Figure 1.13, the term $(J\rho W^1)_W \Delta\zeta$ is approximated as

$$(J\rho W^1)_B.(\Delta\zeta)/z \tag{1.56}$$

and the contravariant component $(J\rho W^1)$ at mid-node B is found as

$$(J\rho W^1)_B = \tfrac{1}{4}(J\rho W^1)_A + \tfrac{3}{4}(J\rho W^1)_{LE} = \tfrac{1}{4}(J\rho W^1)_A \tag{1.57}$$

For the non-orthogonal terms The non-orthogonal terms in the diffusion part of all transport equations, integrated over the hatched computational cell of Figure 1.13, read

$$\left[\frac{\partial}{\partial\zeta}\left(J\mu g^{31}\frac{\partial\Phi}{\partial\xi}\right) + \frac{\partial}{\partial\xi}\left(J\mu g^{13}\frac{\partial\Phi}{\partial\zeta}\right)\right]\Delta\xi\,\Delta\eta \tag{1.58}$$

and are calculated through the following scheme:

$$(J\mu g^{31})_f(\Phi_{fe} - \Phi_{fw}) - (J\mu g^{31})_b(\Phi_{be} - \Phi_{bw})$$
$$+ (J\mu g^{13})_e(\Phi_{fe} - \Phi_{be}) - (J\mu g^{13})_w(\Phi_{fw} - \Phi_{bw}) \tag{1.59}$$

For the particular face (w) lying between points bw and fw, the term $(J\mu g^{13})_w(\Phi_{fw} - \Phi_{bw})$ is approximated by the expression

$$(J\mu g^{13})_{fw}(\Phi_{fw} - \Phi_{LE}) + (J\mu g^{13})_{bw}(\Phi_{LE} - \Phi_{bw}) \tag{1.60}$$

where the first terms contain geometrical information, like the metric g^{13}, which is calculated using coordinates for grid points belonging to the solid wall, while the second term represents information coming only from nodes at the 'periodic boundary' region. In this way, the aforementioned ambiguity of metrics calculation is circumvented.

Low Reynolds turbulence model The turbulence model of the original code was based on the use of wall functions which allow an economy of grid nodes near solid boundaries. This results in more affordable run times for three-dimensional applications. In tip leakage flow cases, two important relative problems are that in the clearance region flow separation and reattachment have been observed, and the model fails to accurately reproduce these. Additionally, the leakage flow is driven by strong pressure gradients. It is generally suggested that in the case of strongly accelerated flows, only low Reynolds versions of the two-equation turbulence model can give accurate predictions; wall functions, which predefine the first node of the wall lying in the logarithmic region, fail to reproduce the flow because viscous effects are now important over a large part of the layer.

In this context, work has been done to include low Reynolds terms in the $k-\varepsilon$ equation (according to the well-known Jones & Launder model) and solving down to the solid walls, where zero Dirichlet boundary conditions are applied to the turbulent kinetic energy and the turbulent energy dissipation respectively. In a two-dimensional counterpart of the present method, the low Reynolds $k-\varepsilon$ model was successfully introduced and the extension of the low Reynolds $k-\varepsilon$ model into three-dimensional space, i.e. in the code that will be used for the tip-clearance studies, could be proposed. This could result in an increase in the number of required grid points, but this seems to be affordable. For more accurate and 'flexible' turbulence modelling one could proposed the introduction of algebraic–Reynolds stress modelling in combination with the $k-\varepsilon$ model, so as to overcome shortcomings of the classical $k-\varepsilon$ model, such as its inability to model separation, curvature and rotation effects.

Faster computer run-times The original TURBO3D code was structured in such a way so as to run on a machine with moderate memory size; this economy in memory resulted in an increase in run times which made the modelling of industrial test cases costly. While implementing the 'clearance grid' in the TURBO3D code, this was restructured so as to take advantage of parallelization and vectorization of a shared memory vector/parallel mini-super-computer. (Alliant FX-80, eight processors, 64 Mbytes of memory). The final run times are of the order of 1 ms/node/iteration. This time is comparable to times reported in the literature by other researchers using standard three-dimensional N-S codes. A speed-up of the order of 3.2 was achieved on a four-processor environment.

3.7 Task 7: demonstration of a three-dimensional laser anemometer

This instrument is a quasi-three-dimensional laser Doppler anemometer, miniaturized to provide ease of use in turbomachines. Its construction has been documented, see Ahmed *et al.* [71], and its potential demonstrated in high-speed centrifugal machines, see Ahmed [72].

To improve the laser power density in the probe volume, the probe was fitted with a new argon-ion laser light source. Also new optical fibre was provided to account for the change in operating wavelength. Both of these changes were successfully engineered.

Problems occurred in areas of probe alignment and signal processing which, given the time available, effectively prevented a successful demonstration. It is concluded that the risk associated with this instrument precludes it from use in any follow-on study.

An alternative is possible however. A three-dimensional laser-time-of-flight system has just become available from a well-known and reputable manufacturer. This system is believed ideal for the intended application and trials are planned.

3.8 Task 8: high-speed cascade facility

This task has been accomplished, according to the technical description provided in the contract [1]. The work undertaken and the achieved targets are described in more detail below.

The annular cascade facility, for detailed measurements in blading representative of modern aeroengine compressors, has been conceived, and its establishment in LTT/NTUA has already started.

Starting from typical values of various design parameters for the annular cascade, a first sizing was proposed for blades, clearances and the straight part of the inlet annulus. The measurements to be performed were specified and consist of two sets, one set of pneumatic measurements and one set of measurements using a three-dimensional laser system. The measurements have been presented in report 5-NTUA-2.

After this phase, information was exchanged between NTUA, Snecma and Turbomeca, in order to define all requirements for both the inlet scroll and the test section. Scroll drawings and manufacturing have already been completed by Turbomeca, while detailed drawings of the rotating hub arrangement have been completed by Snecma. Drawings of the test section layout (casings) as well as the manufacturing of the test section and the rotating hub, will be performed during the next tip-clearance project.

For the design of the ducting connecting the scroll exit to the axial part of the test section, calculations of the flow field were performed (see 7-SN-1 and 8-SN-1 for more details). These calculations were done by means of three-dimensional flow solvers, and they provided the profiles of flow field quantities at the inlet of the test cascade, given the scroll outlet conditions and the geometry of the bend-ducts connecting it to the annular space (details have been given in reports 8-NTUA-1 and 9-SN-1).

A set-up for a 'dummy' facility for testing the scroll outlet conditions was proposed. The overall arrangement was prescribed, the individual pieces were defined with the assistance of Snecma and the final design has been produced, including shop drawings for the parts of this facility. Care was also taken that some of the pieces manufactured for the 'dummy' facility are used in the final facility as well, in order to reduce the overall cost.

A feasibility study for performing three-dimensional measurements with a laser system has been performed. Up-to-date techniques for performing three-component measurements were reviewed and the requirements for performing measurements in the annular cascade facility were established. The features of commercially available systems were examined. A set-up consisting of a one-component and a two-component system, used together, has been proposed. Details are given in report 7-NTUA-1.

A laser system for measuring one velocity component has been specified and procured, with NTUA's own funds. It is already operational at NTUA. Extensive preparation for the procurement of an additional two-component system (which will make possible three-dimensional measurements) has been made, and it is possible to proceed to the procurement of this system as soon as funds become available.

During the conception phase of the facility, it was found that the radial compressor available at the NTUA rig was not suitable for driving the annular facility. A new compressor was found and its implementation in the rig was studied. Procurement and installation of this compressor, which meets the requirements of the test facility, is covered by NTUA's own funds, Snecma and Turbomeca supplying the main parts of a compressor having such characteristics.

At the end of the contract (and according to the initially defined funding for this purpose), some pieces have already been manufactured: the inlet scroll has been manufactured by Turbomeca and has already been transported to NTUA. The underbase of the facility as well as the main supporting brackets have also been manufactured and are available at NTUA. Pictures of these pieces are shown in Section 4.8.

4 Research results

4.1 Task 1

4.1.1 Test cases included in the database

The Compressor test cases currently contained in the database are listed in Table 1.1. The generic name for each case, identification and the source of information are given.

4.1.2 Example of content of database

A sample of the content of the database for one particular test case is shown in Figure 1.13. In this figure a comparison between measurement data and predictions of various flow quantities is shown. A throughflow calculation has been run with experimental data from the database and the results have been introduced into the database.

To get an extensive view of the database one can see the report 8-RR-1; in addition, since each industrial partner has studied database test cases during the project completion, one can find specific industrial contributions in the following meeting reports: 4-SN-2, 6-SN-1, 7-SN-1, 9-SN-1, 2-TM-1, 3-TM-1, 4-TM-1, 6-TM-1, 7-TM-1. A comparison with industrial throughflow software in Kyushu database test cases can also be found in Section 4.5.

4.2 Task 2

Table 1.2 summarizes the list of parameters determining the characteristics of tip-clearance flow and their respective representative ranges of values for multistage high-speed compressors.

Table 1.1 Database test cases

Filefrmt	Brief description	Reference	Contribution
CIT-MK2	CIT 4 stage Axial Comp. 3rd stage rotor studied at 850 rpm and 13.7 kg/s	[2], [3]	Rolls-Royce NTUA
CUTC-1	Cambridge Univ. case 1 single rotor at 432.5 rpm and 31.7 kg/s	[4], [5]	NTUA
CUTC-2	Cambridge Univ. case 2 single stage at 500 rpm and 15.3 kg/s	[6], [7], [8]	NTUA
GELSRC	GE LSRC 4 stage 1st & 3rd stage studied at 824 rpm and 17.4 kg/s	[9] up to [14]	Snecma
IN85GT62	Kyushu Univ. single stage rotor studied at 1300 rpm	[15]	NTUA
PWLSR	P & W low-speed rotor at 510 rpm	[16] up to [20]	Turbomeca
PWTSC	P & W two-stage compressor 2nd stage studied at 650 rpm	[21], [22]	Turbomeca
NGTEC141	Nat. Gas Turbine Est. Research comp. rig C141 all stages studied at 12340 rpm and 24.1 kg/s	[23] up to [26]	Rolls-Royce
LSRCBRIT	CIT LSRC 4 stage 3rd stage rotor studied at 875 rpm and 13.0 kg/s	[27], [28]	Rolls-Royce

4.3 Task 3

The secondary flow calculation code, with the tip-clearance model implemented in it, was used in order to compute the tip-clearance effects in available test cases, where experimental data existed. In view of the level of detail required for validating the present theoretical procedure, appropriate experimental data was scarce, so that all the available experimental results were used to the best of the authors ability. Two linear cascades cases (one compressor and one turbine), a single compressor rotor, and one stage of a multistage compressor were considered. Storer's [42] experiment provided the data for the first case, while Yaras *et al.* [44] experiment was used for the second one. The single rotor case was the one documented by Inoue *et al.* [58], while the multistage compressor case was the one tested in Cranfield [37]. Measurements for various tip clearances are available and concern the kinematic and pressure field at the tip clearance region and for downstream of the trailing edge. However, the available experimental data concern different flow parameters and not all the sets of comparisons can be provided for every case.

Applying the computational procedure described above it was demonstrated

Research results

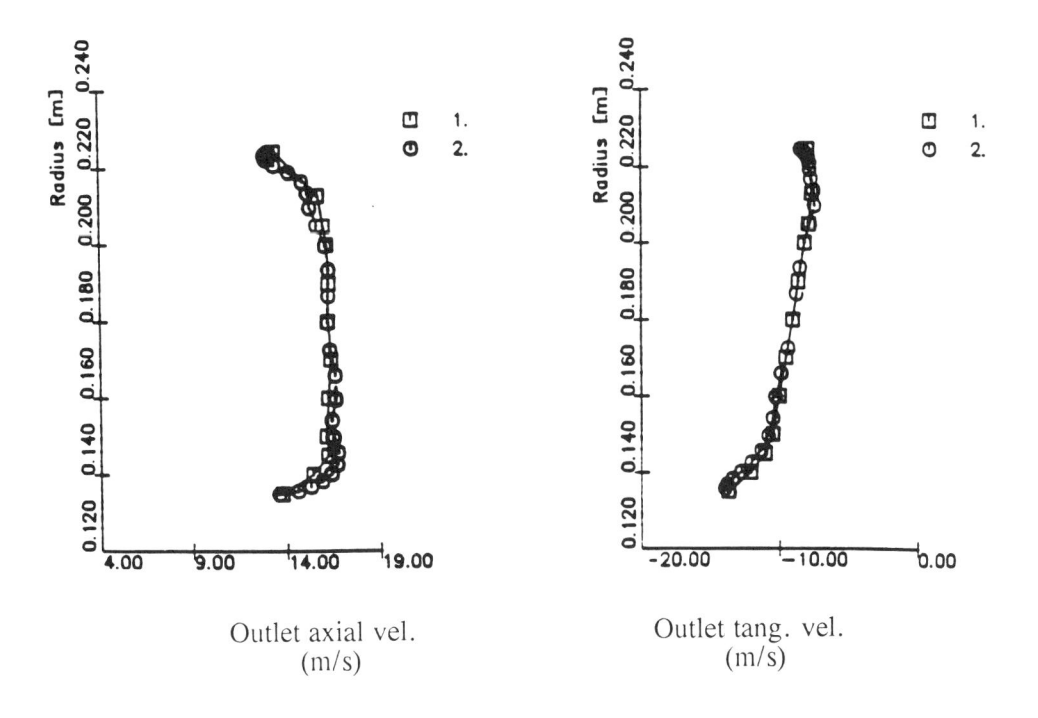

1. Measured, 2. Predicted

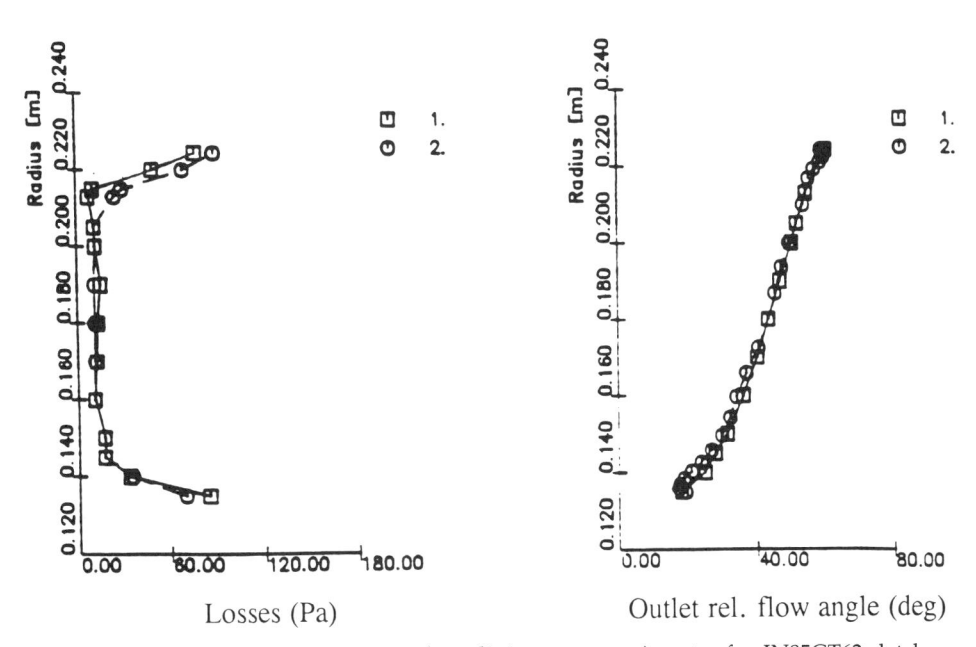

Figure 1.13 Example of comparison of predictions to experiments, for IN85GT62 database test case (tip/chord = 0.004).

Table 1.2 Parameters influencing tip-clearance flow.

Parameters	Snecma	Rolls-Royce	Turbomeca
Tip-clearance gap	0.2–2% of chord	0.5–4% of chord	0–2% of chord
δ^*	1–5% annulus height	1–6% of height 1–6 times tip gap	
Turbulence		4–10%	
Inlet skew	Colateral in previous blade ref. frame	appropriate to rotor tip sect.	
Re(chord)		$> 25 \times 10^5$	
Chordwise blade loading	Included for annular cascade		
$\Delta H/U^2$	0.35–0.6	0.35–0.55	0.34–0.50
Va/U	0.4–0.7	0.45–0.7	0.48–0.61
$s/c = 1/\sigma$	0.6–0.8	0.6–1.0	0.6–1.0
Diffusion factor		0.45–0.55	0.4–0.6
$AR = h/c$	0.8–1.5	1.0–2.0	
t/c	2–5%	2–5%	2–4%
Mach number	0.6–1.3	0.6–1.2	0.9–1.5
Tip speed			300–500 m/s
Casing-blade notion endwall treatment	Not to be included	To be included	
Surface roughness of blades casing		Hydraulically smooth	
Kind of blading		50% reaction or $a° = 0$	
Blade stacking	Radial		

that the calculation results strongly depend upon the value of the leakage mass flow rate. This quantity depends largely upon the pressure difference at the gap height, which is being modified by the presence of the leakage vortex, so that the importance of the correct prediction of this pressure modification was recognized.

In Figures 1.14(b–f) calculation results of the pressure difference at the tip-gap height are compared with experimental data, for various tip clearances (Storer's [42] case). The agreement is good for prediction purposes except in the case of $e/c = 4\%$, where the experimental results cannot be predicted with sufficient accuracy. These calculations were performed, using the pressure

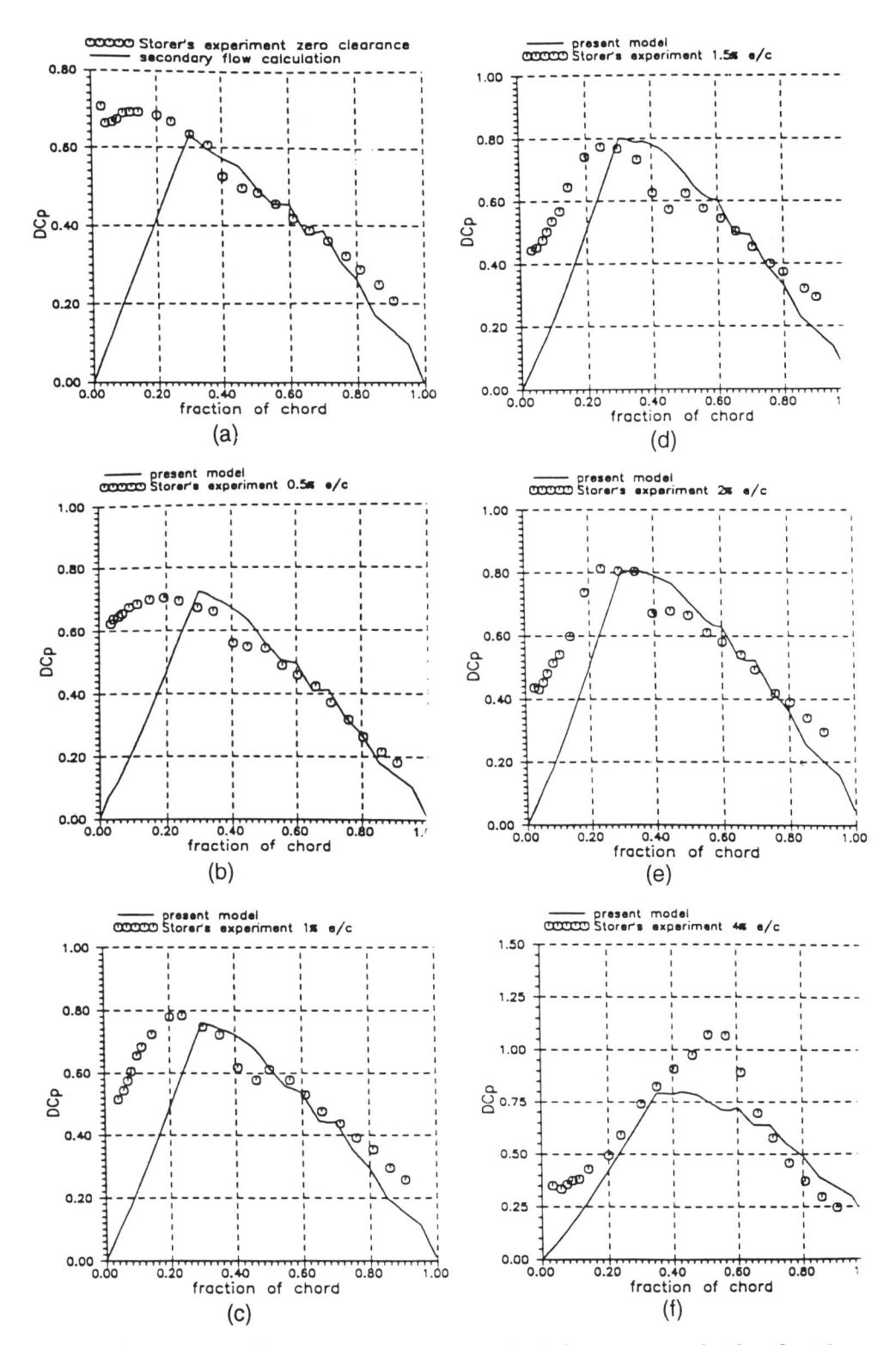

Figure 1.14 Pressure difference at the tip clearance level, for various gap heights (Storer's compressor cascade).

difference obtained by the secondary flow calculation method (Figure 1.14(a)). From these results it may be concluded that the formulation presented in the Appendix may be used for predicting reasonably accurately the pressure disturbance caused by the leakage vortex.

The calculation results for the mean jet velocity component normal to the blade's camber line, for the three available tip clearances, are compared with the experimental data in Figures 1.15(a–c). The corresponding mean flow angles of the leakage jet are presented in Figures 1.16(a–c). The good agreement with the experimental data proves the validity of the assumptions adopted for the gap flow model. The success of the complete theoretical calculation of the discharge coefficient proves the model suitable for prediction purposes and the assumed linear variation of the pressure difference, for the first 30% of the chord, provides satisfactory results for the corresponding area. The validity of the Rains' assumption for the conservation of the longitudinal component of momentum through the gap is proved with the results of the mean jet angle.

Circumferentially averaged distributions of the peripheral components of the vortex flow field, for various tip clearances, are presented in Figures 1.17. They are compared with measured distributions of the mass-averaged peripheral components of velocity, for each tip-clearance value, minus the corresponding distribution for the case with zero clearance (the computation is based on a secondary flow calculation; thus the peripheral variations of the flow quantities are not available). In this way the model itself, as well as the superposition principle adopted in the present work, are validated. The comparisons may be termed as good, if we consider the simplicity of the models used and the assumptions adopted in their development. The fact that a complete theo retical procedure was used for the prediction renders the comparisons more interesting.

Storer's [42] experimental data was used in order to compare calculated and measured total pressure loss through the gap, defined as

$$\int_0^e W_{N_4}(s) \cdot (P_{t_4}(s) - P_{t_1}) \, ds \tag{1.61}$$

Comparisons between theory and experiment are presented in Figure 1.18. The existing differences can be partially explained by the fact that the present model does not include the losses due to the endwall boundary layer inside the gap, which are present in the experimental data. However, the trends are well predicted and the fact that this part of the losses is small compared to the total losses attributed to tip clearance effects, renders this level of comparison adequately accurate for the complete calculation. It should be noted that the coefficient C_D appearing in equation (1.9) has been derived according to (1.9a), while the predictions of other workers seem to have been computed on the basis of the value of C_D, which produced the best fit with experiment.

Similar results were obtained for the third case (Inoue et al. [58]). The circumferentially mean meridional velocity component, as calculated by the

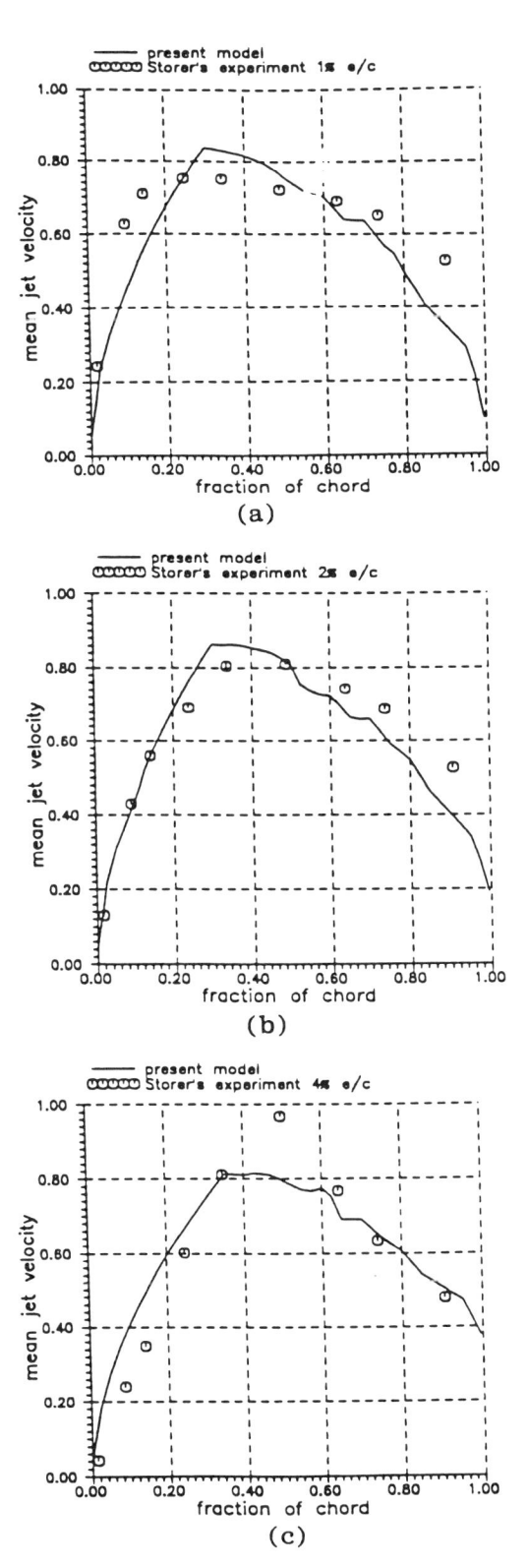

Figure 1.15 Mean jet velocity component normal to the blade camber line (Storer's case).

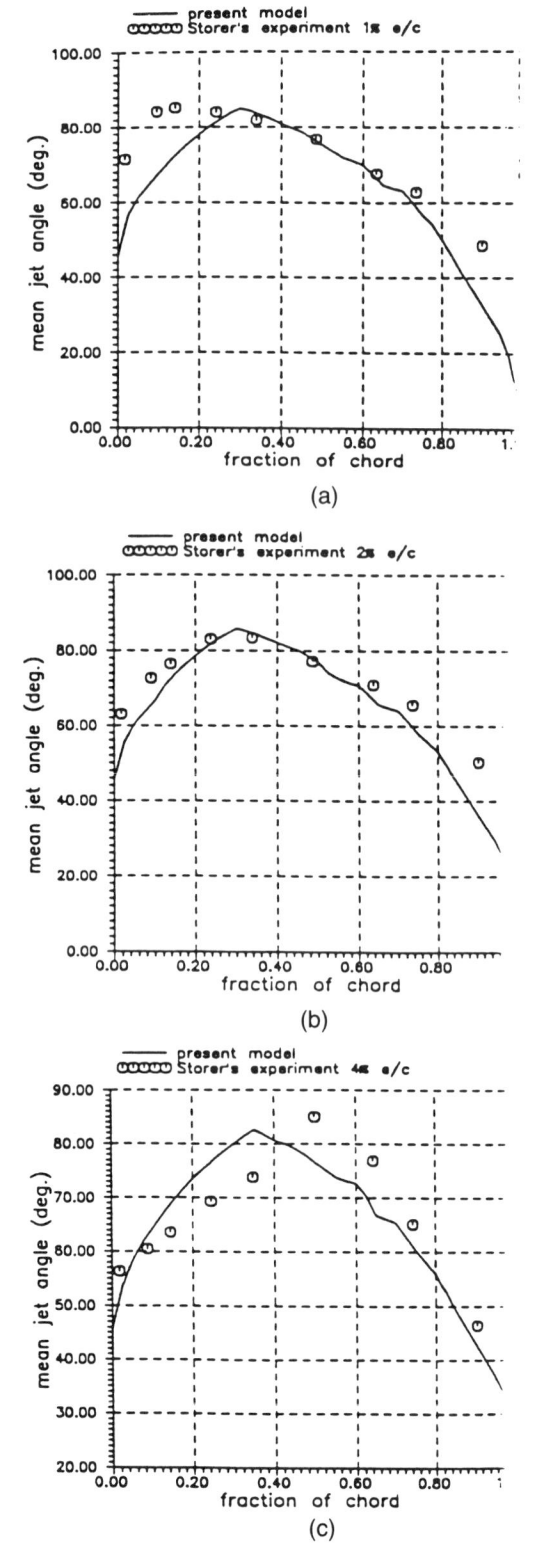

Figure 1.16 Mean jet flow angle at the gap exit for various tip clearances (Storer's case).

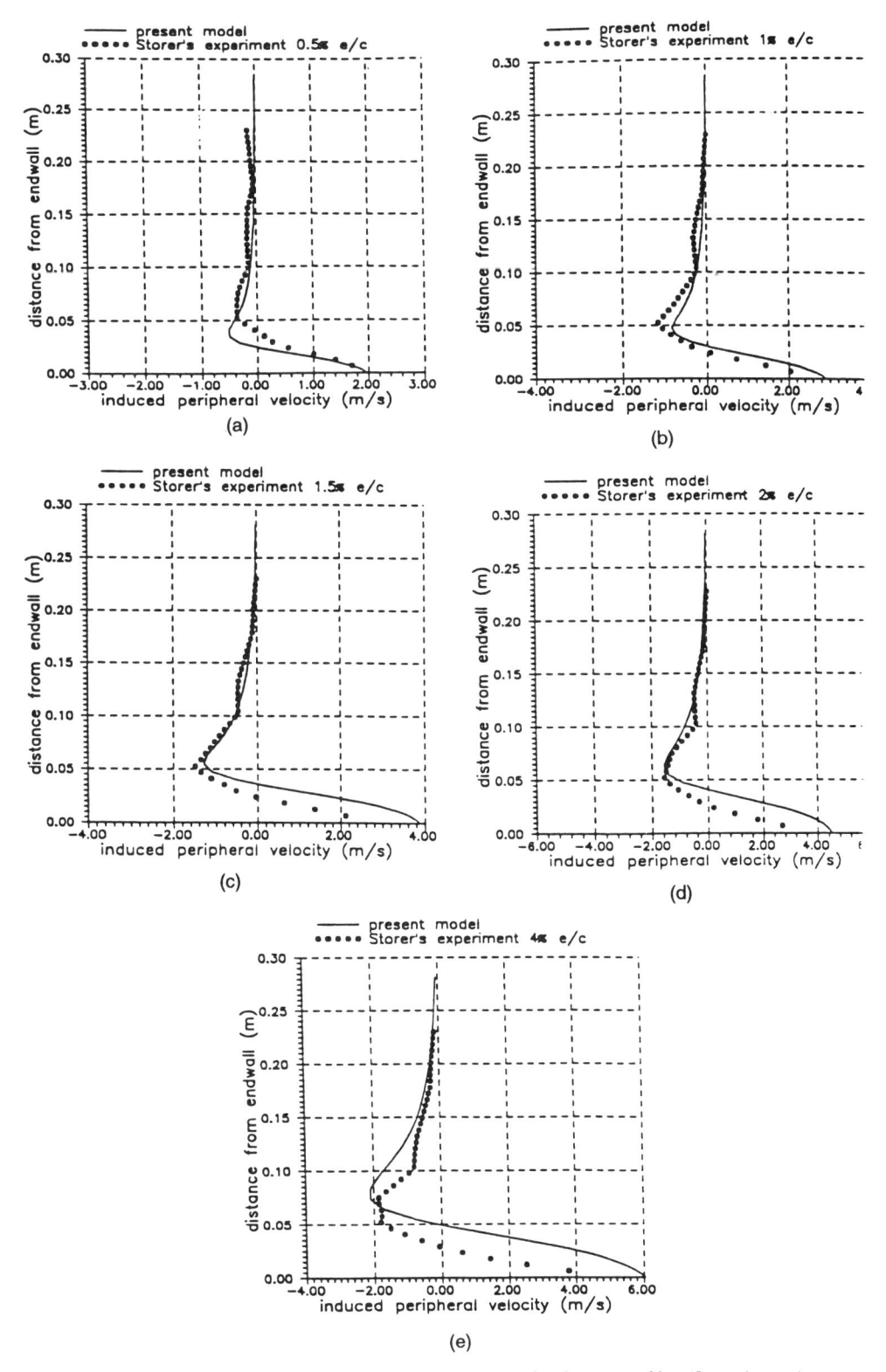

Figure 1.17 Peripherally averaged induced peripheral velocity profiles (Storer's case).

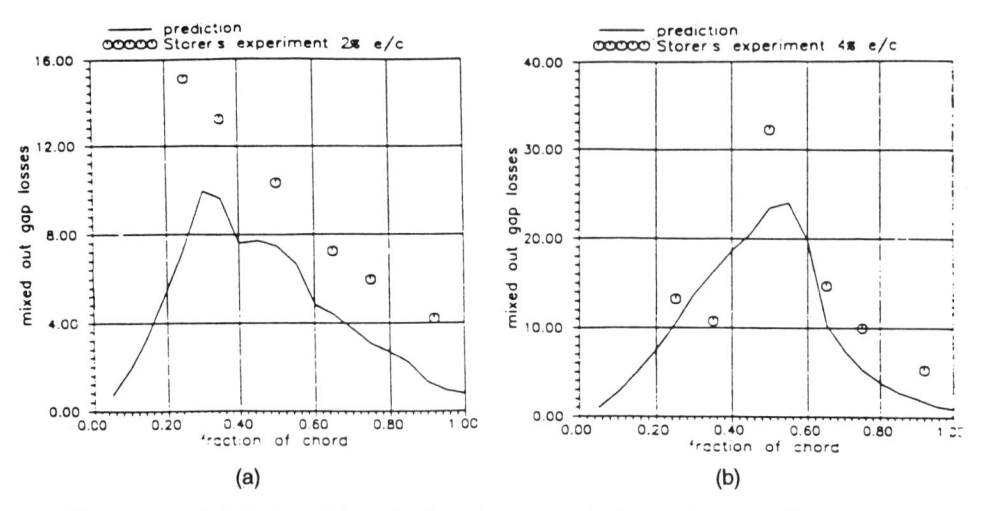

Figure 1.18 Calculation of the mixed out losses inside the tip clearance (Storer's case).

secondary flow calculation method, is presented in Figure 1.19 for a plane 28 mm downstream from the trailing edge at the tip height. In Figure 1.20 the circumferentially averaged relative peripheral velocity component, at the same plane, as calculated by the complete calculation procedure, is compared with experimental results for the case of 0.5 mm clearance. As can be seen, the disturbance caused by the leakage vortex is very small for this minimum clearance case, so that it was treated for comparison purposes as the zero clearance case, in order to calculate the leakage vortex effects for the larger clearance cases.

The circumferentially mean values of the peripheral component of the velocity, minus the one for the minimum clearance case, are presented in Figure 1.21 for the 3.0 and 5.0 mm gap heights respectively. Comparing them with the calculated values of the corresponding component of the induced

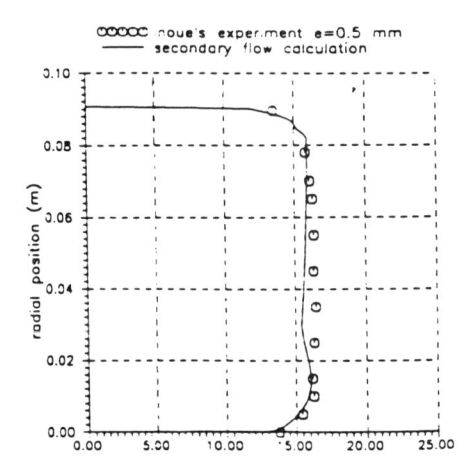

Figure 1.19 Circumferentially averaged meridional velocity profile (Inoue's rotor).

Research results

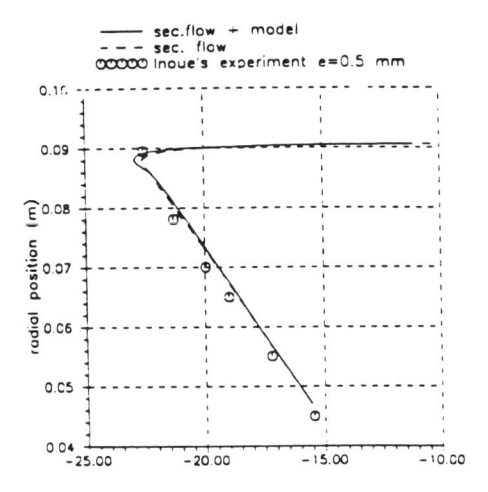

Figure 1.20 Circumferentially averaged relative peripheral velocity with-without tip clearance effects (Inoue's rotor, $e = 0.5$ mm).

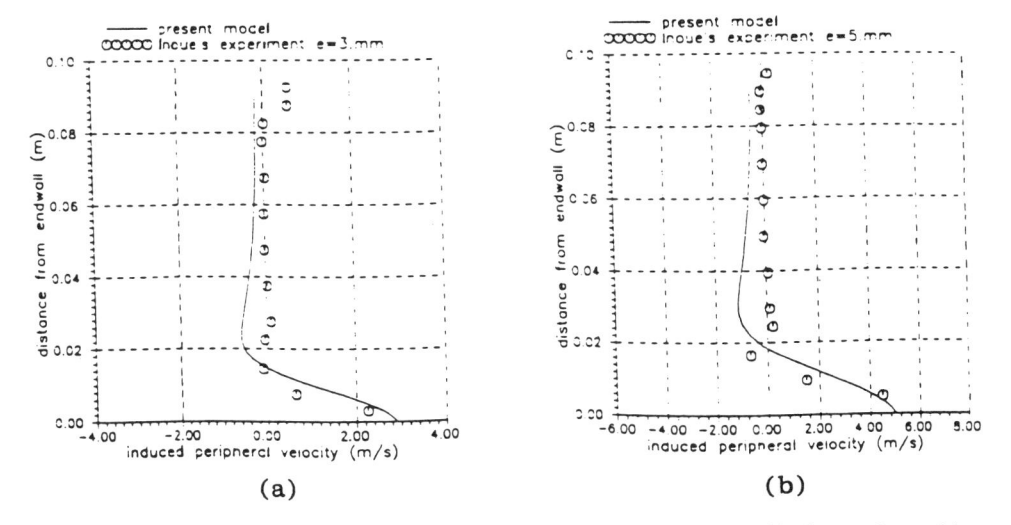

Figure 1.21 Circumferentially averaged induced peripheral velocity profile for $e = 3$ mm (a), and 5 mm (b) (Inoue's rotor).

velocity field, it can be seen that the agreement between theory and experiment is good, if the approximations involved in this comparison are considered.

In order to give indications concerning the validity of the model proposed, comparisons between the complete computational procedure (including the diffusion model) and experiment have been presented in Figure 1.22 for various tip-gap heights. The Inoue et al. [58] experimental results have been used in this case. In the same figure, where the fraction of the bound vorticity shed into the leakage vortex is presented in respect to the gap heights,

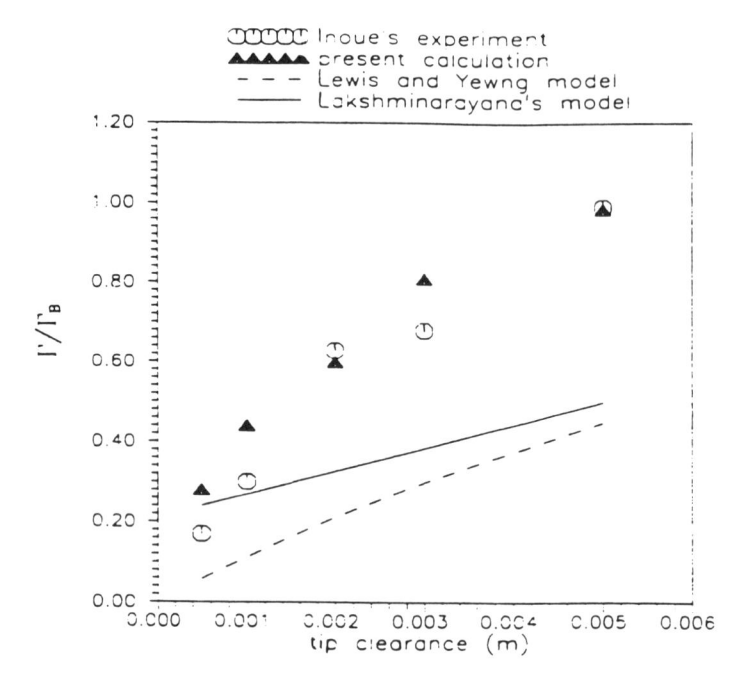

Figure 1.22 Prediction of the leakage vortex circulation for Inoue's rotor, using the present model and the models proposed by Lakshminarayana and Lewis et al.

Lakshminarayana's model [30] and that proposed by Lewis and Yeung [49] appear as well. The agreement between theory and experiment is quite good.

Figure 1.23(a,b) presents comparisons between mass averaged experimental data and peripherally mean predictions for Storer's [42] experiment. The vortex losses were obtained by subtracting the losses for the zero clearance case. Figure 1.23(c,d) presents comparisons between peripherally mean predictions and the circumferentially averaged experimental data of Inoue [58]. The presented values were obtained by subtracting the losses for the smallest clearance of 0.5 mm. The errors introduced by this assumption seem to be negligible, as the losses for the case of 0.5 mm gap are very small compared with those for the large clearances.

Again, the comparisons for both cases indicate good agreement between theory and experiment. Looking at the variation of the total pressure loss it may be also seen that the position and magnitude of the leakage vortex seems to be sufficiently well predicted.

The complete calculation procedure was used also in order to predict the tip-clearance flow effects in the Yaras et al. [44] turbine case. Three available gap heights were considered and the comparison between calculation results for the modified pressure difference at the tip level and the corresponding experimental data are presented in Figures 1.24(a–c). The present model predicts sufficiently well the modified pressure difference due to the leakage vortex except in the first part of the blading near the leading edge region. The strong curvature of the blade near the leading edge and the adopted lineariza-

Research results

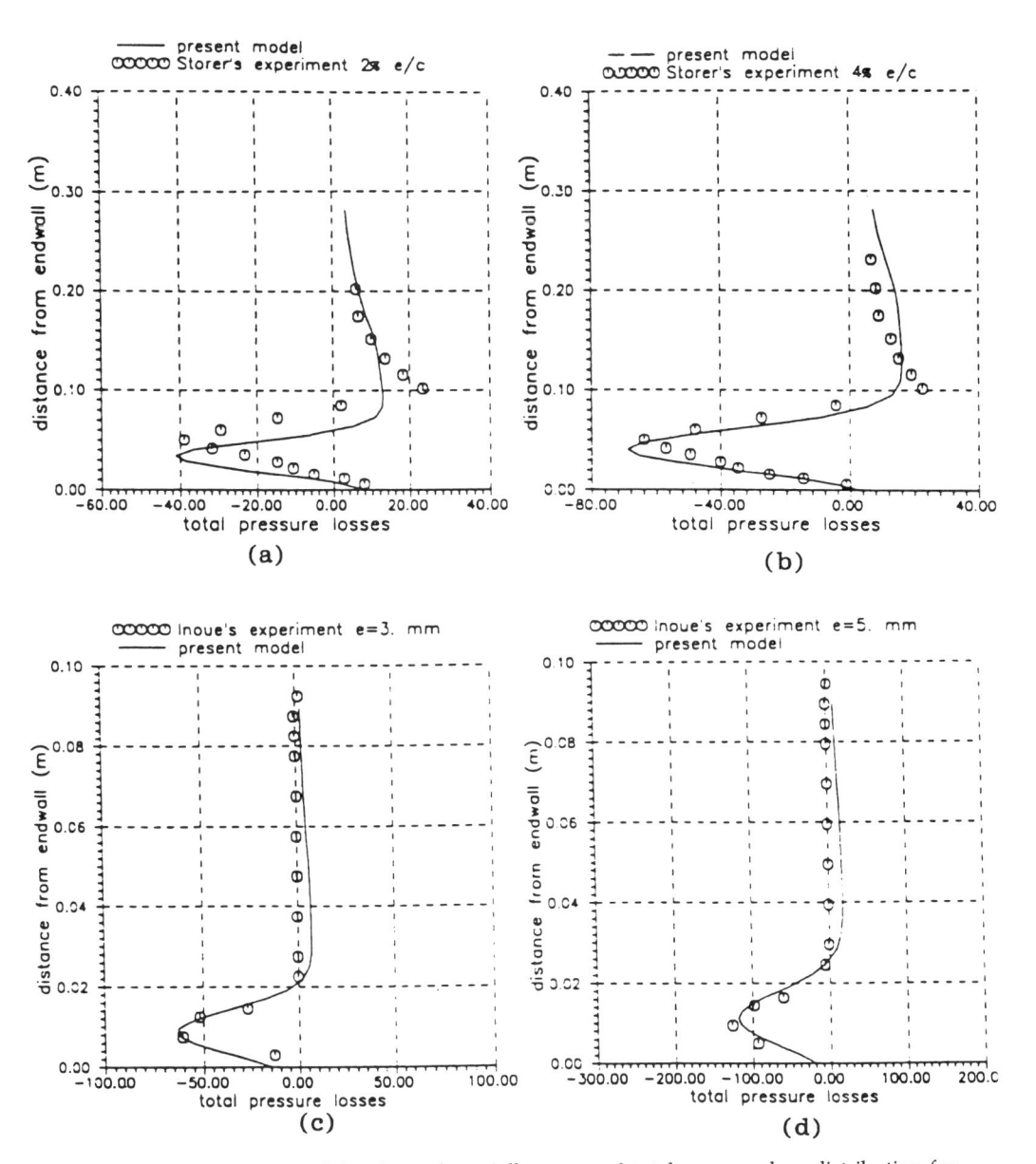

Figure 1.23 Estimation of the circumferentially averaged total pressure loss distribution for Storer's cascade (a,b) and Inoue's rotor (c,d).

tion for the pressure difference of the first 30% of the chord can partially explain the observed discrepancies. However, the calculation results of the mean jet velocity, presented in Figures 1.25(a–c), demonstrate that the corresponding discrepancies in pressure difference have a minor effect upon the calculation of the mass flow rate through the gap. The mechanism of the jet flow in the first 30% of the chord does not seem to depend on the existing

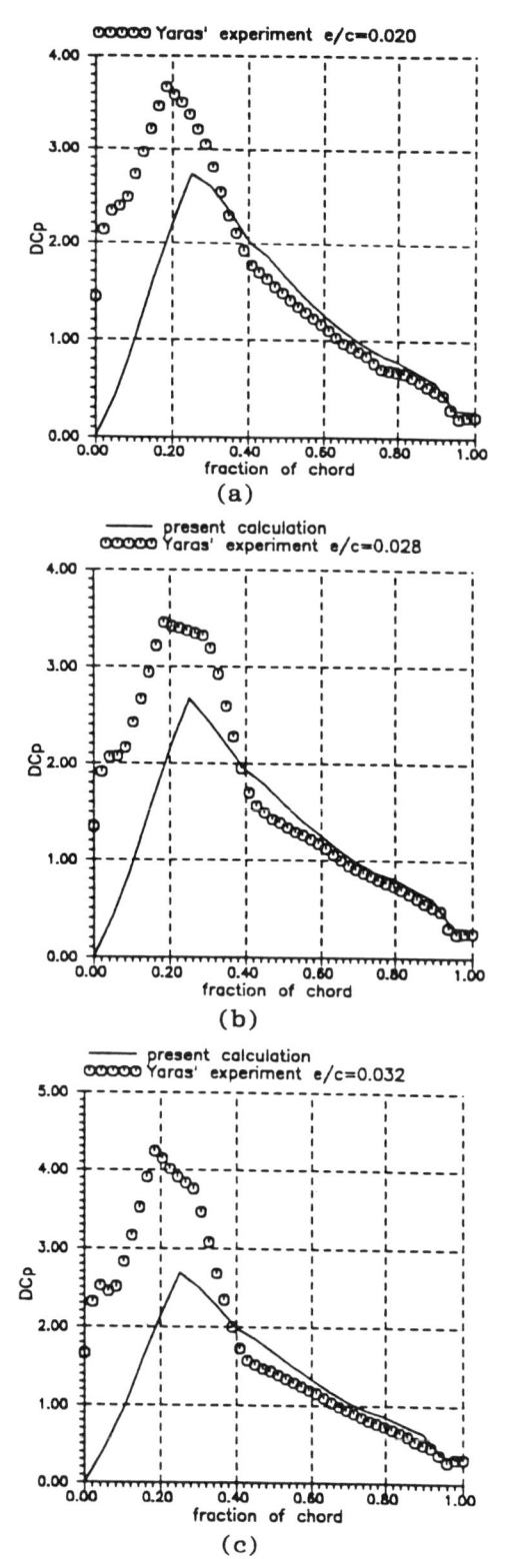

Figure 1.24 Pressure difference at the tip-clearance level for various gap heights (Yaras' turbine cascade).

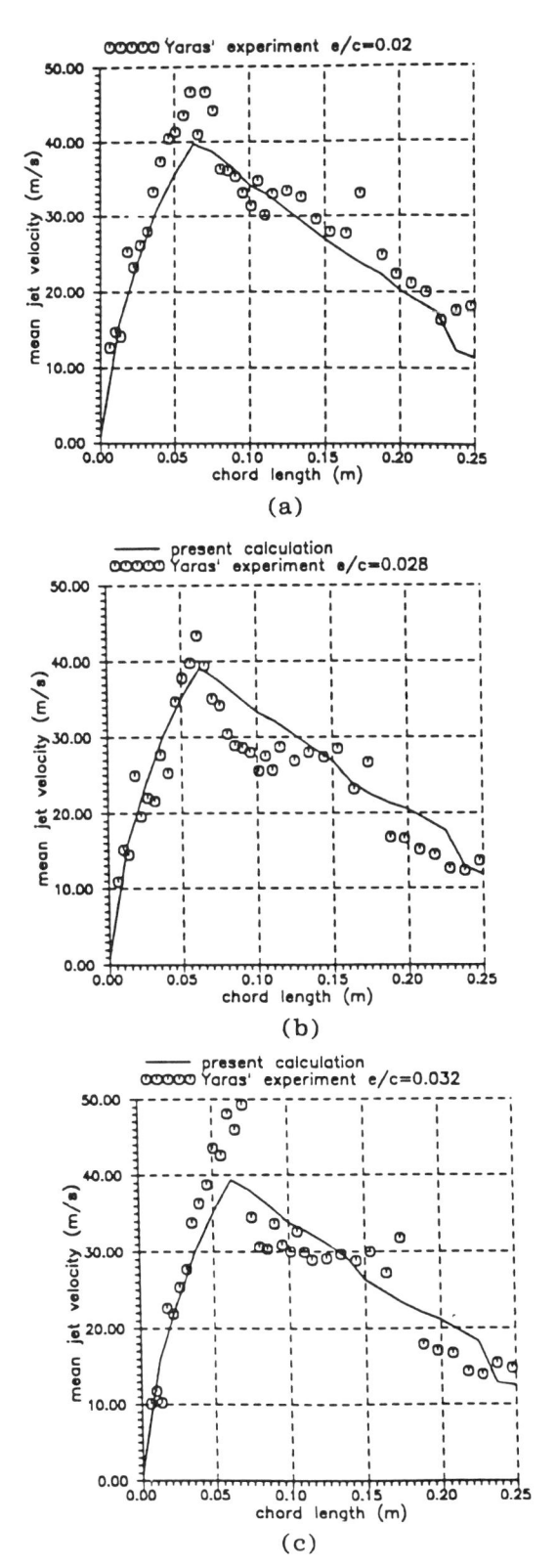

Figure 1.25 Mean jet velocity at the gap exit for various heights (Yaras' turbine cascade).

56 The clearance effects in advanced axial flow compressors

pressure difference and the adoption of a linear variation seems to describe with adequate accuracy the gap exit jet velocity distribution.

The calculation results of the complete calculation procedure for the third stage rotor of the Cranfield case are demonstrated in Figure 1.26. Considering the detail level of the experimental measurements in the tip clearance region the calculation results seem to give a good description of the circumferentially averaged flow field.

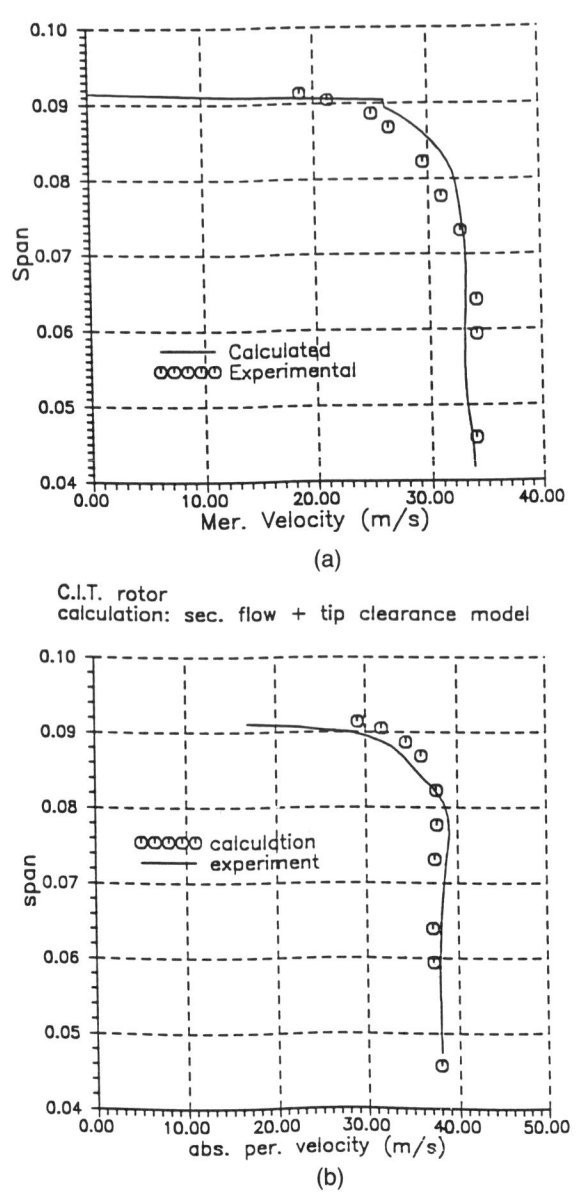

Figure 1.26 Circumferentially averaged (a) meridional and (b) absolute peripheral velocity components for the Cranfield case.

Early versions of the present model had already been provided to the industrial partners, and they were tested against several cases. The computational results were reasonable but not very satisfactory. The method of implementation of the model in the industrial partners' codes is still left open, and could explain some of the discrepances between theory and experiment, while certainly the defects of earlier versions of the code also partly contributed to the unsatisfactory predictions.

Checking the model against several test cases and taking into account the industrial partners' comments and suggestions, a fourth version of the code was developed, which was used to provide the computational results shown in the present report. This version will be provided to the partners during the next phase of the programme.

4.4 Task 4

4.4.1 Overall characteristics

Figures 1.27 and 1.28 show the characteristics for this construction of the low-speed compressor with 1.8% and 0.7% tip clearance/height ratio, respectively. The variation of compressor efficiency, pressure rise and work coefficient with flow coefficient are presented. The efficiency calculation includes an allowance for so-called tare torque, see Ivey [73]. Detailed traverse data were obtained for $Va/U = 0.63$ and 0.595 (so-called peak efficiency or design and near surge, respectively).

It is interesting to assess the effect of tip clearance on compressor efficiency and pressure rise coefficient. Reducing the tip-clearance ratio from 1.8% to 0.7% has the effect of increasing peak efficiency by approximately 0.6% and increasing the peak pressure rise coefficient from 0.43 to 0.45. Such effects are consistent with expectations. Note also that variations are shown for pre- and post-traverse. The variation of efficiency confirms the contention that resolution of efficiency with this facility is possible to within 0.3%.

4.4.2 Blade static pressure profiles

Figure 1.29 shows a schematic of the location of static pressure tappings. No single aerofoil can support all theses tappings and consequently they are located conveniently across 6 rotors and 6 stators. 'Reference' tappings are employed to ensure repeating conditions within blade rows.

Figure 1.30 and 1.31 shows static pressure profiles at hub, mid and tip for rotor and stator at 1.8% and 0.7% tip clearance, respectively. Note, the stator tip is adjacent to the rotating drum and the rotor tip is adjacent to the casing.

Figure 1.32 shows a comparison between the two constructions' different tip-clearance ratios of rotor tip static pressure distribution. Interpretation of these results is not yet complete.

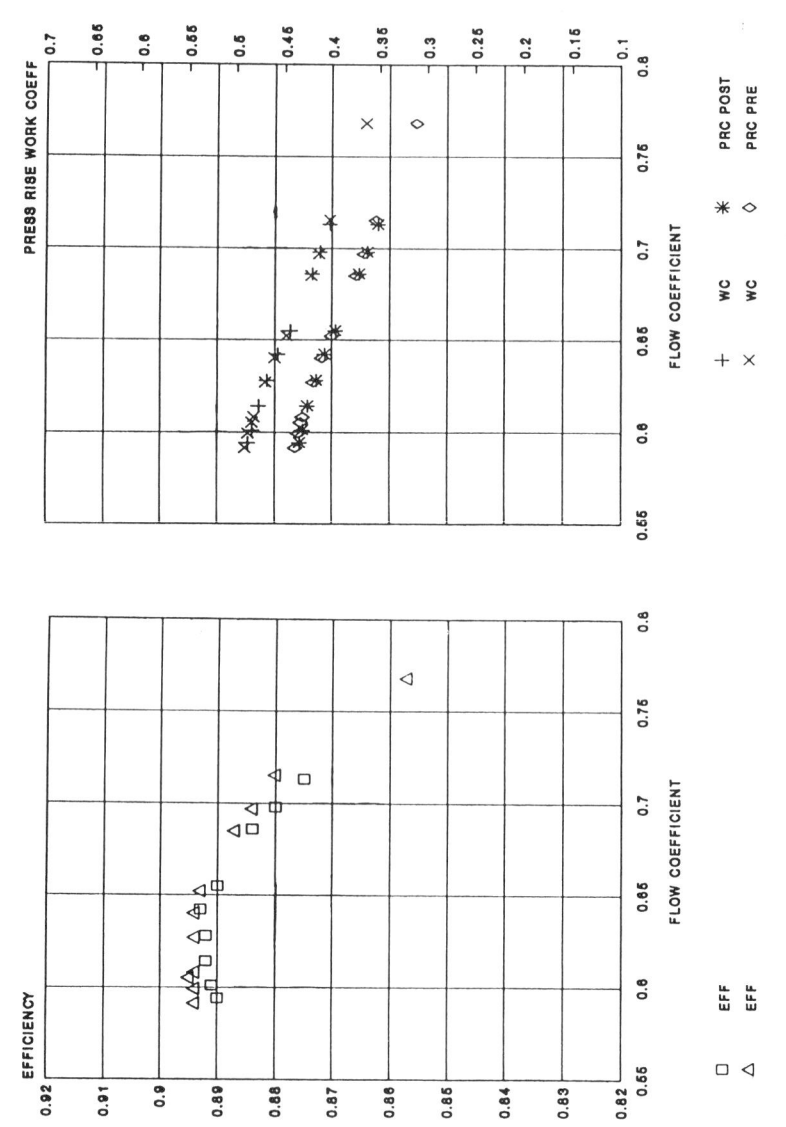

Figure 1.27 Pre- and post-traverse characteristics for 1.8% tip-clearance/height ratio.

Research results

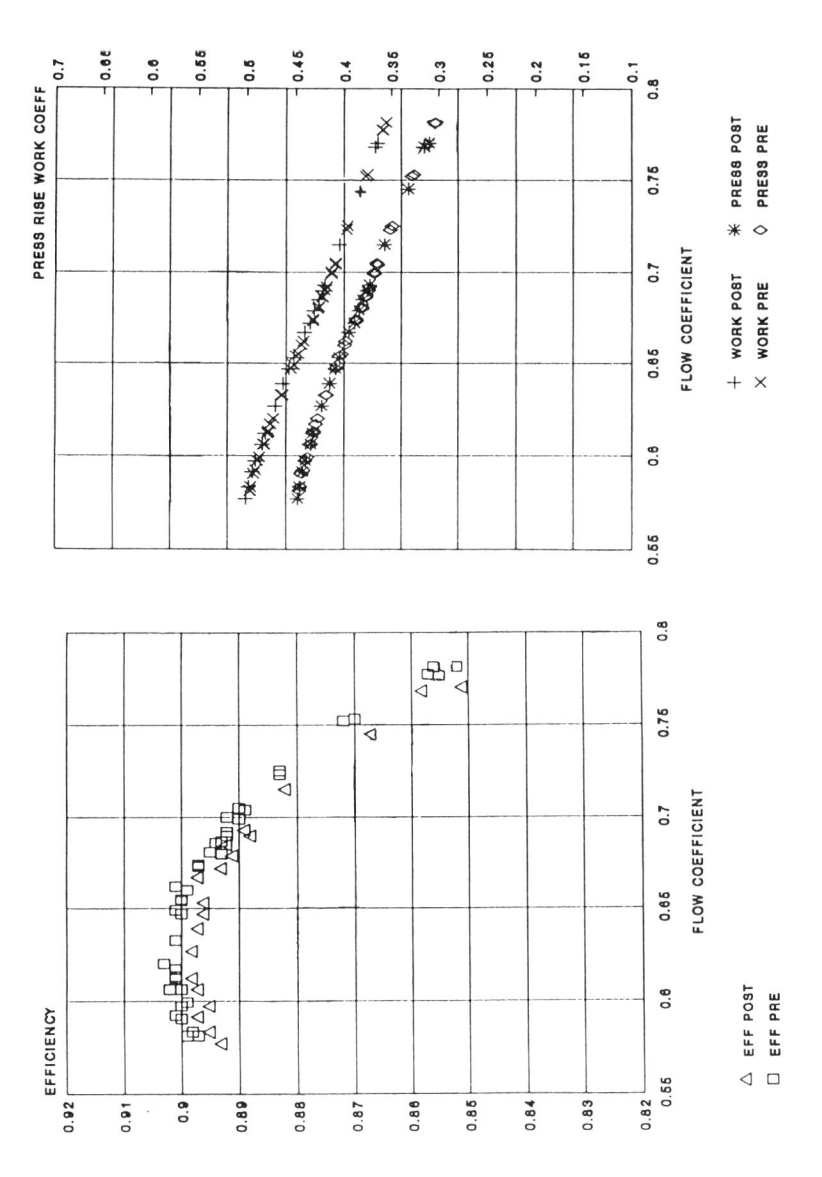

Figure 1.28 Pre- and post-traverse characteristics for 0.7% tip-clearance/height ratio (tare torque corrected).

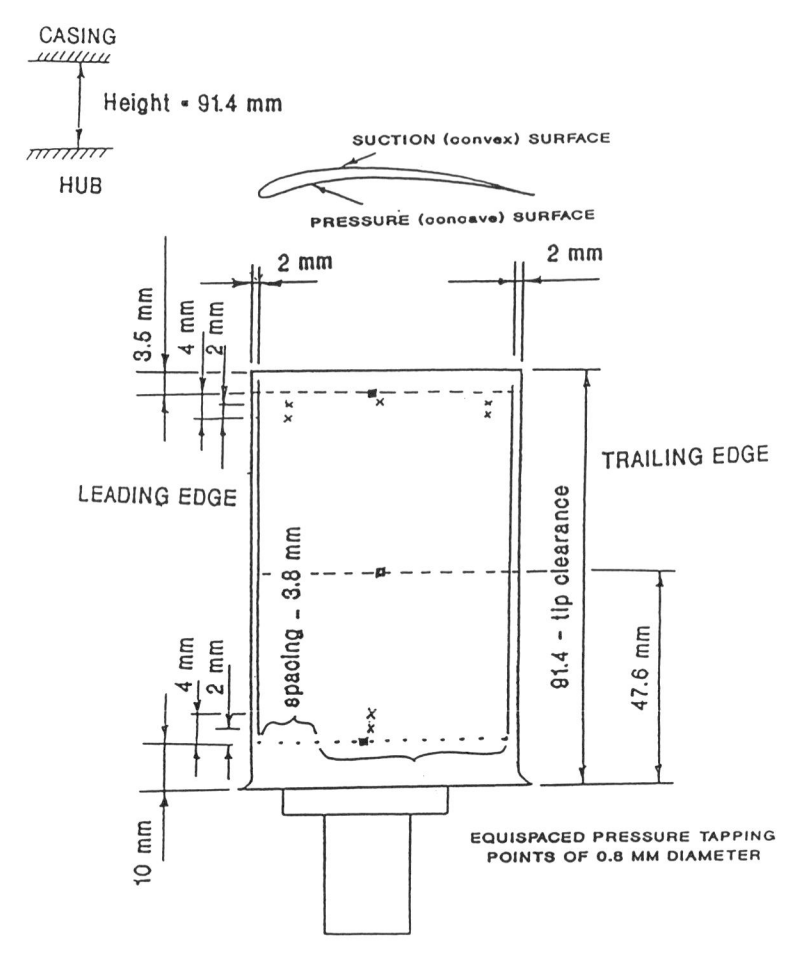

Notes :

(1) Not drawn to scale

(2) The dotted lines represent the centreline for pressure tapping points at two other heights (on separate blades)

(3) There are 14 tapping points on the suction surface and 12 tapping points on the pressure surface at each height

(4) The crosses represent locations for additional tapping points - 6 nearer the tip and 2 nearer hub heights

(5) The boxes represent reference tapping points on the surface opposite to those mentioned in note (3)

Figure 1.29 Instrumented blade.

Research results

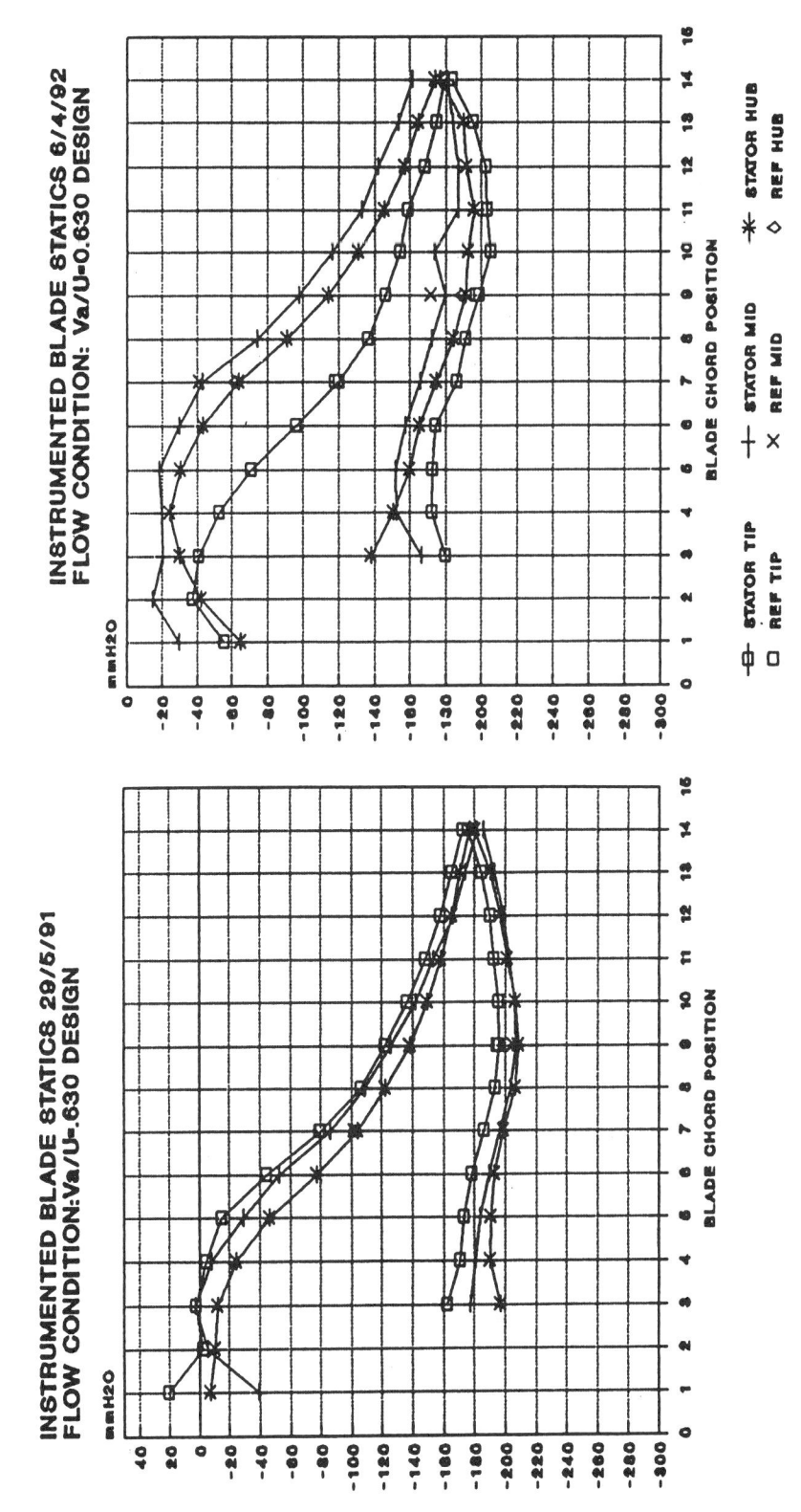

Figure 1.30 Static pressure profiles for 1.8% tip clearance/height ratio.

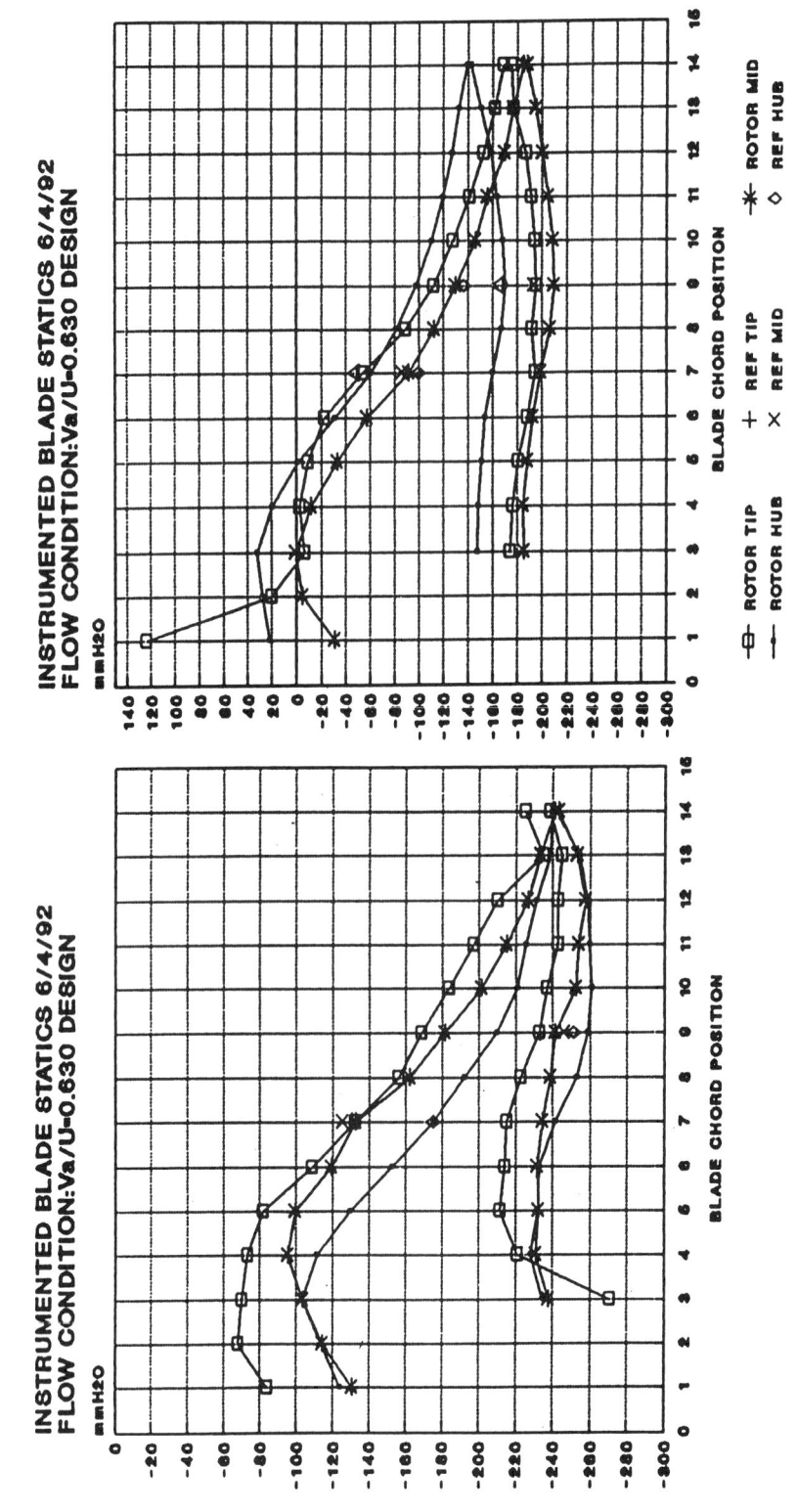

Figure 1.31 Static pressure profiles for 0.7% tip clearance/height ratio.

Research results 63

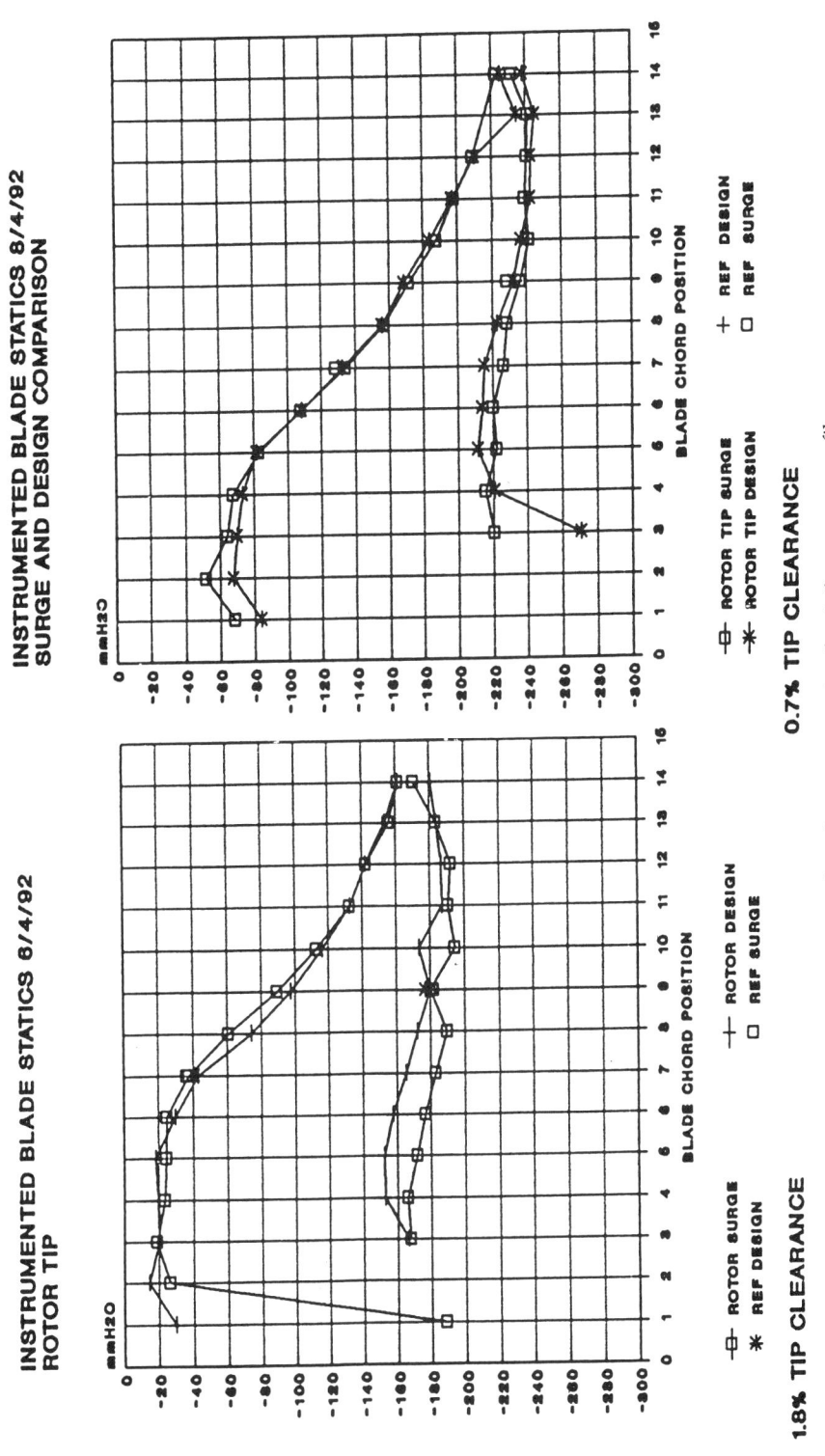

Figure 1.32 Effect of tip clearance on rotor tip static pressure profile.

4.4.3 Streamline curvature analysis

It is useful to document the responsibilities associated with the production of this analysis. Cranfield's contribution was to collect data from the compressor and to satisfy certain quality criteria. Only limited processing and analysis were conducted at Cranfield.

Traverse data was circumferentially averaged and input to an advanced streamline curvature code by Rolls-Royce. This analysis could be found in two reports [74], [75]. This details the output from this analysis together with relevant comments.

For completeness the conclusions from these reports are re-stated below.

1.8% tip-clearance/height ratio (J.P. Barton)

(1) The use of cobra probes has been demonstrated for the first time on the Cranfield rig and useful results have been obtained, particularly in the near endwall regions.

(2) Tip-clearance effects and other secondary flow phenomena have been well defined by a very detailed set of measurements at the maximum efficiency operating point.

(3) Both stage 1 and stage 3 of the four-stage rig have been traversed. There are significant differences between stages 1 and 3. This demotrates that a single stage is not likely to be representative of embedded stages of a multistage compressor.

(4) A clearance vortex has been observed, produced by the stator hub clearance flow both in stage 1 and in stage 3.

0.7% tip-clearance/height ratio (K.F. Young)

(1) The radial extent of the influence of tip-clearance flows decreases as the tip clearance is decreased.

(2) The peak level of loss near the wall becomes less as tip clearance is reduced.

(3) Increased rotor tip clearance 'spoils' the inflow to the stator in the casing region and consequently increases the stator losses there.

(4) The radial variations of loss and deviation in the near-wall region are consistent with expectations.

(5) Resolution of these near-wall effects has been possible only as a result of the high radial density of traverse locations in these regions.

Research results

It is apparent that the attempt to produce a comprehensive and accurate set of data extremely close to both hub and casing has been successful. The analysis can support a detailed interpretation of tip-clearance flow. Finally, the decision to adopt the use of cobra probes has been justified.

4.4.4 Laser anemometry

This section presents velocity measurements in the region of a tip clearance flow in a compressor having a 1.8% tip clearance/height ratio. A preliminary analysis of these measurements has been conducted and contour plots of axial velocity and rotor-relative flow angle are presented. Although interesting features of the flow are revealed, it is too early to draw definite conclusions from these measurements.

L2F measurements

Figure 1.33 shows a matrix for the measurements obtained using the L2F technique. The axial stations correspond to 4, 33, 66, 96 and 125% of chord. At each radial location up to ten circumferentially discrete measurements were obtained.

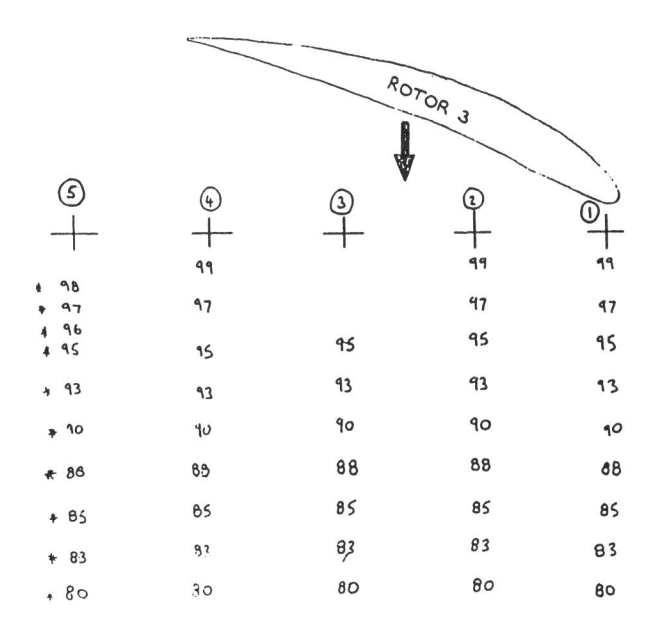

Figure 1.33 Measurement matrix for L2F anemometry.

Figure 1.34 shows velocities measured at 83% height and at the axial stations indicated above. At this height the flow appears ordered and follows the blade passage. There is no evidence of over-turning.

Figure 1.35 shows similar measurements but at 95% height. At 4% chord the flow is ordered ; however at 33% chord and greater, there is evidence of flow disturbances within the passage. (Indeed, in the mid passage region at 66% chord, measurements were not possible.)

Figure 1.36 shows velocity measurements within the tip clearance region at 99% annulus height. Tip clearance flow is clearly evident at 33% chord. Measurement difficulties at 66% were such that no data is available.

Preliminary analysis
Preliminary analysis of the results described above has been conducted. Figure 1.37 shows a contour plot of axial velocity at 33% chord. The mid-passage region appears ordered having higher velocities towards the suction side of the blade and lower velocities towards the pressure side. On either side of the rotor tip, axial velocities are slightly lower on the suction side than on the pressure side.

Figure 1.38 shows contours of rotor relative flow angle at 33% chord. As before for heights below approximatively 97% the flow is ordered. Closer to the casing there is evidence of increased underturning on the suction side.

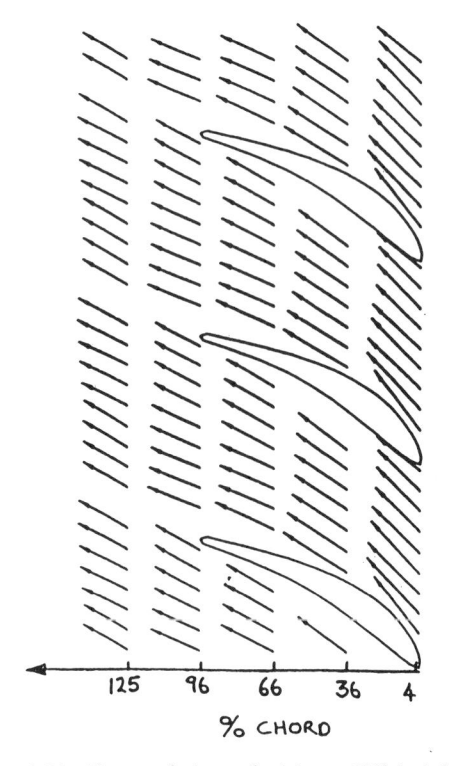

Figure 1.34 Rotor relative velocities at 83% height.

Research results

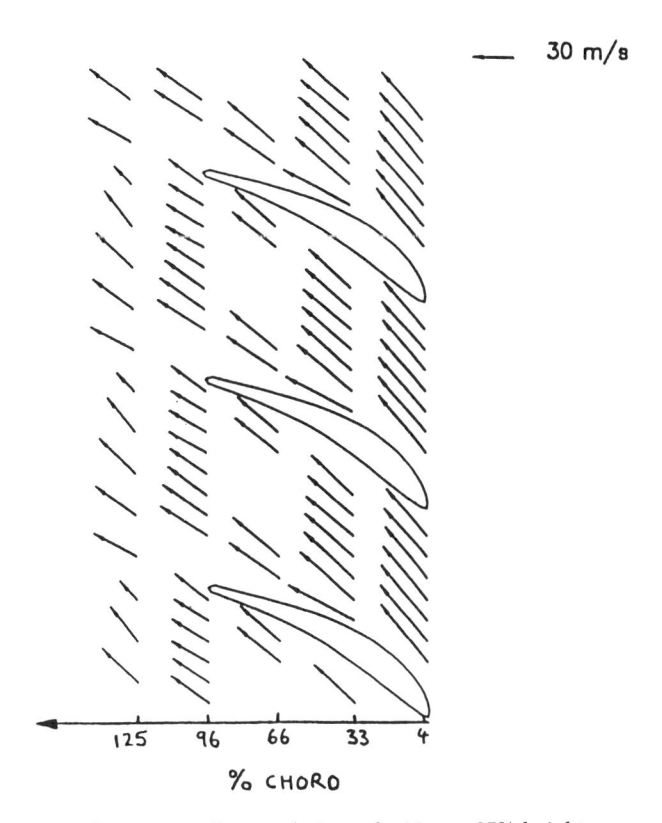

Figure 1.35 Rotor relative velocities at 95% height.

Figure 1.39 shows a contour of axial velocity at 125% chord. The location of the wake region is clear and given at a pitch location of approximately 12. Note, however, a second region of low axial velocity at approximately 95% height and a pitch location of 14–15.

Figure 1.40 shows contours of rotor-relative flow angle at 125% chord. At a pitch location of 12 and 87% height, there is a region of low tangential flow which corresponds to the low axial velocity wake region. Note that the second low axial velocity region (see Figure 1.40) shows a region of high tangential flow. It is interesting to speculate that this second region may indicate evidence of a tip-clearance flow.

4.5 Task 5

Since the model computes the deficit quantities between the non-disturbed flow (with zero clearance) and the actual flow field with the real tip clearance, and while the required flow quantities are only given at the tip height, the computed data and especially the outlet axial velocity as well as the peripheral

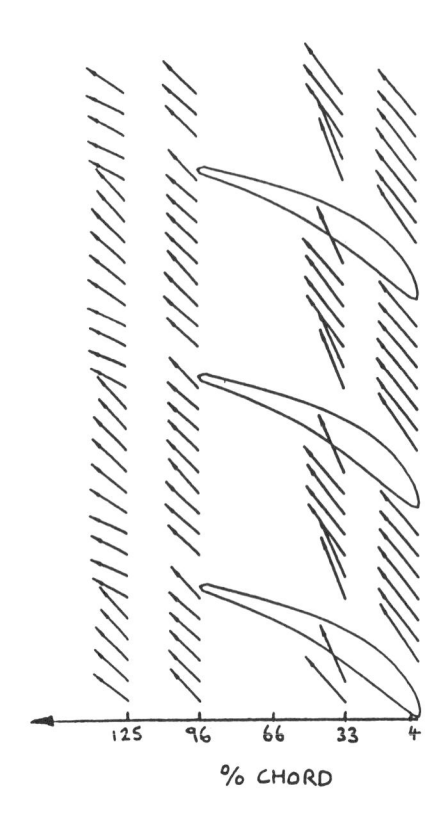

Figure 1.36 Rotor relative velocities at 99% height.

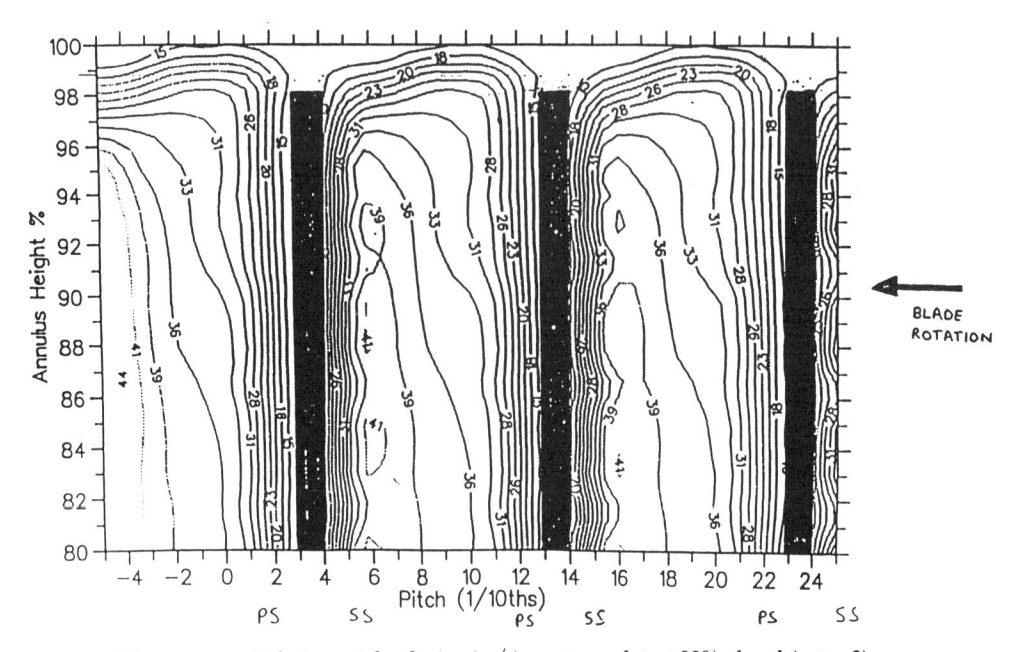

Figure 1.37 Relative axial velocity (m/s) contour plot at 33% chord (rotor 3).

Research results

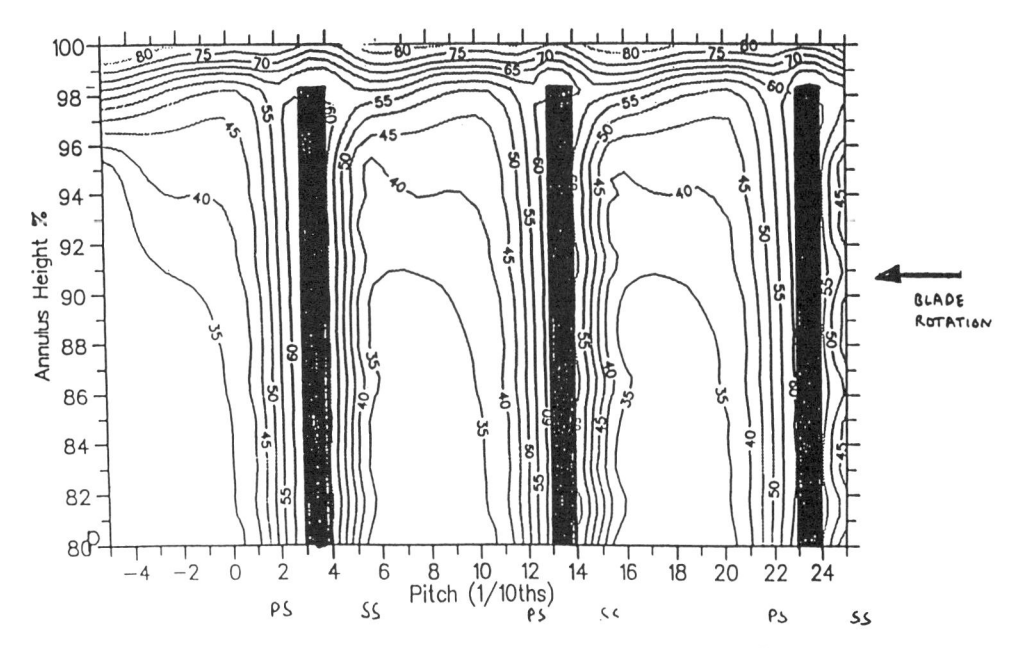

Figure 1.38 Rotor relative flow angle (degrees) at 33% chord (rotor 3).

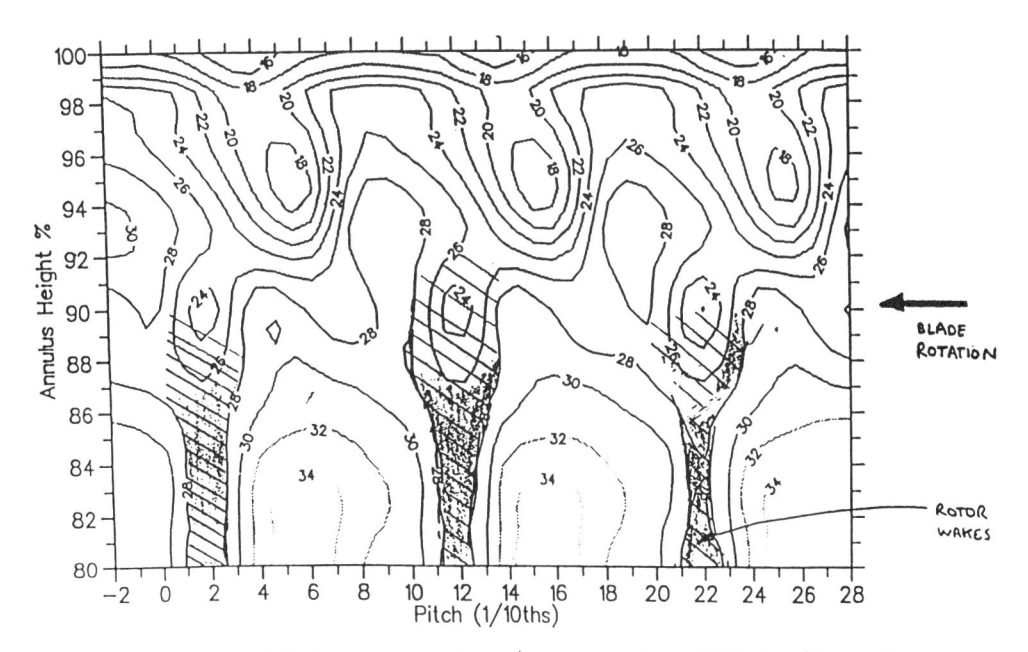

Figure 1.39 Relative axial velocity (m/s) contour plot at 125% chord (rotor 3).

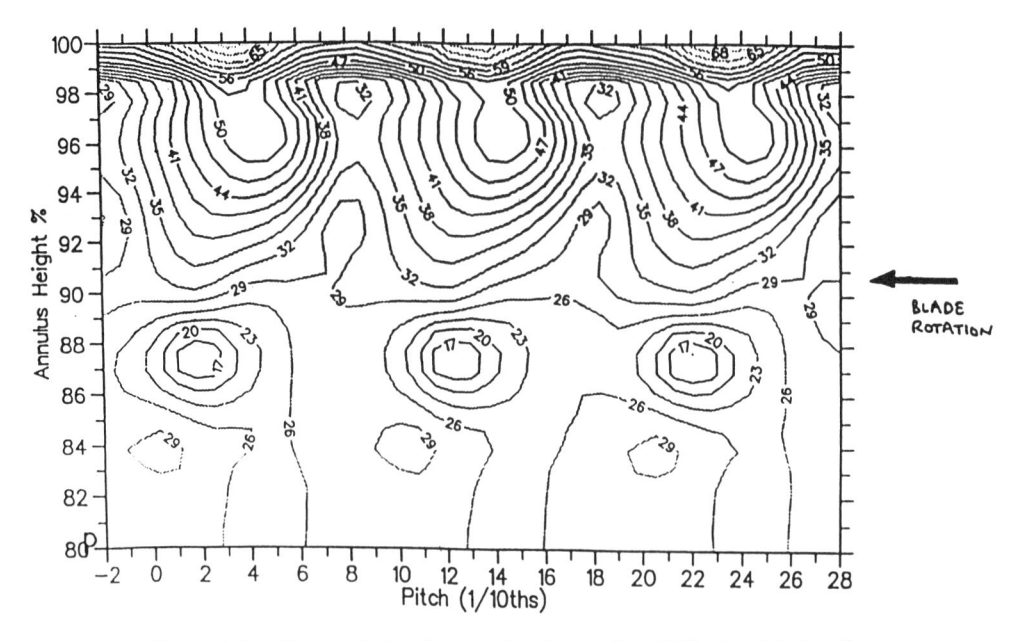

Figure 1.40 Rotor relative flow angle (degrees) at 125% chord (rotor 3).

velocity have been directly added to the meridional throughflow once to get the outlet air angle radial profile.

Several test cases of the database have been used; the meaningful case is that of the Kyushu low-speed compressor, since three tip clearance sizes have been studied. Figures 1.41–1.43 show the corresponding results for the largest tip size respectively in Snecma, Rolls-Royce and Turbomeca environmental software. Results with the other database test cases are also achieved [8-RR-1, 10-RR-2, 10-SN-6].

The validation of the present model has been also achieved against experimental measurements on the Cranfield Low Speed Research Compressor obtained during task 4 completion. This has been done in NTUA environmental software, see Figure 1.26.

4.6 Task 6

4.6.1 Linear compressor cascade of storer

The tip clearance flow is studied in a linear cascade, for which measurements exist in the literature (Storer and Cumpsty [66]). The geometry of the cascade and the kinematic data are listed below:

Chord	300.0 (mm)
Pitch	180.0 (mm)
Span	435.0 (mm)

Research results

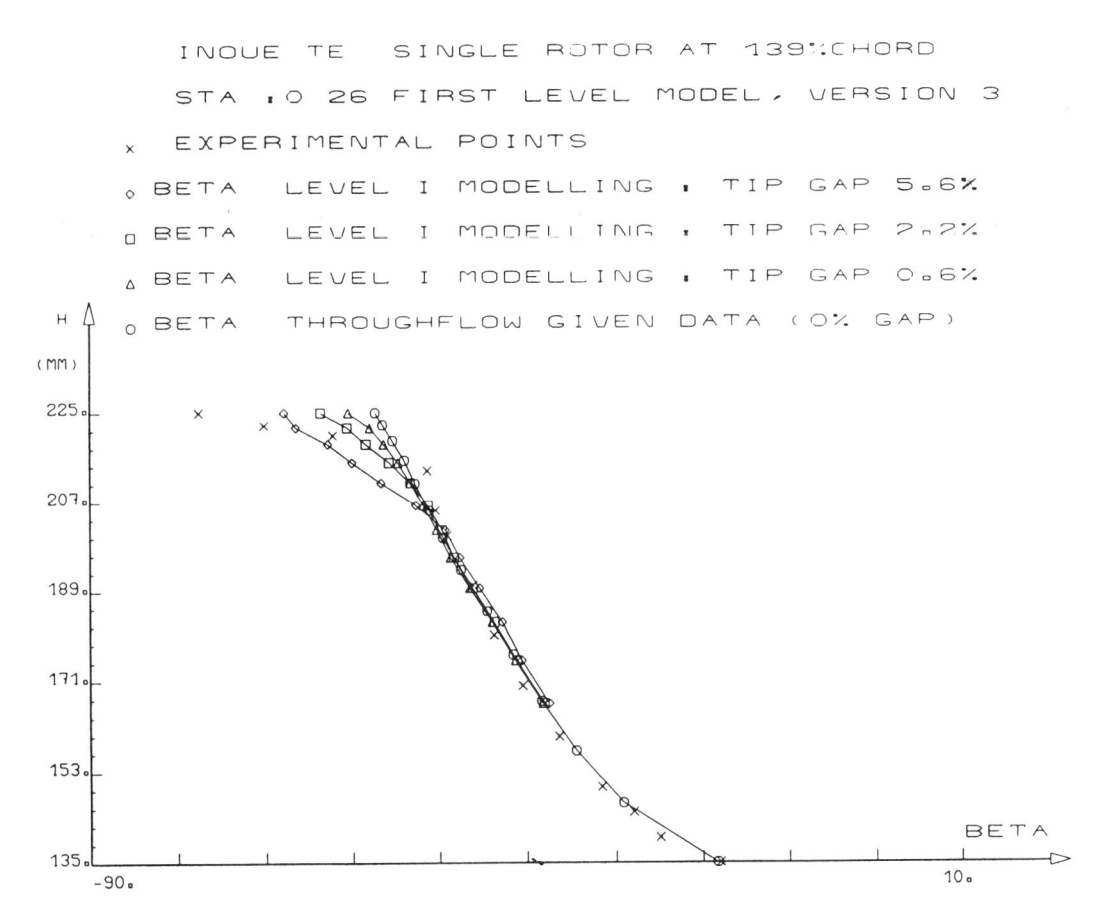

```
INOUE TE  SINGLE ROTOR AT 139% CHORD
STA : 0 26 FIRST LEVEL MODEL, VERSION 3
 x    EXPERIMENTAL POINTS
 ◊    BETA   LEVEL I MODELLING , TIP GAP 5.6%
 □    BETA   LEVEL I MODELLING , TIP GAP 2.2%
 △    BETA   LEVEL I MODELLING , TIP GAP 0.6%
 ○    BETA   THROUGHFLOW GIVEN DATA (0% GAP)
```

Figure 1.41 Present model results on Kyushu's rotor obtained in Snecma environmental software.

Maximum thickness/chord	0.05
Camber (circular arc)	45.5°
Stagger	22.2°
Inlet flow angle	45°
Inlet Mach No.	0.07
Reynolds No. based on chord	5×10^5

Inlet end wall boundary layer 70% upstream 2.66% of span

Three different cases were studied, namely :

Case 1.1 tip clearance/chord = 0, grid size $25 \times 31 \times 50$
Case 1.2 tip clearance/chord = 0.02, grids size $25 \times 40 \times 50$, $16 \times 14 \times 26$
Case 1.3 tip clearance/chord = 0.04, grids size $25 \times 34 \times 50$, $10 \times 8 \times 26$

It should be noted here that in the experiment, inlet boundary layer displacement thickness is quoted at 140% chord upstream and was equal to 2.9 mm.

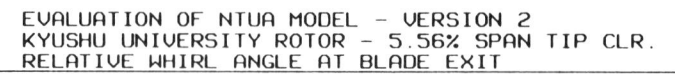

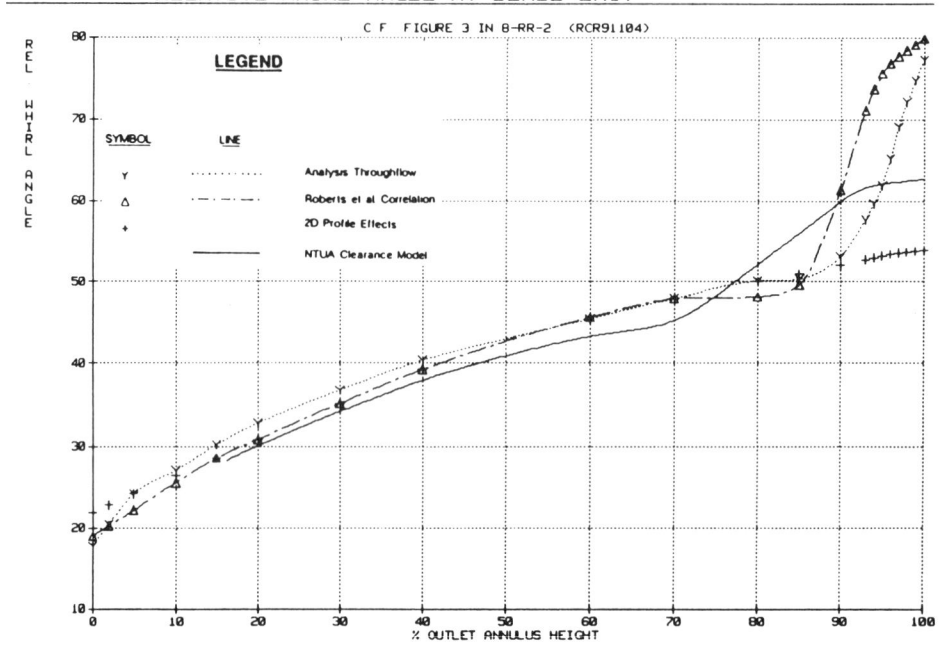

72   The clearance effects in advanced axial flow compressors

Figure 1.42 Present model results on Kyushu's rotor obtained in Rolls-Royce environmental software.

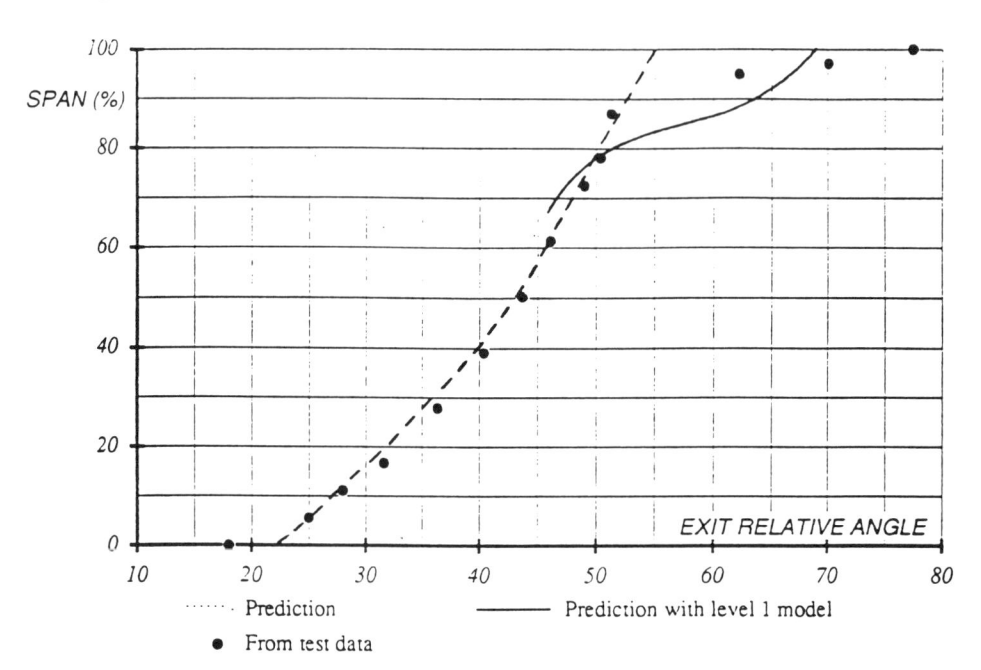

Figure 1.43 Present model results on Kyushu's rotor obtained in Turbomeca environmental software.

The calculation inlet boundary was not placed that far upstream for economy reasons.

Figure 1.44(a) shows the basic grid which was stacked in order to produce the three-dimensional grid. Figure 1.44(b) shows the hub, suction side and pressure side grids together with the magnification of the grid used to fill the tip-clearance region. The predicted values of c_p are compared with experimental measurements in Figure 1.45. Figure 1.46 compares the predicted and measured values of c_p on the shroud for 0 and 4% of chord clearance. In contrast to the other test cases examined within this project, it can be seen that the predicted distribution of c_p is different from the measured one. LTT/NTUA has contracted the authors of this work in order to check that the geometry and kinematic data used in the calculation agreed with those of the experiment. Our efforts, however, did not lead to a reduction of this difference in c_p, which is quite critical with respect to tip-clearance effects, since the pressure difference between the suction and pressure sides of the blade is the driving force of the leakage flow. Consequently, andy underestimation of this pressure difference will lead to a reduced leakage flow, something which was observed in our case; qualitatively the main characteristics of the leakage flow have been captured as can be seen in Fig. 1.46.

4.6.2 Linear compressor cascade of Flot

The flow in a compressor linear cascade, for which experimental investigations were performed by Flot (Leboeuf and Flot [67], Flot [68]), in the wind tunnel of the Ecole Centrale de Lyon, were studied for a zero and a non-zero tip clearance. The profile of the blade was a NACA-65-12-A10-10, with a chord length $c = 0.13$ m and a solidity equal to 1/0.8. The stagger angle was 30° and the span/chord ratio was equal to 2.1. The two cases examined correspond to tip clearances equal to:

Case 2.1   zero, grid 25 × 31 × 48
Case 2.2   1.1% of the chord length, grids 25 × 32 × 48, 10 × 8 × 24

The grid was constructed by stacking a two-dimensional H-type grid. Kinematic and thermodynamic data used in the calculation are listed below:

(a) the inlet stagnation conditions were

        total pressure   101.325 Pa
        total temperature   288.1 K

(b) precise velocity profiles, in both the axial and the 'peripheral' directions, were given in the original references. These were used as inlet conditions, determining the value of the incoming mass flow rate. Concerning inlet flow angle, it is to be said that the above-mentioned velocity profiles introduce a swirling flow corresponding to a mean inlet velocity angle

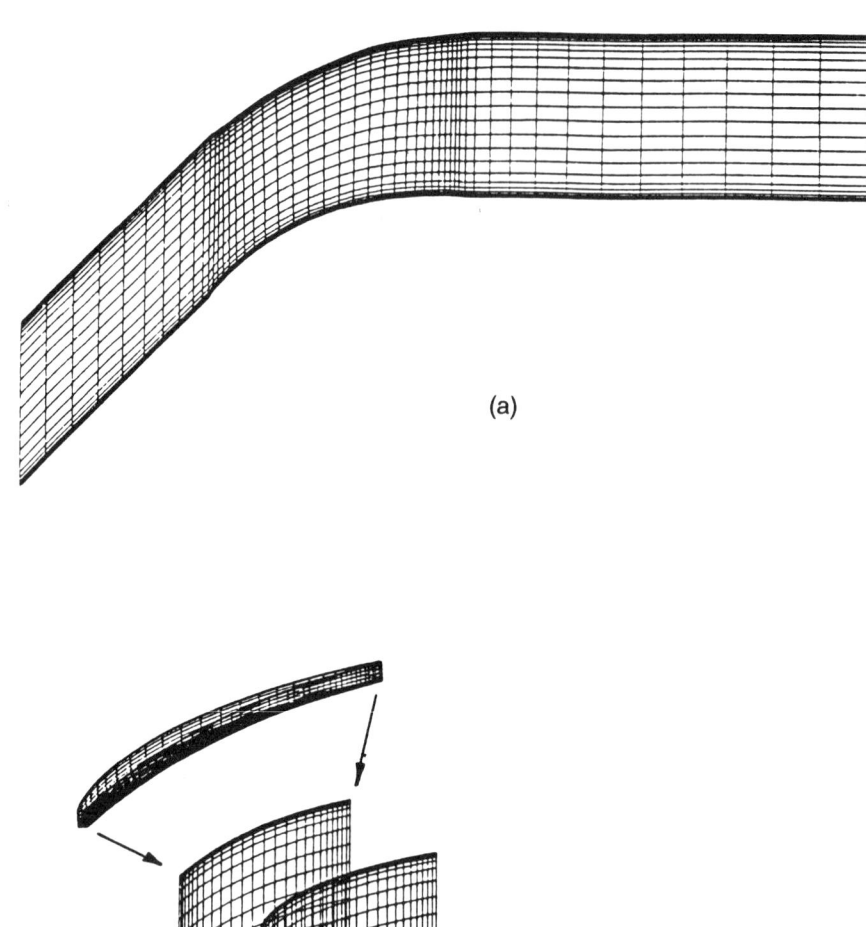

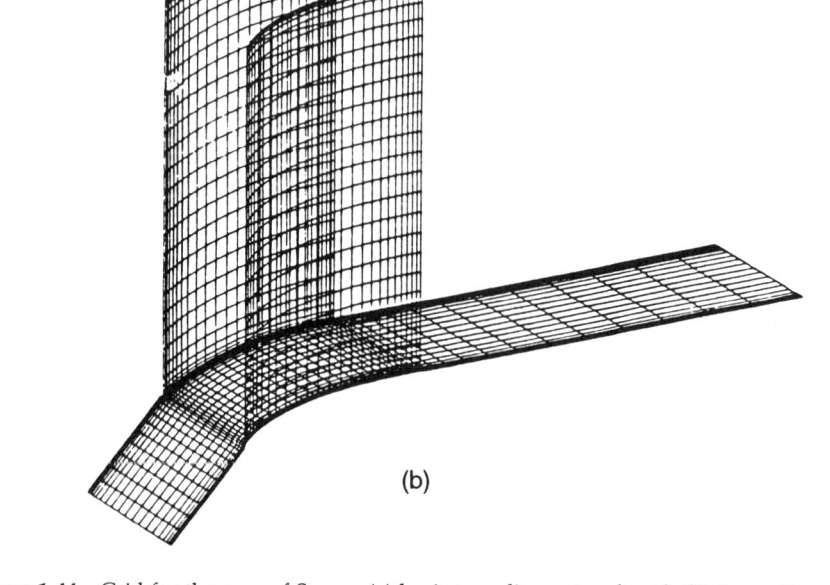

Figure 1.44 Grid for the case of Storer: (a) basic two-dimensional grid; (b) three-dimensional grid.

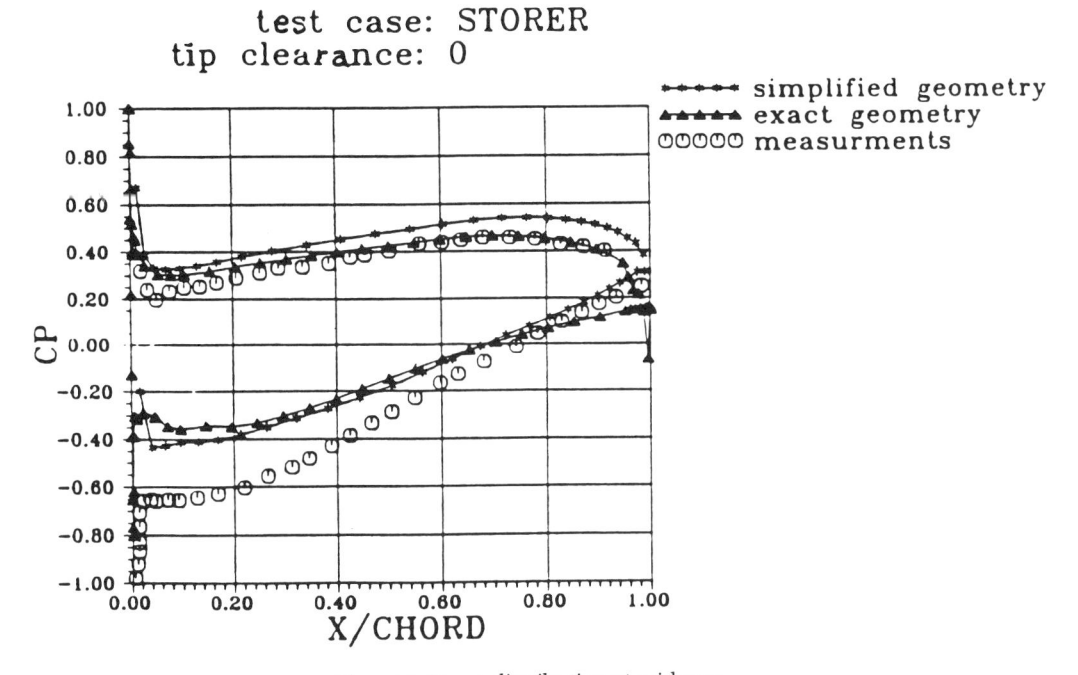

Figure 1.45 c_p distribution at midspan.

$\beta_1 = 56.2°$. The inlet velocity profiles were symmetric with respect to the mid-span plane.

(c) the inlet turbulence intensity was set equal to $T_u = 0.5\%$

Figures 1.47(a) and (b) show the basic grid which was stacked in order to produce the three-dimensional grid and the 3-D grid itself respectively. The distribution of velocity at midspan is shown in Figure 1.48. Figure 1.49(a) shows the measured velocity vectors at the mid-span and at a plane lying at a distance of 0.5 mm from the shroud. The latter field could be identified as those vectors that deviate from the 'throughflow' direction and generally have smaller magnitudes. These could be directly compared to Figures 1.49(b) and 1.49(c) showing predicted velocity vectors at the corresponding blade-to-blade planes. A comparison between Figures 1.49(a) and 1.49(c) demonstrates that the accurate modelling of the blade tip geometry has allowed the reproduction of the characteristic patterns of clearance flows. Firstly, strong secondary flow effects can be observed in Figure 1.49(c), where flow is directed from the pressure to the suction side. This flow strongly interacts with the jet emerging from the tip clearance near the leading edge at the suction side; this jet results from the large pressure difference between the pressure and suction sides of the blade, close to the leading edge. The 'collision' of the secondary flow with the leakage jet forms a stagnation line that starts at the leading edge and develops downstream halfway between the pressure and suction sides. Upon

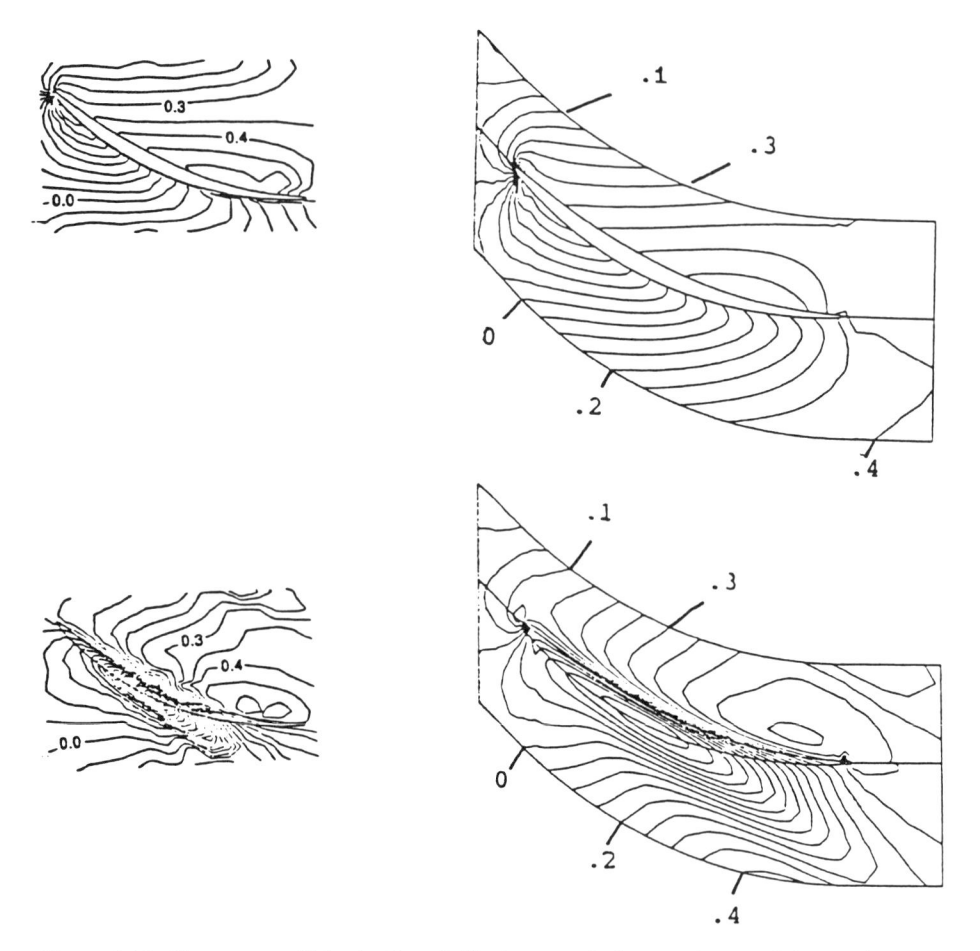

Figure 1.46 Pressure coefficient at 0 and 4% clearance Left: experiment, right: prediction.

exiting the blade section, this line curves and forms a spiral, near the suction side.

Figure 1.50 presents the pressure coefficient contours on the shroud for the cases of no-tip clearance (Figure 1.51(a)) and $\tau = 0.011$ C (Figure 1.50(b)). This dimensionless pressure coefficient is defined as the pressure rise from the inlet static pressure divided by the dynamic pressure corresponding to the mean inlet velocity. In Figure 1.50(b), a low pressure through may be observed at approximately 10% of the chord from the leading edge. This through corresponds to the generation of the tip leakage vortex and has been observed by other researchers as well (for example Kang and Hirsch [78]).

4.6.3 Axial compressor rotor of Inoue

The tip clearance flow is studied in a low speed axial compressor rotor, for which measurements exist in literature (Inoue *et al.* [69], [70]). A 12 blade

Research results

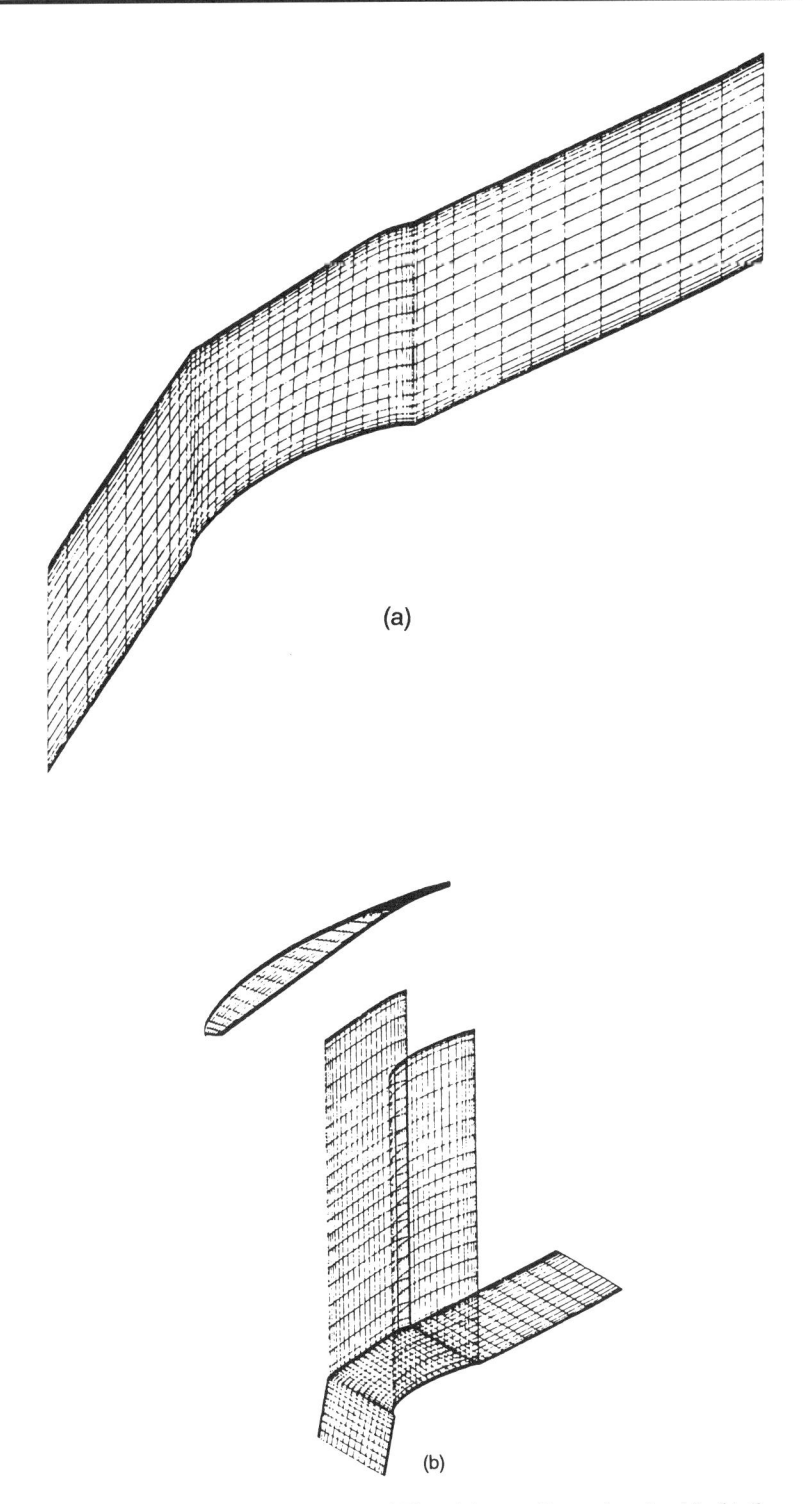

Figure 1.47 Calculation grid for the case of Flot: (a) two-dimensional grid; (b) three-dimensional grid.

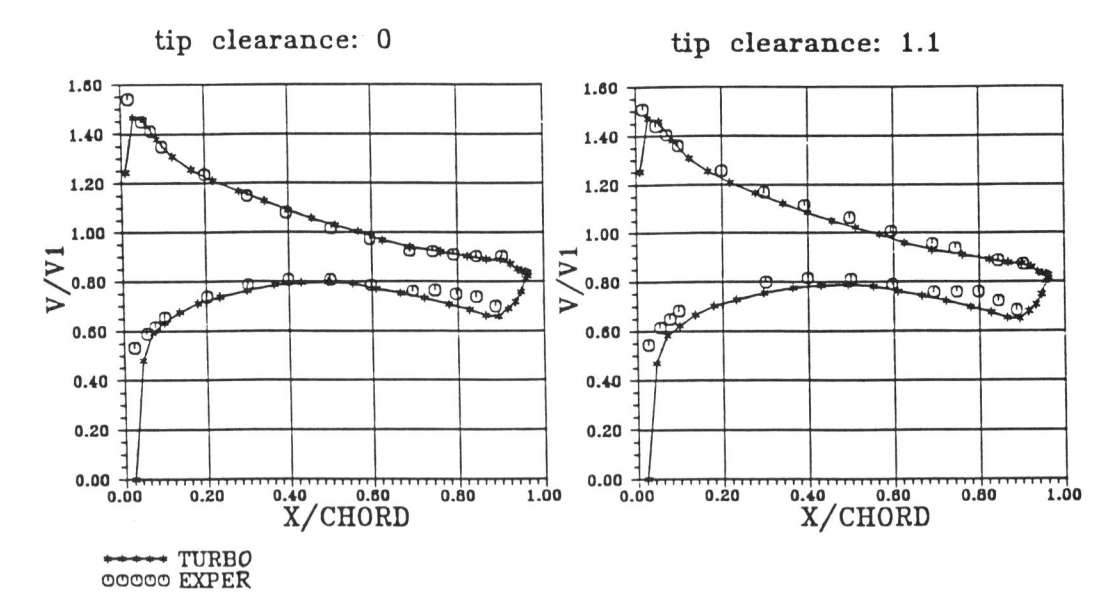

Figure 1.48 Midspan velocity distribution, Flot case.

isolated rotor, of 449 mm casing diameter and a hub-to-tip-ratio equal to 0.6, is used. The rotor blade was built from the information found in the above references, where geometrical data were given for three radial positions, namely at the hub, at midspan and at the tip. The geometrical data and the cases examined are summarized below:

(a) geometrical data for three radial positions

	Root	Mid-span	Tip
Radius	135.0	180.0	224.5 (mm)
Chord length	106.1	117.8	117.5 (mm)
Camber (c_{li})	1.30	0.62	0.32
Solidity	1.50	1.25	1.00
Stagger	31.5	47.2	56.2 (deg)
Max.thickness/chord	0.10	0.08	0.06

(b) the camber line was determined from the value of c_{li}

(c) cross-sections of this blade with cylindrical surfaces of any radius create aerofoils belonging to the NACA65 series. For example, the blade geometry at the tip was NACA65

(d) a fourth-order interpolation scheme was used to calculate the radial distribution of the above quantities

(e) the size of the rotor was kept constant and the various tip-clearance sizes

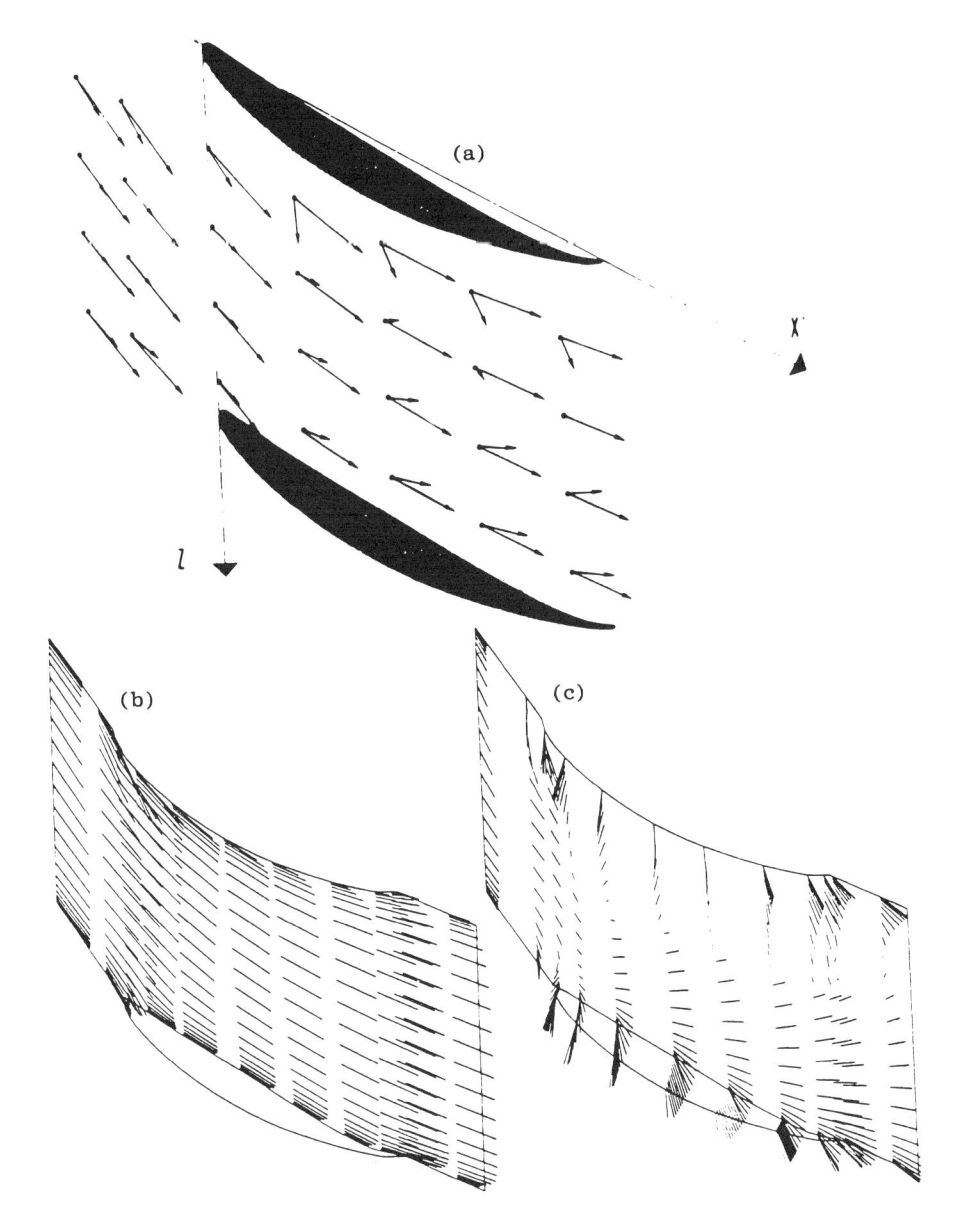

Figure 1.49 Velocity vectors at midspan and at plane lying between the blade tip and the shroud, Flot case: (a) measured velocity vectors at midspan and 0.5 mm from shroud; (b) predicted velocity vectors at midspan; (c) predicted velocity vectors 0.5 mm from shroud.

are studied by changing the diameter of the casing wall. Three different cases were studied, namely:

Case 3.1: tip clearance = 0.5 mm, grids 25 × 34 × 58, 12 × 10 × 25
Case 3.2: tip clearance = 3.0 mm, grids 25 × 39 × 58, 12 × 15 × 25
Case 3.3: tip clearance = 5.0 mm, grids 25 × 39 × 58, 12 × 15 × 25

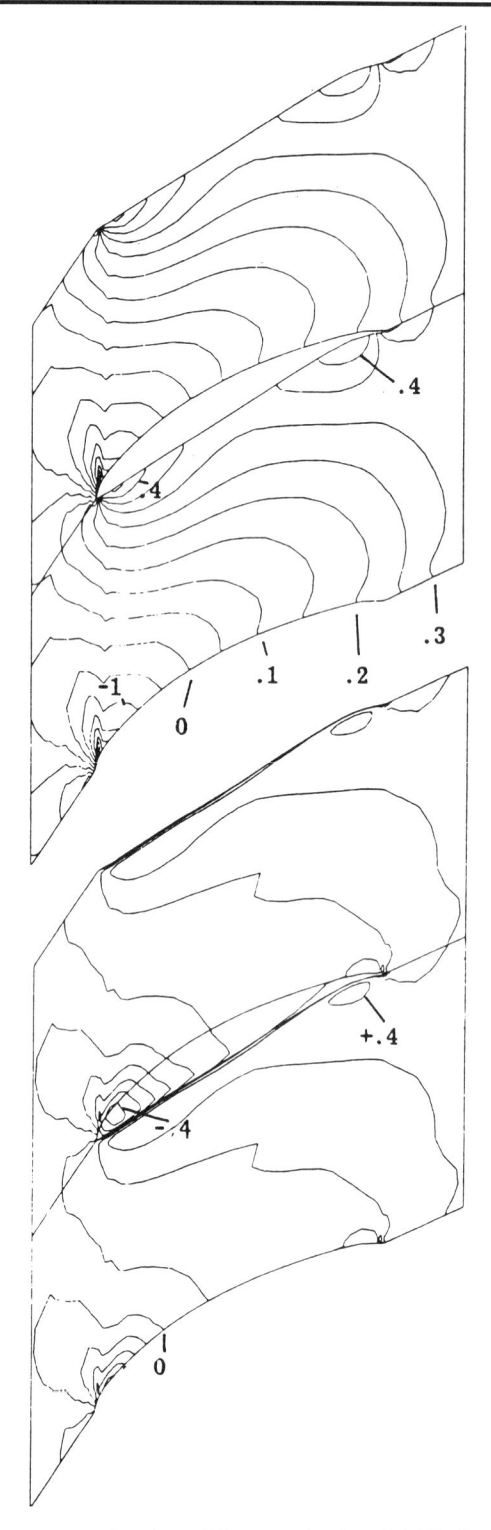

Figure 1.50 c_p contours on the shroud for zero (top) and 1.1% tip clearance, Flot case.

Research results

It should be noted that the casing diameter was constant from inlet to exit in the calculation, while in the experiment this diameter varied in the region near the rotor due to the different sleeves that were used for obtaining different tip clearances with the same rotor.

Kinematic and thermodynamic data used for these three calculations are listed below:

(a) the rotor rotates with a constant rotational speed $N = 1300$ r.p.m.

(b) the inlet stagnation conditions were

$$\begin{aligned} \text{Total Pressure} &= 10^5 \text{ Pa} \\ \text{Total Temperature} &= 288.1 \text{ K} \end{aligned}$$

(c) the mass flow rate for the whole rotor passage was 22.44 kg/s

(d) the working fluid was air

(e) the inlet turbulence intensity was 4%, while equal hub and shroud boundary layer thicknesses (0.0078 span) were set at the inlet.

Figure 1.51 shows two views of the grid that was used in the calculation. The highly stretched grid lines at the upper part of both the pressure and suction sides match the extra grid used for meshing the tip-clearance region. Figure 1.52 shows the axial velocity, relative peripheral velocity and relative exit angle

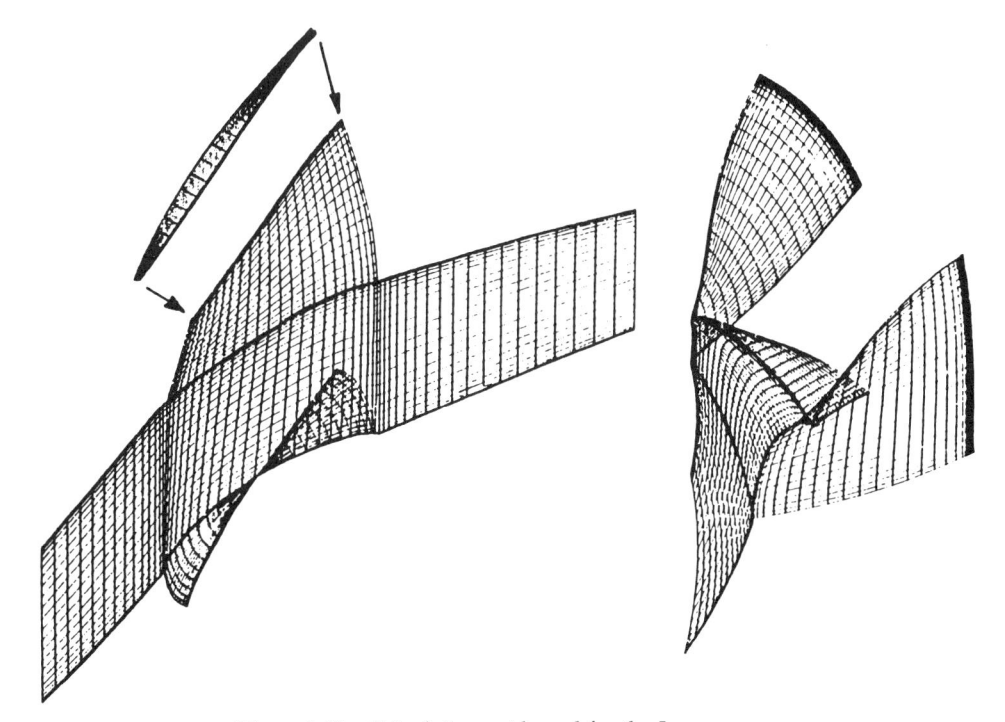

Figure 1.51 Calculation grid used for the Inoue case.

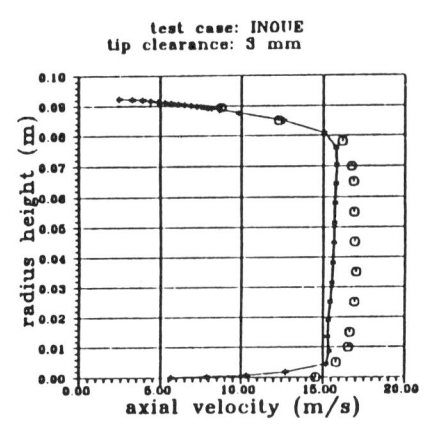

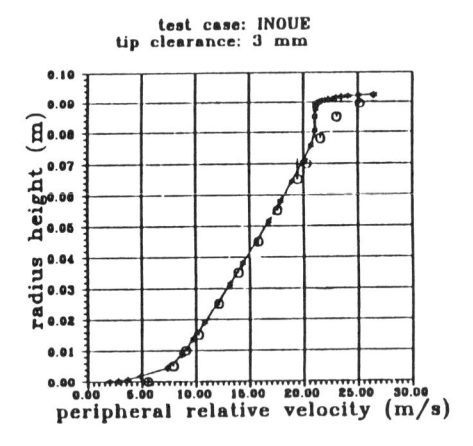

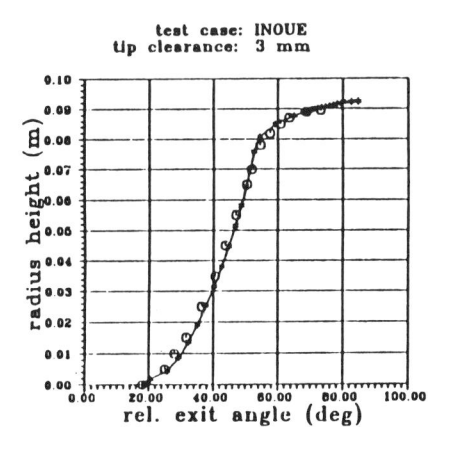

Figure 1.52 Relative exit angle, axial and peripheral relative velocity: 3 mm of tip gap.

Research results

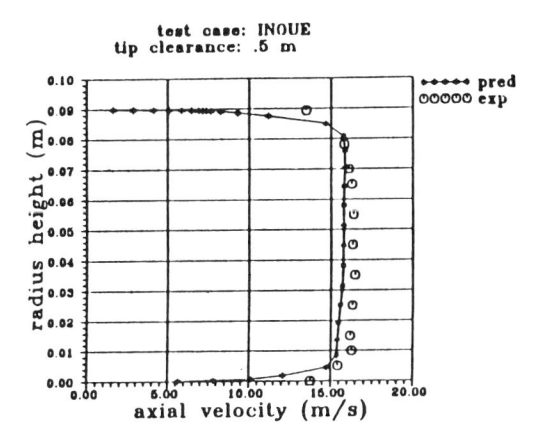

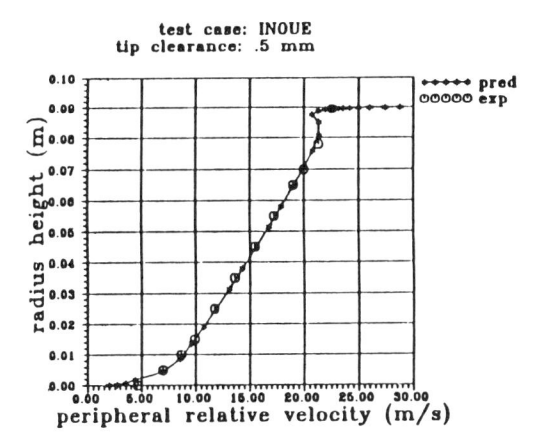

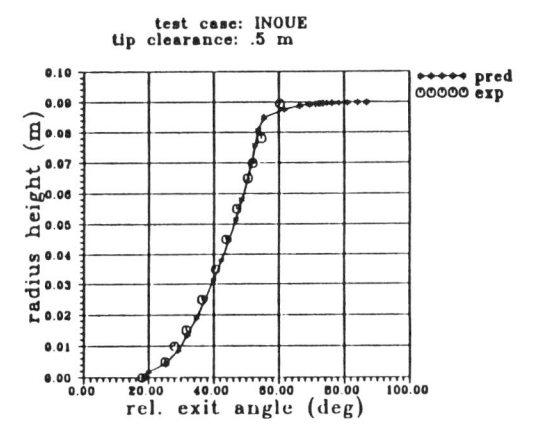

Figure 1.53 Relative exit angle, axial and peripheral relative velocity: 0.5 mm of tip gap.

for the 3 mm tip clearance, while Figure 1.53 shows the same quantities for 0.5 mm clearance. The agreement between prediction and experiment is very good. Figure 1.54 presents relative velocity profiles at planes normal to the rotation axis halfway between the leading and trailing edges for three different tip clearances, $\tau = 0.5$ mm, $\tau = 3$ mm and $\tau = 5$ mm. Positive radial velocity components near the pressure side at blade height can be observed. Within the tip gap, the flow in the case of the small clearance ($\tau = 0.5$ mm, Figure 1.54(a)) closely resembles a Couette type flow with little mass flowing through the gap. At the suction side it can be observed that the leakage flow interacts with the secondary flow.

Figures 1.55, 1.56 and 1.57 compare pressure coefficient contours between experiments and prediction on the shroud, for $\tau = 0.5$ mm, $\tau = 2$ mm and $\tau = 3$ mm respectively. The pressure coefficient is defined as the pressure rise from the inlet stagnation pressure divided by the dynamic pressure corresponding to the blade tip speed. It can be observed that the minimum pressure trough moves away from the leading edge with increasing tip clearance, a fact which has also been observed in the experiments. Figures 1.58 and 1.59 compare relative velocity vectors for $\tau = 3$ mm at two radii lying in the clearance gap. It can be seen that the predicted flow field has the same trends as the measured one.

4.6.4 Compressor rotor of Cranfield

The flow in a low-speed compressor rotor stage was studied. Experimental measurements for this particular rotor have been taken at Cranfield on their four-stage research compressor (Robinson *et al.* [72]). The particular rotor stage consisted of examining 79 blades and two tip clearances, one of 0.7% of span and another of 1.8% of span. The grid dimensions were $25 \times 39 \times 53$ and $12 \times 15 \times 28$ for the main and clearance grids respectively. The blade profile was generated by considering the blade as consisting of four curves which were given.

(a) geometrical data for three radial positions

	Root	Mid-Span	Tip
Radius	517.0	564.0	610.0 (mm)
Stagger	−22.60	−28.18	−33.69 (deg)
Chord length	60.27	59.20	58.68 (mm)
Solidity (s/c)	0.6822	0.7577	0.8267
max. thickness/chord	0.1236	0.1095	0.1048

(b) a fourth-order interpolation scheme was used to calculate the blade profile at different radial positions.

Kinematic and thermodynamic data used for the calculation

Research results

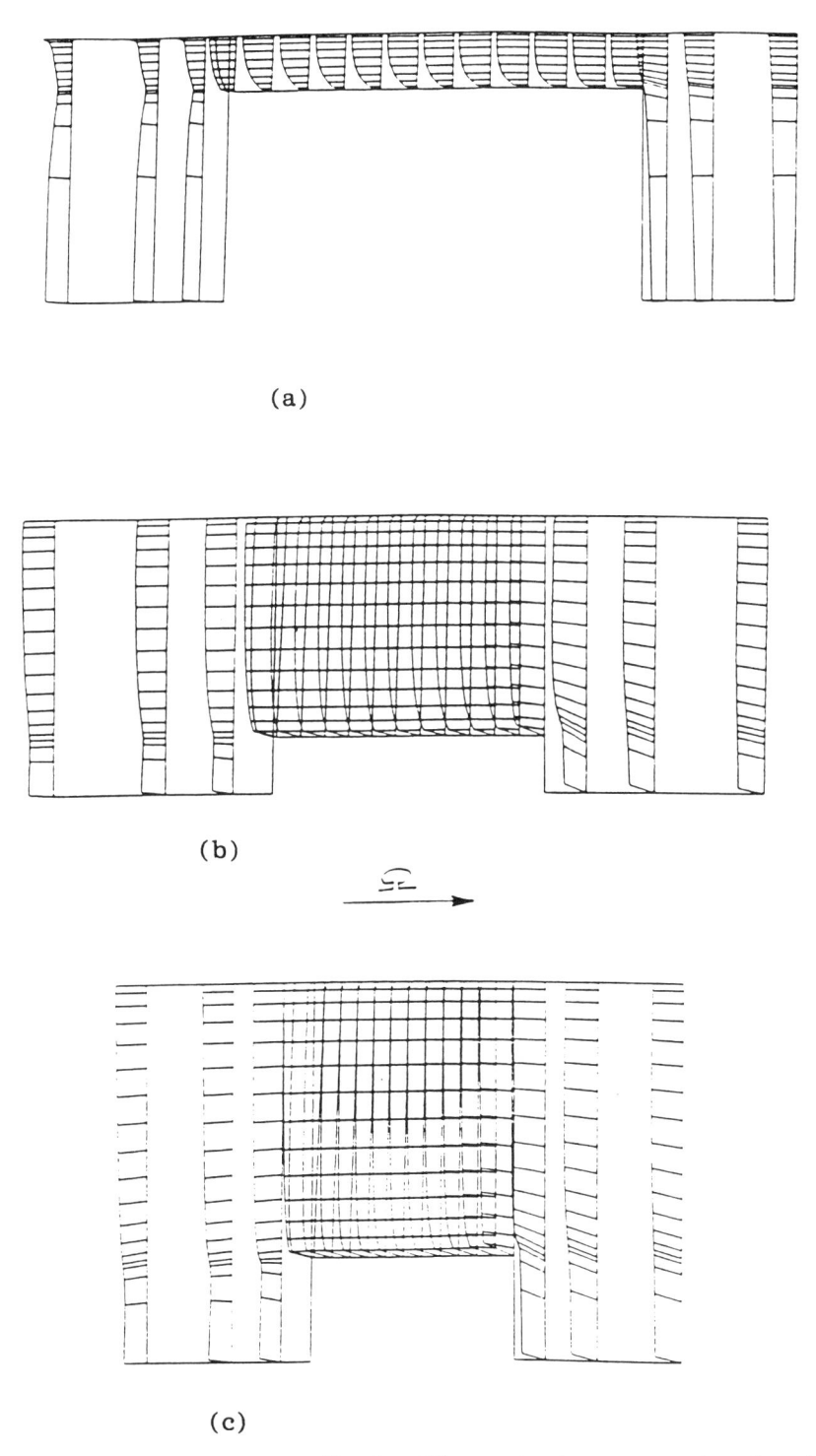

Figure 1.54 Relative velocity profiles for different tip clearances χ at a plane half-way between the leading and trailing edges: (a) $\tau = 0.3$ mm; (b) $\tau = 3$ mm; (c) $\tau = 5$ mm.

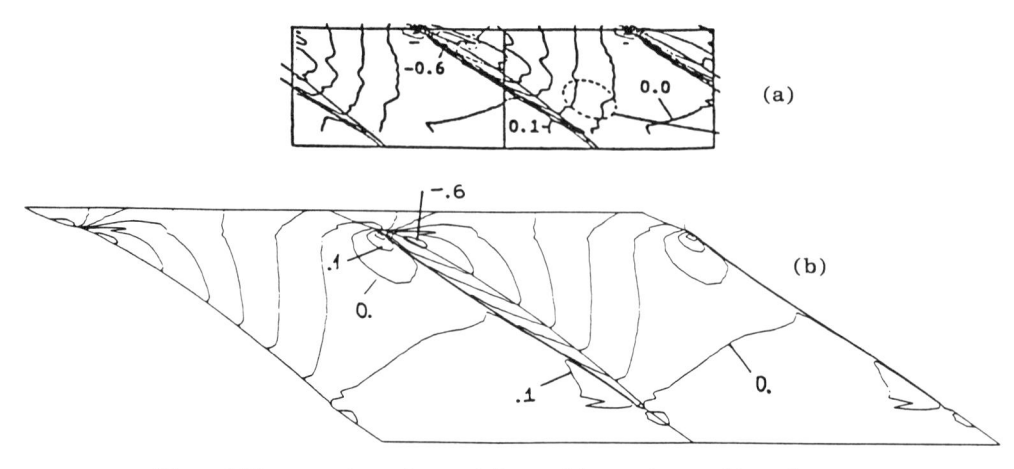

Figure 1.55 c_p contours for $\tau = 0.5$ mm: (a) experiment; (b) prediction.

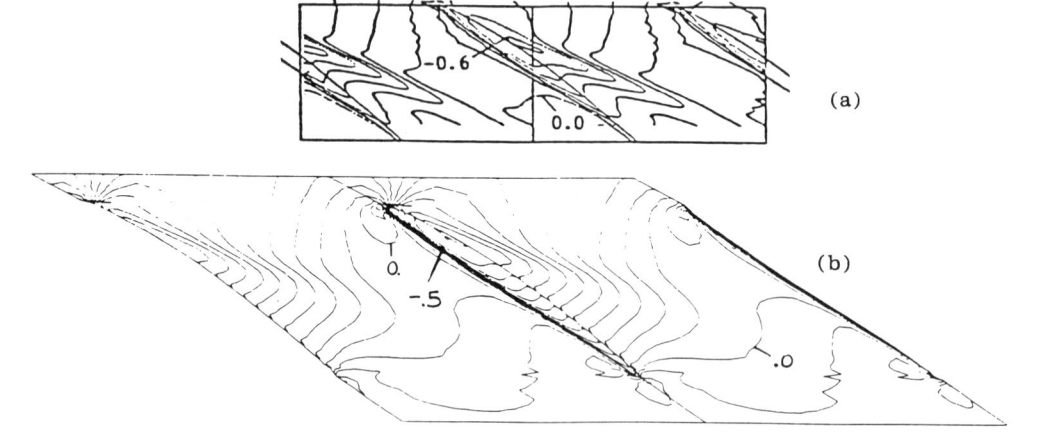

Figure 1.56 c_p contours for $\tau = 2$ mm: (a) experiment; (b) prediction.

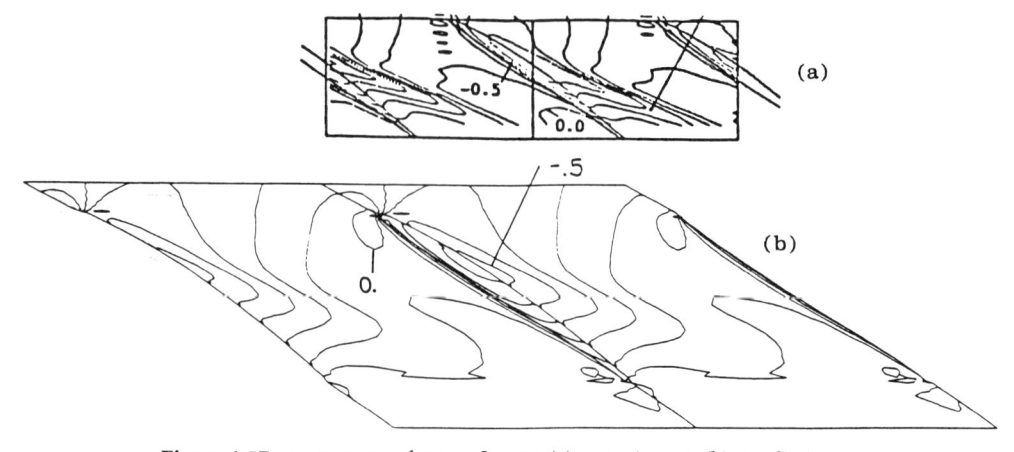

Figure 1.57 c_p contours for $\tau = 3$ mm: (a) experiment; (b) prediction.

Research results

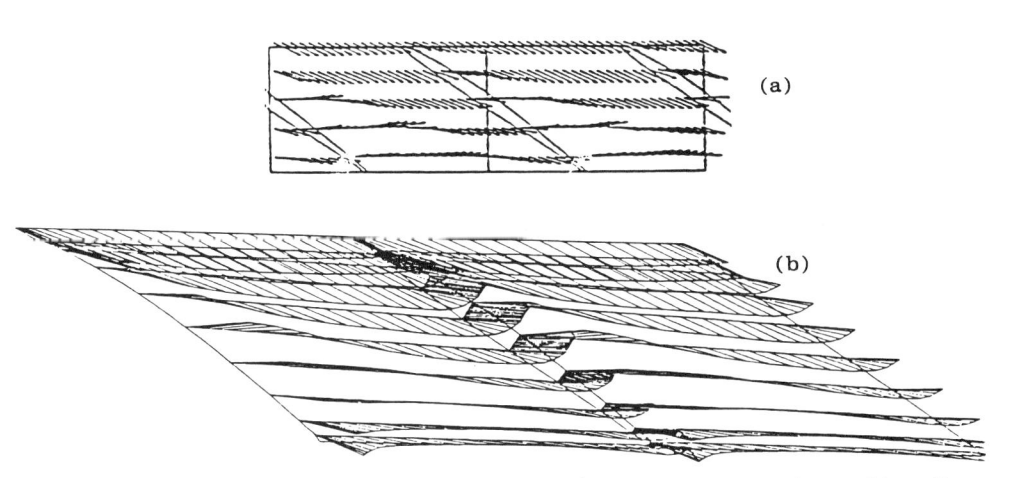

Figure 1.58 Relative velocity vectors for $\tau = 3$ mm, $r/r_{tip} = 1.006$: (a) experiment; (b) prediction.

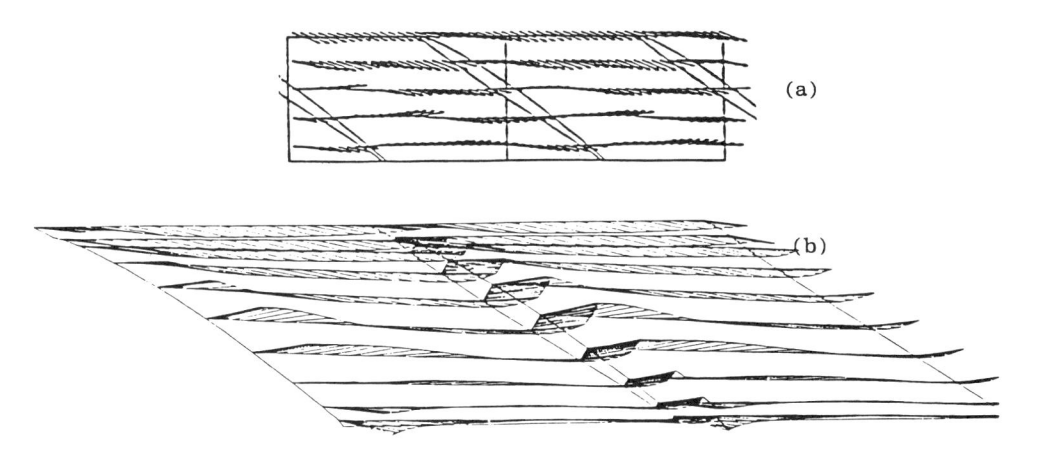

Figure 1.59 Relative velocity vectors for $\tau = 3$ mm, $r/r_{tip} = 1.012$: (a) experiment; (b) prediction.

(a) the rotor rotates with 876.0 r.p.m.

(b) the inlet stagnation conditions were

$$\text{Total Pressure} = 103\,400 \text{ Pa}$$
$$\text{Total Temperature} = 288 \text{ K}$$

(c) The velocity distribution at rotor inlet was obtained from the results of a meridional solver from Cranfield and the resulting mass flow rate was 13 kg s^{-1}.

Figure 1.60 shows two views of the calculation grid used. Figures 1.61 and

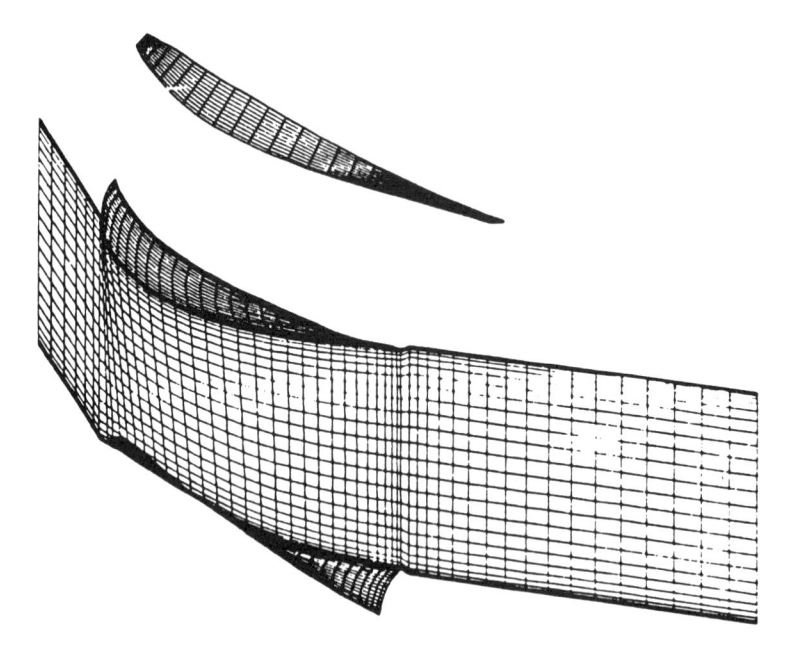

Figure 1.60 Computational grid used for the Cranfield rotor.

1.62 show radial distributions of axial velocity, relative peripheral velocity and relative exit angle for the 0.7 and 1.8% of span clearances. The agreement between prediction and experiment is good. Figure 1.63 shows relative velocity vectors at three positions down the blade.

4.7 Task 7

At the time of writing, successful use of this instrument is only just beginning on bench tests and its employment in a turbomachine is not yet realised. Note that the potential of this instrument remains; however, development of this technique is recognized as having high relative risk.

4.8 Task 8

Figure 1.64 displays the high-speed cascade rig (HSCR) overall arrangement as well as the responsibility for manufacturing.

As soon as the new compressor has been found and therefore their main characteristics known, outlet scroll conditions are provided by means of an existing computation. Then the inlet scroll has been designed and manufactured by Turbomeca: Figure 1.65 displays it in the NTUA laboratory.

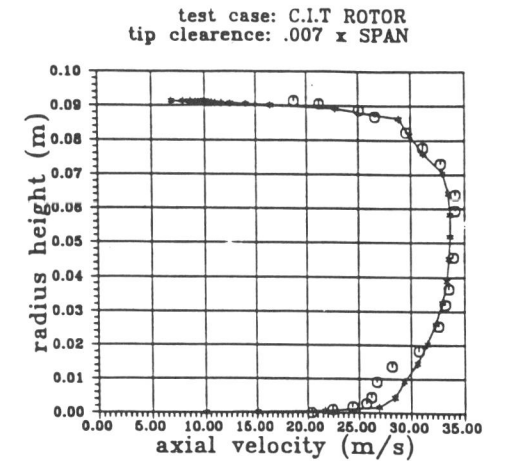

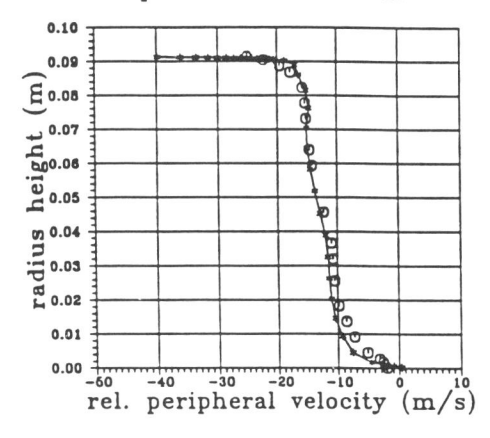

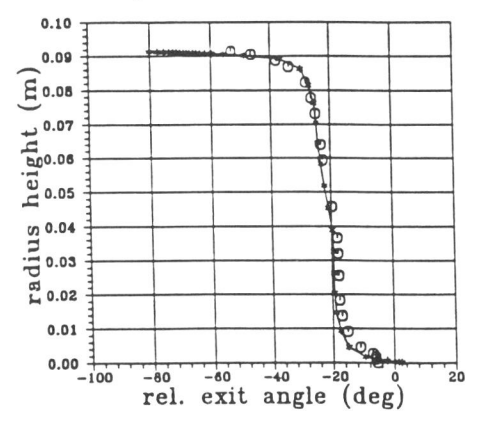

Figure 1.61 Rotor exit relative exit angle, axial and peripheral relative velocity: $\tau = 0.7\%$ span.

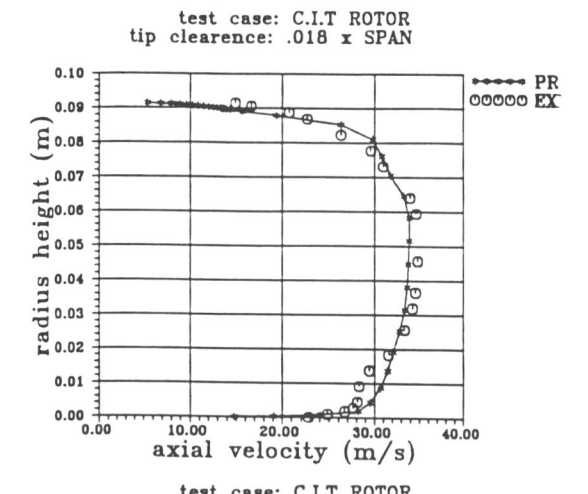

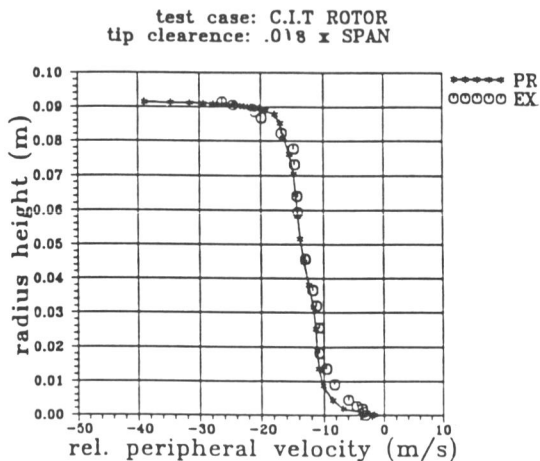

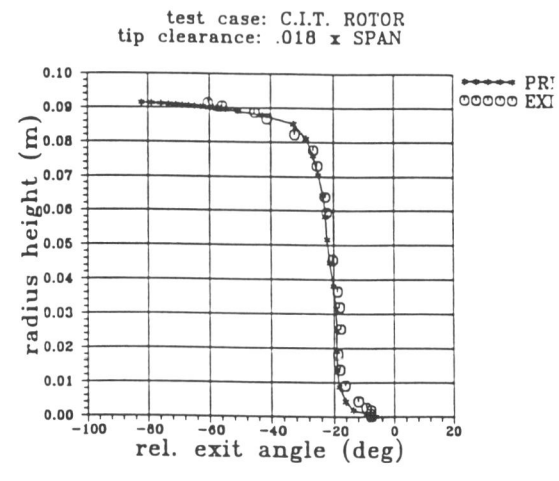

Figure 1.62 Rotor exit relative exit angle, axial and peripheral relative velocity: $\tau = 1.8\%$ span.

Research results 91

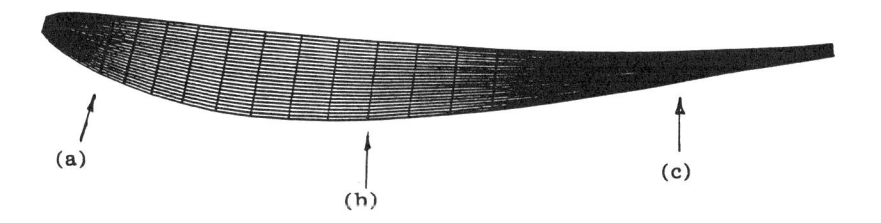

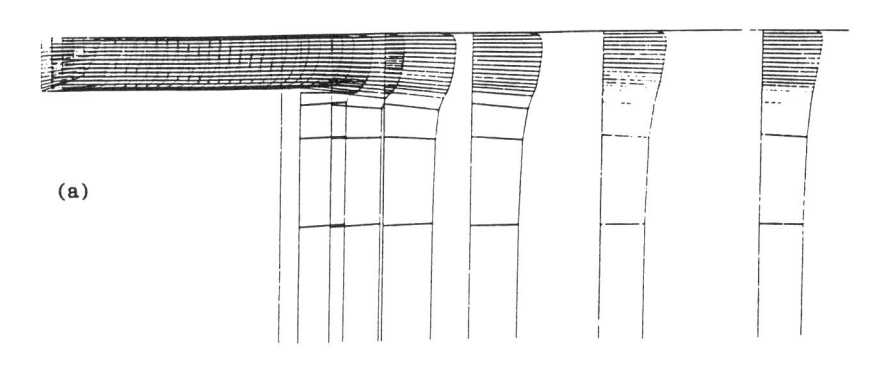

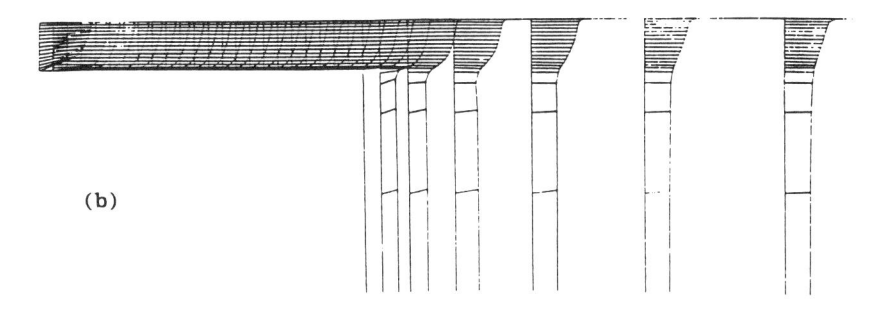

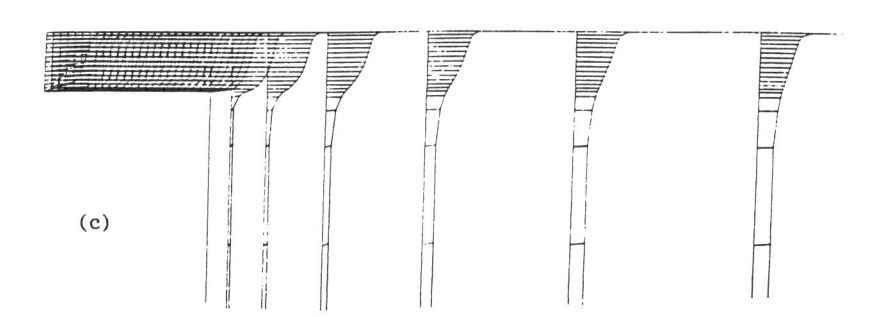

Figure 1.63 Relative velocity vectors at planes normal to the axis.

BMW-RR

NTUA

SNECMA

TURBOMECA (I)

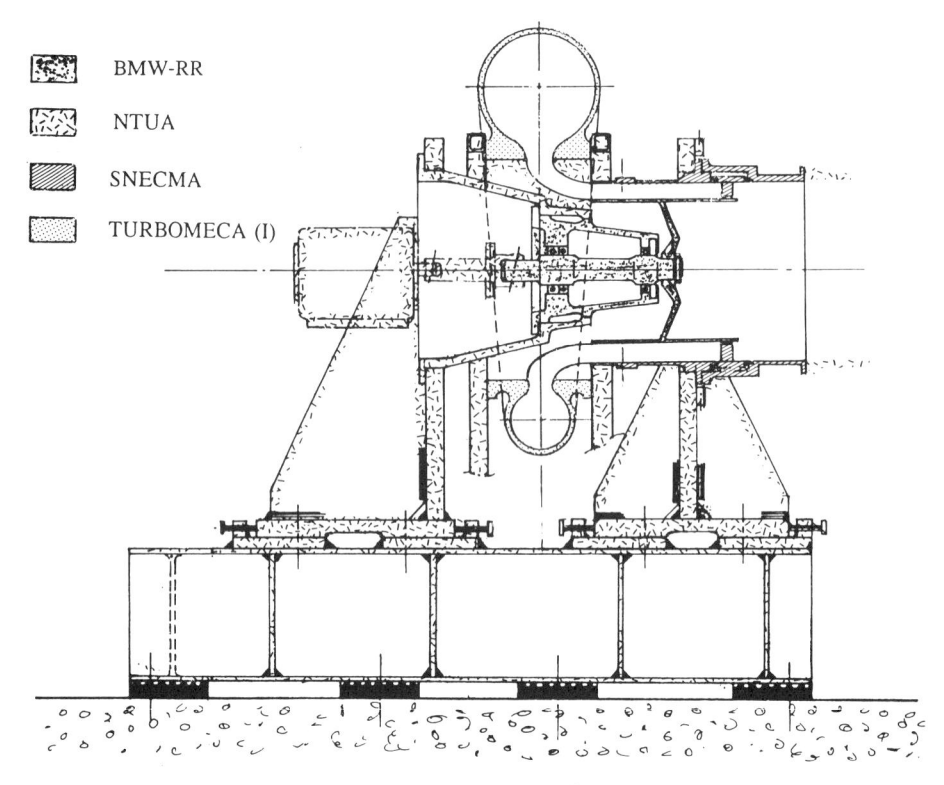

Figure 1.64 Overall view of HSCR.

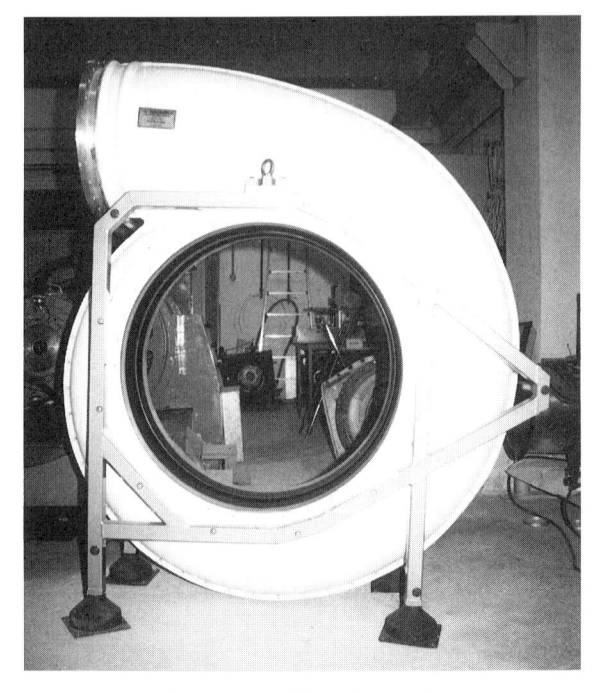

Figure 1.65 The inlet scroll.

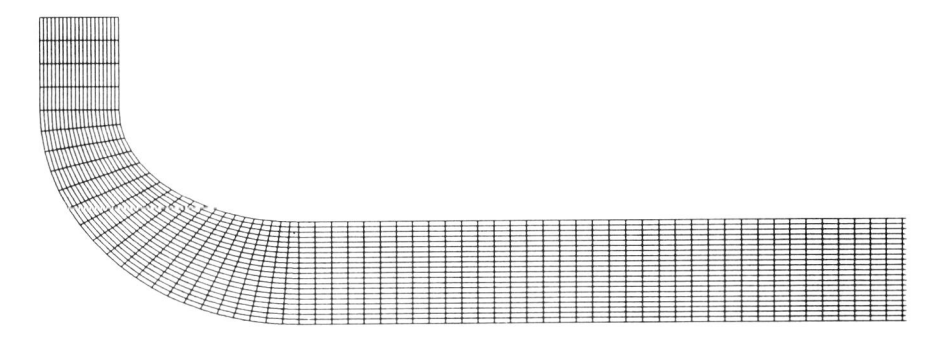

Figure 1.66 Bent duct grid.

Calculations of the bend-duct connecting the inlet scroll to the test section have been completed to ensure a reasonable flow field near the outer casing, that is without separation. Figure 1.67 displays such meridional velocity contours while Figure 1.66 shows the grid used.

A draft of the test blade has been done (see Figure 1.68) and will be finalized as soon as actual inlet scroll characteristics will be achieved.

Figure 1.67 Meridional velocity contours.

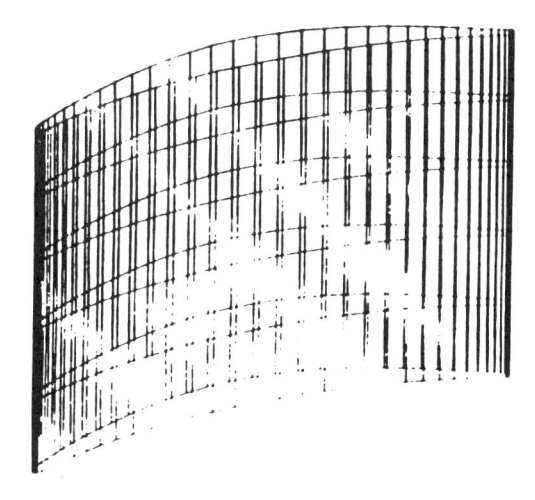

Figure 1.68 Test blade profile.

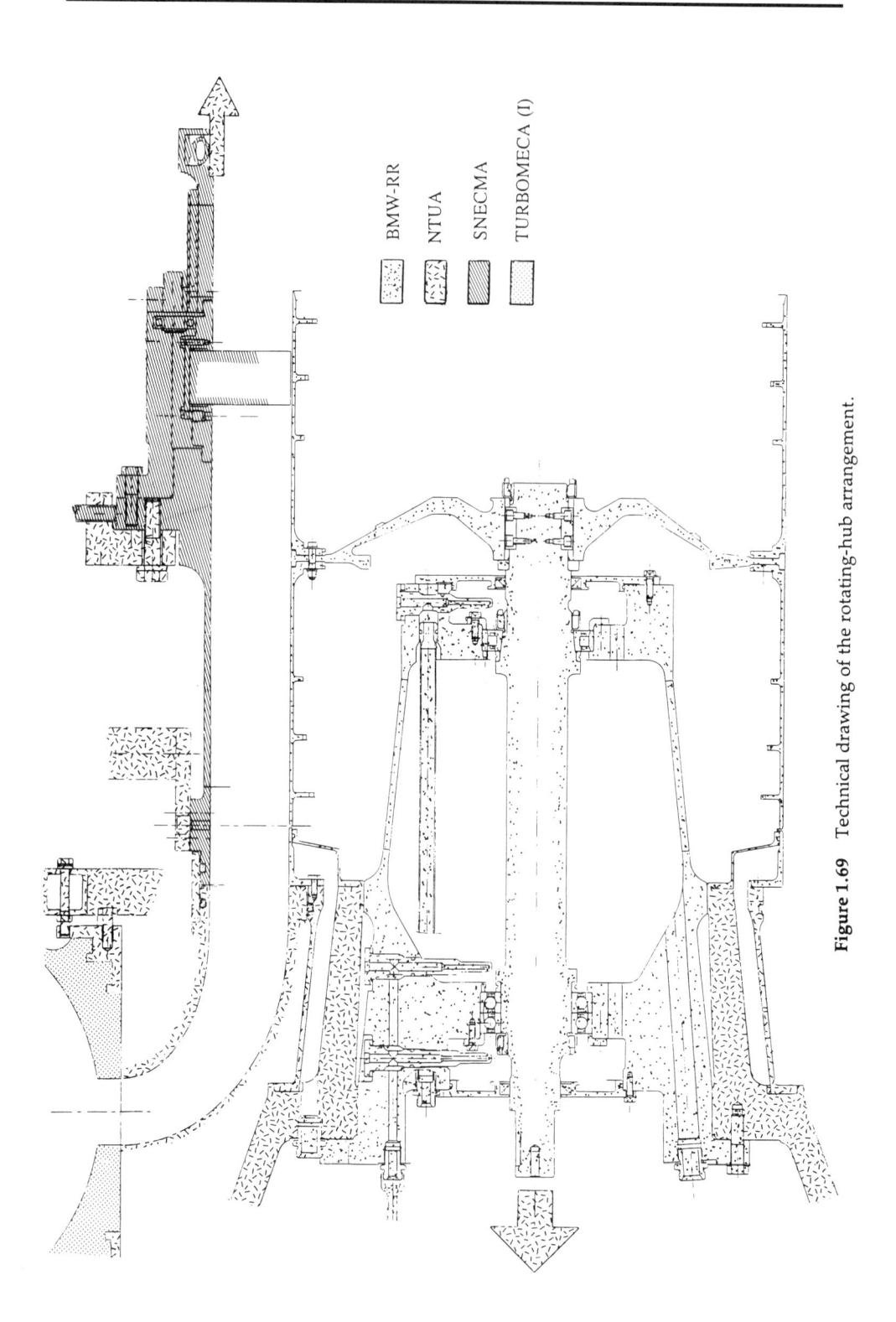

Figure 1.69 Technical drawing of the rotating-hub arrangement.

Figure 1.70 Underbase and supporting brackets.

The technical drawing of the rotating hub arrangment is displayed in Figure 1.69; detailed drawings are available.

Also, the underbase as well as the main supporting brackets have been manufactured (Figure 1.70).

5 Conclusions

5.1 Task 1

A database using existing open literature experimental results has been achieved and augmented by present experiments on the Cranfield four-stage LSRC.

It contains for each case, full geometrical data as well as measurements and/or predictions of flow field in a format agreeable to the industrial partners.

The database has been used for theoretical validations both by NTUA and industrial partners.

5.2 Task 2

A list of important parameters that determine the environment in which tip-clearance flows develop, in the rear stage of axial flow compressors has been achieved. That will settle the parameters range of values for relevant and meaningful experiments or theoretical assumptions.

5.3 Task 3

From the theoretical developments presented in this report it may be seen that the two models are able to describe the basic tip-clearance mechanisms with sufficient accuracy for the cases considered, essentially without any empirical information, other than the theoretical assumptions underlying them. Although the present work was performed for the compressor case, the validity of both models was tested against all available experimental results, including those for turbines. The basic reason for this was the scarcity of experiments with the level of detail required for validating the discussed models.

Comparisons of the complete theoretical procedure with available experimental results were successful and the superposition principle was proven to be adequate for setting up the complete calculation procedure. The interaction of the models within an integrated external flow plus a secondary flow calculation procedure introduced no appreciable error to the prediction results.

The basic assumptions of the gap flow model were tested, using also the second-level tool to reconstruct the flow inside the tip clearance. The formation of the *vena contracta* at the gap entrance, the two dimensionality of the flow inside the gap and the simple velocity profiles adopted at the gap exit were proven to be adequate for the description of the flow inside the tip clearance.

The new model adopted for the calculation of the leakage vortex strength and development seems to give satisfactory results, providing also a radial distribution of the total pressure losses inside and downstream of the blade row. However, a further development and testing of the loss calculation procedure is needed in order to be able to obtain higher reliability of the model.

The targets of this task, for the development of a theoretical tool compatible with the existing meridional codes, have been achieved successfully. However, the present calculation algorithm concerns only one-stage machines and further development is needed for multi-stage machines. The further testing of the proposed procedure in multi-stage cases and the refinement of the adopted models can be the subject of future work.

5.4 Task 4

This report has presented significant data which will assist development of the level one and two models of tip-clearance flow. The effect of tip-clearance on overall compressor properties (efficiency and pressure rise coefficient) has been determined. The chordwise distribution of static pressure for both rotors and stators has been measured; data in the rotor tip region will prove to be particularly useful. Detailed area traverses have been obtained for stages 1 and 3 using both wedge and cobra pneumatic probes. This data is more accurate and of greater density than previously obtained, and the resulting analysis has generated a better understanding of tip-clearance flow axial compressors.

Two versions of the Cranfield multistage compressor have been achieved, the first at a tip clearance/height ratio of 1.8%, the second at a ratio of 0.7%. With hindsight, all the pneumatic data collected appears to be perfectly consistent with the change in tip clearance. The full benefit of this data is yet to be realized, however, two factors require reiteration .

Firstly Cobra type probes are required if detailed and accurate measurements on multistage turbomachines are essential, or if meaningful comparisons with engine core compressors are to be drawn.

Laser anemometry has been employed successfully on the Cranfield low-speed research compressor. The laser two-focus technique has detailed the blade-to-blade flow in the tip-clearance region and using contour-plotting techniques it may prove possible to infer the three-dimensional structure and the location of the tip-leakage vortex.

Finally, the capacity of the Cranfield large-scale low-speed research compressor to undertake detailed project-type investigations of physical phenomena has been demonstrated. The further application of advanced instrumentation, particularly in the rotation frame of reference, to this compressor will reveal the structure of tip-clearance flows more fully.

5.5 Task 5

First and second versions of the first-level model have been introduced into the framework of industrial throughflow software and tested against a selection of the database test cases.

It is well known that the influence of the tip-clearance jet may extend even up to midspan for low aspect ratio. This particular behaviour is reproduced by the code: the area of tip effects influence increases from 16% of immersion up to 48% while the tip gap comes from 0.56% to 5.56% respectively. However, in the present state, the prediction accuracy does not seem reliable enough to use the model. Some further improvements are expected.

Up to now (second version of the code), gradients of the throughflow profiles are not taken into account, while only meridional throughflow quantities within tip clearance are used as input data to the model. In other words, the state of the flow within which the tip-clearance flow develops is supposed to be radially uniform, which is not realistic. This, in fact, should change with the last version, where spanwise tip-clearance mixout losses are computed. Thus a second iteration of the meridional throughflow code with these additional losses should provide better flow profiles.

The model is too time consuming to offer the possibility of direct implementation in industrial proprietary throughflow software for a strong coupling between the tip-clearance jet and the meridional flow.

This model needs further development to be incorporated as a design tool in industrial environmental software, but it is hoped that subsequent versions of the code should further improve the accuracy in predicting tip-leakage effects.

5.6 Task 6

For this Task, the three-dimensional Navier–Stokes code 'TURBO3D', developed in Cranfield, was provided to the LTT/NTUA, which undertook modifications so that the code could be used as a second-level tool for the validation of the first-level tool developed by the same group within this project.

Listed below are the modifications which have been implemented in the original code, improving its modelling performance in the respective areas.

(1) Accurate modelling of the blade tip geometry through the introduction of a square tip.

(2) Accurate modelling of the leading and trailing edges of the blade by retaining the blade's original shape in these regions.

(3) With respect to turbulence modelling, work was done on the alternative use of a low-Reynolds $k-\varepsilon$ model, instead of the wall function formulation built in the original version of the code.

(4) Restructuring of the code so as to improve its computer run-times, which were reduced to values smaller than 1 ms/node/iteration.

The examined cases, selected in cooperation with the other partners, covered linear and peripheral stationary or spinning compressor cascades, and a qualitative and quantitative comparative study for different sizes of tip gap was performed. The investigation of the various flow situations occurring in tip-clearance flows have not been exhaustively covered, but quite a few relevant problems were identified and some of them have been successfully tackled.

Grid generation is always the first step, and is far from being negligible in terms of engineering cost. The H-grid topology was retained for the following three reasons:

(a) When dealing with compressor cascades, the H-type grid is a couvement gridding technique. Problems usually arise close to the leading edge, where an ambiguity occurs, but this was discussed and a remedy was proposed herein;

(b) Parabolic solvers, based on a space marching algorithm, tend to make use of a primitive direction that must approximate the main flow direction. H-type grids respect this primitive direction;

(c) The extra grid used in the clearance region is easily 'blended' in the H-type grid.

Depending on the geometry of each particular test case, two different meshing approaches could be identified. The examined linear cascades were built by stacking a single two-dimensional grid, with appropriate clustering in the spanwise direction, in order to take into account endwall effects. On the other hand, peripheral cascades required the construction of a different two-dimensional mesh at each radius (in the case of twisted of blades) that was wrapped on the respective cylindrical surface. The resulting grid depends on how efficiently the two-dimensional domain is meshed, but for more stiff cases a further insight to local grid generation (in the tip region) will be required. In the case of peripheral cascades, the repeated generation of high-quality two dimensional meshes makes inevitable the use of a three-dimensional 'automatic' grid generation process. The problem will be even more acute when low Reynolds turbulence models will be in use.

The task of accelerating the whole code, by making use of the vector and parallel capabilities of our mini-super-computer, was not clearly foreseen but it proved to be extremely valuable in developing and using the code, since it allowed affordable computational times with the existing resources.

Under the existing computer limitations turbulence modelling was restricted to the use of the very economical wall function and an effort to pass to a low Reynolds approach was initiated, first for a two-dimensional counterpart of the code. The completion of this work in the three-dimensional code, with the probable incorporation of algebraic Reynolds stress models, seems to be a necessary step in order to obtain more accurate results, even at the expense of computer memory and computing time. Of course the further assessment of the present second-level tool, and as a consequence of the first-level one, in well-documented experimental results will highlight the advantages and disadvantages of such an approach.

5.7 Task 7

Demonstration of the miniature three-dimensinal laser Doppler anemometry system was not achieved. An alternative is possible, however. A three-dimensional laser-time-of-flight system has just become available from a well-known and reputable manufacturer.

5.8 Task 8

To get detailed three-dimensional experimental data, a high-speed cascade rig has been studied and partially built. It will be finalized and used during the next IMT AC3A project.

The inlet scroll has been studied, manufactured and is already in the NTUA laboratory; underbase and brackets supporting the test stand are achieved as well. Detailed drawings of the rotating-hub arrangement are available; they will be used in the next phase for manufacture.

To anticipate the next phase, provision of a laser system for one velocity component measurements has been achieved with private funds. It will be used connected with a two-component system to provide three-dimensional detailed measurements.

6 References

[1] BRITE/EURAM Contract AERO-0021-C(CD), Project PL 1036 on Tip Clearance Effects in Advanced Axial Flow Compressors, from 1.4.1990 to 30.6.1992.
[2] Robinson, C.J., Northall, J.D. and McFarlane, C.W.R., Measurements and Calculation of the Three Dimensional Flow in Axial Compressor Stators, With and Without End Bends, *ASME Paper* 89-GT-6.
[3] Private communication with Gallimore, S.J.(Rolls-Royce) who provided Rolls-Royce data.
[4] Hunter, I.H and Cumpsty, N.A., *Journal of Engineering for Power*, **104**, 1982.
[5] Smith, G., *Ph. D. Thesis*, Cambridge University, 1970.
[6] MacDougall, N.M., Stall inception in axial compressors, *PhD Thesis*, Cambridge University.
[7] Goto, A., Measurements of the flowfield behind a compressor rotor, Cambridge University internal report (unpublished), 1989.
[8] Storer, J.A., 1987, An investigation of the use of hub clearance to reduce stator losses, Rolls-Royce internal report.
[9] Wisler, D.C., *Core Compressor Exit Stage Study*. Volume I, *Blading design*, NASA CR 135391.
[10] Wisler, D.C., *Core Compressor Exit Stage Study*. Volume II, *Data and performance*, NASA CR 159498.
[11] Wisler, D.C., *Core Compressor Exit Stage Study*. Volume III, *Data and performance*, NASA CR 159499.
[12] Wisler, D.C., *Core Compressor Exit Stage Study*. Volume IV, *Data and performance*, NASA CR 165357.
[13] Wisler, D.C., *Core Compressor Exit Stage Study*. Volume V, *Data and performance*, NASA CR 165358.
[14] Wisler, D.C., *Core Compressor Exit Stage Study*. Volume VI, *Final Report*, NASA CR 165554.
[15] Inoue, M., Kuroumaru, M., and Fukuhara, M., Behaviour of tip leakage flow behind an axial compressor rotor, *ASME Journal of Engineering for Gas Turbines and Power*, **108**, 7–13, 1986.
[16] Dring, R.P., Joslyn, H.D. and Hardin, L.W., *Compressor Rotor Aerodynamics: Analytical and Experimental Investigation*, United Technologies Research Centre, UTRC 80-15, March 1980.
[17] Dring, R.P., Joslyn, H.D., and Hardin, L.W., *Experimental Investigation of Compressor Rotor Wakes*, AFAPL-TR-79-2107, Air Force Aero Propulsion Laboratory, Technology Branch, Turbine Engine Division (tbx), Wright-Patterson Air Force Base, OH.
[18] Dring, R.P., Joslyn, H.D. and Hardin, L.W., An investigation of compressor rotor aerodynamics, *ASME Journal of Engineering for Power*, **104**, 84–96, January 1982.
[19] Dring, R.P., Joslyn, H.D. and Wagner, J.H., *Axial Compressor Middle Stage Secondary Flow Study*, Final Report submitted to NASA-Lewis Research Centre for contract no. NASA-23157, December 1982.
[20] Dring, R.P. and Serovy, G.K. Test Cases for Computation of Internal Flows in Aero Engine Components, AGARD Advisory Report no. 275, p. 152, July 1990.
[21] Dring, R.P., Joslyn, H.D. and Hardin, L.W., *Compressor Rotor Aerodynamics: Analytical and Experimental Investigation*, United Technologies Research Centre, UTRC 80–15, March 1980.
[22] Dring, R.P. and Serovy, G.K. *Test Cases for Computation of Internal Flows in Aero Engine Components*, AGARD Advisory Report no. 275, p. 286, 1990.

[23] Dransfield, D.C. and Calvert, W.J. *Detailed Flow Measurements in a Four Stage Axial Compressor*, ASME Paper 76-GT-46, 1976.
[24] Dransfield, D.C. and Calvert, W.J. *Detailed Flow Measurements in a Four Stage Axial Compressor*, NGTE Note no. Nt.965, 1976.
[25] Roberts, W.B., Serovy, G.K. and Sandercock, D.M. Modelling the 3-D flow effects on deviation angle for axial compressor middle stages. *ASME Journal of Engineering for Gas Turbines and Power*, **108**, 131–137, 1986.
[26] Roberts, W.B., Serovy, G.K. and Sandercock, D.M. *Design Point Variation of 3-D Loss and Deviation for Axial Compressor Middle Stages*, ASME Paper 88-GT-57, 1988.
[27] Barton, J.P. *Results of the First Test at Datum Rotor Tip Clearances carried out on the Cranfield Low Speed Research Compressor for the Brite/Euram Tip Clearance Project*, RCR91120, Brite/Euram Report 10-RR-2, June 1992.
[28] Young, K.F. *Brochure ECC1/157D. Interim Brite/Euram Tip Clearance Investigation Report on Second CIT Compressor Build*, RCR 91122, Brite/Euram Report 10-RR-3, June 1992.
[29] Rains, D.A. *Tip Clearance Flow in Axial flow Compressors and Pumps*, California Institute of Technology, Hydrodynamics and Mechanical Engineering Laboratories Rep. No 5, 1954.
[30] Lakshminarayana B. Methods of predicting the tip clearance effects in axial turbomachinery, *ASME Journal of Basic Engineering*, D., **92**(3), 467–482, 1970.
[31] Ohayon G. Une Méthode de Prédiction des Effets du Jeu Radial dans les Compresseurs Axiaux, *Journal de Mécanique Appliquée*, **5**(3)
[32] *Tip Clearance Effects in Axial Flow Compressors*, Report on the Tip Clearance Model, Report: 4-NTUA-2, January 1991.
[33] *Tip Clearance Effects in Axial Flow Compressors*, Report on 1st Year Progress. Report: 5-NTUA-1, March 1991.
[34] *Tip Clearance Effects in Axial Flow Compressors*, Report on Intermediate Progress, Report: 6-NTUA-1, June 1991.
[35] *Tip Clearance Effects in Axial Flow Compressors*, Report: 7-NTUA-2, September 1991.
[36] *Tip Clearance Effects in Axial Flow Compressors*, Report: 8-NTUA-3, January 1992.
[37] *Tip Clearance Effects in Axial Flow Compressors*, Report: 9-NTUA-2, May 1992.
[38] *Tip Clearance Effects in Axial Flow Compressors*, Final Technical Progress Report. Report: 10-NTUA-1, June 1992.
[39] Wadia, A.R. and Booth, T.C. Rotor tip leakage: Part II—design optimization through viscous analysis and experiment, *ASME Journal of Engineering for Power*, **104**, 162–169, 1982.
[40] Booth, T.C., Dodge, P.R. and Hepworth, H.K. Rotor-tip leakage: Part I—basic methodology, *ASME Journal of Engineering for Power*, **104**, 154–161, 1982.
[41] Moore, J. and Tilton, J.S. Tip leakage flow in a linear turbine cascade, *ASME Journal of Turbomachinery*, **110**, 18–26, 1988.
[42] Storer, J.A., Tip clearance flows in axial compressors, *Ph.D. Thesis*, Dept. of Engineering University of Cambridge, 1991.
[43] Sjolander, S.A. and Amrud, K.K., *Effects of Tip Clearance on Blade Loading in a Planar Cascade of Turbine Blades*, ASME Paper 86-GT-245, 1986.
[44] Yaras, M.I., Yingkang, Z., and Sjolander, S.A., Flow field in the tip gap of a planar cascade of turbine blades, *ASME Journal of Turbomachinery*, **111**, 276–283, 1989.
[45] Moore, J., Moore, J.G., Henry, G.S. and Chaudhry U., Flow and heat transfer in turbine tip gaps, *ASME Journal of Turbomachinery*, **111**, 301, 309, 1989.

[46] Milne-Thompson, L.M., *Theoretical Hydrodynamics*, (5th edn) Macmillan, London.
[47] Lakshminarayana, B., Effects of a chordwise gap in an aerofoil of finite span in a free stream, *Journal of the Royal Aeronautical Society*, **68**, 276–280, 1964.
[48] Lakshminarayana, B., and Horlock J. H., *Leakage and Secondary Flow in Compressor Cascades*, ARC R&M, 3483.
[49] Lewis, R.I., and Yeung, E.H.C., *Vortex Shedding Mechanisms in Relation to Tip Clearance Flows and Losses in Axial Fans*, ARC R&M No. 3829, 1977.
[50] Yamamoto, A., Endwall flow/loss mechanisms in a linear turbine cascade with blade tip clearance, *Transactions of the ASME, Journal of Turbomachinery*, **111**, pp. 264–275, July 1989.
[51] Inoue, M. and Kurumaru M., Structure of tip clearance flow in an isolated axial compressor rotor, *ASME Journal of Turbomachinery*, **111**, 250–256, 1989.
[52] Yaras, M.I., Sjolander, S.A. and Kind, R.J., *Effects of Simulated Rotation on Tip Leakage in a Planar Cascade of Turbine Blades, Part II: Downstream Flow Field and Blade Loading*, ASME paper 91-GT-128, 1991.
[53] Yaras, M.I. and Sjolander, S.A., Development of the tip leakage flow downstream of a planar cascade of turbine blades: vorticity field, *ASME Journal of Turbomachinery*, **112**, 609–617, 1990.
[54] Lamb H., *Hydrodynamics* (6th edn), Dover Publications, London.
[55] Newman, B.G., Flow in a viscous trailing vortex, *Aeronautical Quarterly*, 149–162, 1959.
[56] Owen, P.R., The decay of a turbulent trailing vortex, *Aeronautical Quarterly*, **XX**, 69–78.
[57] Chen, G.T., Greitzer, E.M., Tan, C.S. and Marble, F.E. *Similarity Analysis of Compressor Tip Clearance Flow Structure*, ASME paper 90-GT-153, 1990.
[58] Inoue, M., Kuroumaru, M. and Fukuhara, M., *Behaviour of Tip Leakage Flow Behind an Axial Compressor Rotor*, ASME paper 85-GT-62, 1985.
[59] Kunz, R.F., Lakshminarayana, B. and Basson, A.H., *Investigation of Tip-Clearance Phenomena in an Axial Compressor Cascade Using Euler and Navier–Stokes Procedures*, ASME Paper 92-GT-299, 1992.
[60] Moore, J. and Moore, J.G., *Shock Capturing and Loss Prediction for Transonic Turbine Blades Using a Pressure Correction Method*, ISABE Paper 89-7017, 1989.
[61] Hah, C. and Reid, L., *A Viscous Flow Study of Shock-Boundary Layer Interaction, Radial Transport, and Wake Development in a Transonic Compressor*, ASME Paper 91-GT-69, 1991.
[62] Choi, D. and Knight, C.J., *A Study of H and O-H Grid Generation and Associated Flow Codes for Gas Turbine 3D Navier–Stokes Analyses*, AIAA Paper 91-2365, 1991.
[63] Watanabe, T., Nozalei, O., Kikuchi, K. and Tamura, A., Numerical simulations of the flow through cascades with tip clearance, *ASME, Numerical Simulations in Turbomachinery*, 131–138, 1991.
[64] Patankar, S.V. and Spalding, D.B., A calculation procedure for heat, mass and momentum transfer in three-dimensional parabolic flows, *Int. J. Heat Mass Transfer*, **15**, 1787–1806, Pergamon Press, 1972.
[65] Rhie, C.M. and Chow, W.L., Numerical study of the turbulent flow past an airfoil with trailing edge separation, *AIAA Journal*, **21**(11), 1525–1532, 1983.
[66] Storer, J.A. and Cumpsty, N.A., *Tip Leakage Flow in Axial Compressors*, ASME Paper 90-GT-127, 1990.
[67] Leboeuf, F. and Flot, R., *Low Subsonic Compressor Cascade NACA 65*, AGARD-AR-275, 1975.

References

[68] Flot, R., Contribution a l'Etude des Ecoulements Secondaires Dans les Compresseurs Axiaux, *Ph.D. Thesis*, Lyon, 1975.

[69] Inoue, M., Kuroumaru, M. and Fukuhara, M., *Behavior of Tip Leakage Flow Behind an Axial Compressor Rotor*, ASME Paper 85-GT-62, 1985.

[70] Inoue, M. and Kuroumaru, M., Structure of tip clearance flow in an isolated axial compressor rotor, *ASME Journal of Turbomachinery*, **111**, 250–256, 1989.

[71] Ahmed, N.A., Elder, R.L., Forster, C.P., and Jones, J.D.C. Miniature laser anemometer for 3d measurements, *Measurement Science Technologie*, **I**, 272–276.

[72] Ahmed, N.A. Flow studies in impeller passages, *PhD Thesis*, Cranfield Institute of Technology, 1988.

[73] Ivey, P.C., *A Review of Progress to Date in a BRITE/EURAM Investigation of Tip-Clearance Effects in Advanced Axial Compressors*. Technical Note. N°5.CIT.2, Cranfield Institute of Technology, 1996.

[74] Smout, P.D., The problem of static pressure measurement in turbomachinery annuli using traversable instrumentation. *Xth Symposium of Aerodynamics. Measurements Techniques*, VK1, 1990.

[75] Kang, S. and Hirsch, Ch., *Experimental Study on the Three Dimensional Flow within a Compressor Cascade with Tip Clearance: Part I—Velocity and Pressure Fields*, ASME Paper No 92-GT-215, 1992.

2 Low emission combustor technology (LOWNOX I)

F. Joos and G. Pellischek*

This report, for the period February 1990 to May 1992, covers the activities carried out under the BRITE/EURAM Area 5: 'Aeronautics' Research Contract No. AERO-CT89-0016 (Projet : AERO-P1019) between the Commission of the European Communities and the following:

*MTU Motoren- und Turbinen-Union München GmbH (coordinator)
BMW-Rolls Royce GmbH, BMW-RR
Rolls Royce Plc, RR
Société Nationale d'Etude et de Construction de Moteurs d'Aviation, Snecma
Société Turboméca, TM
Volvo Flygmotor AB, Volvo
Deutsche Forschungsanstalt für Luft- und Raumfahrt, DLR
Defence Research Agency, DRA
Office National d'Etudes et de Recherches Aérospatiales, ONERA
CERT/ONERA
Cranfield Institute of Technologies, CIT
Instituto Superior Técnico Lisboa, IST
University of Karlsruhe
University of Rouen-CORIA
Contact: Mr. G. Pellischek, MTU München GmbH, Abt. ETW, Dachauer Strasse 665, D-80995 München, Germany (Tel : +49/89/1489-2413, Fax : +49/89/1489-6301)

Advances in Engine Technology. Edited by R. Dunker
© 1995 John Wiley & Sons Ltd.

0 Abstract

To provide an assessment of the pollutant emissions procedure by future aero engines, five tasks were reviewed.

Task 1. Definition of combustor conditions for future aero engines

Formation of pollutants within gas-turbine combustion chambers will be dependent upon the combustor design technology employed and the combustor operating conditions. In order to assess future engine emission levels with both conventional and low-emission combustors, the combustor conditions presumed for engines to be certified at the end of the decade were specified. Data from seventeen engines, supplied by the industrial partners of the project, were summarized under the six engine categories of propfans, turbofans, turboprops, turboshafts, APUs and supersonic turbofans.

Task 2. Assessment of pollution levels using present technology

The combustor operating conditions were used in subsequent tasks to predict the expected pollution levels using future engine cycles and current combustor technologies as well as using low NO_x combustor technologies.

In the near future, aero engines will be subject to the present ICAO recommendations, but due to the reduction of emission levels of recommendations to come, improved combustors with reduced NO_x emissions will be required.

Task 3. Summary of pollution reduction methods

A clear reduction in the emission of nitrogen oxides should be possible by the careful design of the combustion chamber. As combustion chambers have already been optimized with regard to the emission of unburnt hydrocarbons and carbon monoxide, futher modifications to the combustors are not likely to bring about a further reduction in these emissions.

Task 4. Fundamental investigations

Task 4.1. Study on atmospheric atomization with air-blast atomizers

A prediction model for spray formation was assessed and developed by using advanced measurement techniques during tests and the numerical simulation of an air-blast injector. Advanced measurement techniques were applied to study the flow field downstream of an air-blast injector. The prediction model for spray formation was developed and validated with measurements carried out in this BRITE/EURAM project to simulate the phenomena downstream of this atomizer and to compare them with the measured results.

Task 4.2. Demonstration of limiting capabilities of laser diagnostics

Traditional combustor exit plane measurements of pollutant concentrations do not provide any information about their origin or the detailed mechanism of their formation. Application of non-intrusive laser-based diagnostics will provide such information but is not practical near engine operating conditions. As part of the fundamental experimental programme, the range of conditions over which non-intrusive techniques can be applied was explored. Phase Doppler anemometry (PDA) proved to be best for the fuel drop size and velocity measurements, which were demonstrated successfully up to 12 bar in an isothermal spray and up to 5 bar in a burning model combustor. Coherent antistokes Raman scattering (CARS) measurements of gas temperatures within similar model combustors were demonstrated up to 6 bar but finally a more conservative limit of 2.5 bar was suggested to maintain spatial and temporal resolution.

In the associated cross-correlation exercise on spray diagnostic instruments, surprisingly good agreement of drop size and velocity measurements was demonstrated on an isothermal kerosene spray from a circular fuel injector. During the programme, significant confidence was generated in the comparability of test bed standards and in the reliability of the instruments employed. The comparison of two CARS systems showed significant discrepancies but also proved to be invaluable in identification of problems associated with turbulent, liquid-fuelled combustion which could not be reproduced with a conventional calibration device.

Task 4.3. Influence of primary zone architecture and stoichiometry on pollution formation in a tubular combustor

The influence of primary zone air-flow distribution upon pollutant emissions and lean blow-out was studied and could be understood. Five liner configurations were tested and investigated numerically. Air-flow distribution, gaseous emission and smoke measurements were carried out with the liners mentioned up to 40 bar. These combustors were submitted to lean blow-out tests, down to 0.6 bar. In order to design an optimum combustor and to enable its pollutant emission level assessment, three-dimensional calculations were performed. The phenomenon of lean blow-out was studied and modelling results were compared with experimental measurements. The last configuration tested, a rich quench lean combustor showed a NO_x emission index reduction of 50% compared with the baseline tubular combustor.

Task 4.4. Influence of fuel distribution and droplet size on homogeneity and pollution formation

The influence of fuel distribution and droplet size on homogeneity and pollution formation was investigated experimentally, using a lean-burning, rectangular combustor. To realize high air flow through the combustor head, generic tests of several fuel air nozzles were carried out. The final version consists of 14 nozzles instead of the three-nozzle conventional sector design. The results show 50% lower NO_x emission compared with a conventional combustor design.

Task 4.5. Influence of air distribution on homogeneity in a tubular primary zone combustor

The effect of a variation of combustor primary zone entry geometries on gas turbine emissions was studied. Isolated primary zones were applied for these experiments so that complications due to continued combustion and the production of pollutants in the intermediate and the dilution zones could be avoided. Favoured by the relative simplicity of isolated primary zones, a wide range of operating conditions could be varied parametrically. The results indicate that combustor designs can be influenced to an extent so that excellent idling emissions, smoke and combustion stability can be achieved and NO_x emissions are reduced by 30% to 40%. The NO_x reduction mechanism was identified.

Task 4.6. Smoke production in rich burning primary zones

The increased importance of gas turbine emissions initiated very specific activities in combustor design and in the evaluation of alternative emission reduction strategies. Particulate soot generated in fuel rich zones within the combustor is both the precursor of exhaust smoke emission and the principal source of thermal radiation to the combustor liner. Therefore, it is an unacceptable pollutant as well as a significant determinant for combustor durability and life. Whilst lean-burning strategies, developed for reduced NO_x emissions, will also tend to reduce smoke levels, other approaches—i.e. those incorporating rich-burn–rapid-quench designs—very substantially enhance soot production in the richer-burning zones and therefore have subsequently to incorporate a highly effective burn-out.

Smoke measurements are traditionally restricted to the engine exhaust and involve techniques of sample extraction and their subsequent analyses. For purposes of detailed and more mechanistic combustor design analyses, however, spatially resolved measurements are required from within the combusting zones, employing techniques constrained to minimize the perturbations to both the flow field and the measured sample. It is an objective of the research undertaken to investigate options for mapping soot concentration in the rich-burning zones of a combustor liner and to demonstrate suitable techniques via measurement in a model tubular combustor over a range of operating conditions (varying, in particular, air/fuel ratios and pressures). The particular combustor selected for this investigation features also in other diagnostic studies within the low emission combustor technology programme and incorporates elements of the rich-burn–rapid-quench design also investigated.

Task 4.7. Modelling and optimization of an existing premix duct, computational and experimental research

To obtain more information about the air flow, detailed fuel–air mixing and fuel-evaporation tests were carried out in the premix duct of an existing annular combustor. In spite of intensive efforts, the adaptation of an existing combustor code failed because of the very complex geometry of the combustor. A simplified combustor model was introduced and a new mixing tube was derived from the computer calculations. A segment combustor already available with the conventional premix duct configuration was tested first at atmospheric pressure without combustion and afterwards at a 3.5 bar pressure level with fuel burning.

Velocity and droplet measurements at the mixing tube exit applying phase Doppler particle anemometry (PDPA) as well as emission-rate measurements were accomplished. Additional atmospheric and emission tests were run with the improved mixing-tube configuration without the annular combustor.

The computational results show that more work has to be invested before a reliable calculation for a given combustor arrangement can be performed. The test results indicate that the fuel–air mixing and the fuel vaporization require further investigation to define new suitable premixing tubes. Therefore, more detailed work is necessary in a follow-on research programme.

Task 5. Selection of pollution reduction methods

Summarizing, three different promising combustion concepts for use in low-emission aero engines were identified. These three methods are the double annular combustor, the lean premixed and prevaporized combustor, and the rich-burn, quick-quench and lean-burn combustor. They differ in terms of both the potential for nitrogen oxide reduction and the development effort and risk involved.

- 20–40% reduction can be obtained by a radially or axially staged combustor in which the fuel is burned lean without premixing and prevaporizing. The development risk of this combustor is modest.

- The rich-lean combustion method comes with a lower development risk than the lean premixed concept. Its reduction potential is between 80% and 90%. Since in the main reaction zone the fuel must be burnt with minimal oxygen present, cooling for this zone must necessarily be convective leading to a thermal problem of the combustor walls. Sooting is a further problem, and another is that of rapidly, homogeneously admixing the air to the reaction gases in the first zone.

- The effectiveness of lean combustion can be enhanced by premixing and prevaporizing the fuel. The reduction potential here is 80–90%. This concept, however, is embarrassed by largely pressure- and temperature-related auto-ignition and flashback problems. Unlike the rich-lean concept the lean premixed concept requires further in-depth background research to resolve the inherent problems.

All three concepts showed potential in advanced low-emission aero engines, although they differ in the amounts of development investments they still require. For this reason, further investigations into emissions reduction methods are envisioned for the next project phase.

Assuming that requisite fundamental work is carried out and a start is made on the development of appropriate combustors, referred to present levels, a 30% reduction in the emission of nitrogen oxides by engines of the mid-nineties would appear feasible, and a reduction by 80% should be possible by the end of the century. The achievement of these targets calls for investigation of both the combustion and combustor control aspects.

1 Introduction

Pollutants produced by aero engines are CO, unburnt hydrocarbons (UHC), nitrogen oxides and smoke. During the last thirty years, the emission level of these pollutants widely changed. The emission index of CO and UHC decreased significantly. Smoke emission initially decreased as well, but nowadays, an increasing tendency to smoke production can be stated. Since the development of gas-turbine engines, the emission index of NO_x has increased steadily.

The decrease in smoke emission was achieved by the introduction of new fuel-atomizing systems. These new systems, however, incurred problems related to relight at high altitudes and the lean flame-out limit.

At present, considerable effort is being made to lower the smoke emissions of the combustion chambers of new large civil engines (Rolls-Royce Derby, SNECMA, MTU) and small engines (BRR, Turboméca). However, advanced technology for lowering the NO_x emissions has not been incorporated into current production engines.

Low-pollution combustion chambers, in which attempts were made to reduce the NO_x emissions, were investigated in the United States in a number of research programmes (ECCP, PRTP, SCERP) in the mid-seventies. Similar investigations were also carried out by Rolls-Royce Derby and by SNECMA. Combustion chambers suitable for production were not forthcoming from these programmes. Typically, the decrease in NO_x emission was accompanied by an increase in the CO and UHC emissions. Furthermore, safety problems (autoignition, flashback) could not be resolved satisfactorily.

As a consequence of this situation, more generic research in this area is needed.

A reassessment of the possibilities of low-pollution combustion has to be undertaken since improved tools are available for the analysis of the physical relationships leading to the formation of pollutants. Innovative methods were developed for high-precision analysis of two-phase flow and for non-intrusive measurement of temperature and concentration. These methods of measurement, which represent the most advanced state of the art, were applied to this project.

2 Research objectives

The major point of interest of this work is to provide an assessment of the pollutant emissions produced by future aero engines. This task is considered an urgent one since increased pollutant emissions from aircraft are expected for two reasons:

- The volume of air traffic will increase significantly (up to twice today's level).
- New civil aero engines emit more NO_x because of higher pressure ratios.

An assessment of the pollutant emissions will be provided when the project is completed.

At the end of the two-year pilot phase, methods of pollutant reduction suitable for application under the operating conditions of future aero engines were selected. For the selection of these methods, various fundamental investigations had to be performed, providing the following benefits:

- The systematic investigation and comparison of advanced measurement methods allowed the assessment of their quality and their limits of application.
- Supported by detailed measurement, the understanding of liquid atomization had to be improved in favour of the specific design of future atomizers.
- Numerous possibilities for mixing fuel and air could be investigated, and this allowed the comparison of their results with regard to the achieved mixture homogeneity.
- Calculation methods for the description of the atomization process and the formation of NO_x in combustion chambers had to be tested.

Although these investigations were aimed specifically at the requirements of aero engines, the results and methods derived can be applied to many areas of combustion technology, in particular to industrial gas turbines.

The main objectives of the tasks investigated are listed below.

Task 1. Definition of combustor conditions of future aero engines

Important combustor conditions which result from the engine performance are combustor inlet pressure and temperature and combustor outlet temperature. The influence of these parameters is also important for the pollution formation process as well as for the applicability of different pollution reduction methods. Combustor conditions of engines certified at the end of the nineties had to be determined separately for large and small civil engines. MTU, Rolls-Royce Derby and SNECMA performed this task for large engines whereas Turboméca, BRR and also MTU estimated combustor conditions for small engines.

Task 2. Assessment of pollution levels using present technology

Data of pollution levels of present aero engines under five different ratings (idle, take-off, climb, cruise, approach) had to be collected and compared with overall correlations for the different pollutants (CO, UHC, NO_x, smoke). The best-fitting correlation had to be selected to carry out the assessment of pollution levels of future combustors comprising conventional combustor technology. This task had to be performed for small and large engines separately.

Task 3. Summary of pollution reduction methods

It was required to set up a summary of different pollution reduction techniques applicable to aero engines as well as to industrial gas turbines. Specific aspects of each technique, e.g. degree of reduction, limits of applicability, critical elements of the relevant method, had to be considered. Therefore, it was necessary to summarize and critically discuss examples of low-emission combustors described in the literature.

Task 4. Fundamental investigations

A coordinated series of studies had to be undertaken by each of the partners to investigate fundamental aspects of combustion with the following objectives:

(a) to quantify the achievable reduction potential in pollution level by various technically feasible low-emission combustor technologies and generate confidence in the selection of the most promising methods for Task 5.

(b) to provide an improved understanding of the primary zone processes which have a strong influence on emissions levels by use of advanced instrumentation techniques and mathematical modelling.

The quality of fuel preparation within a combustor primary zone together with the uniformity of subsequent fuel–air mixing are two key parameters influencing the production of gaseous emissions. The first had to be studied in Tasks 4.1, 4.2, 4.4 and 4.6, whereas the second was dealt with in Tasks 4.3, 4.4, 4.5 and 4.7.

Task 4.1. Study on atmospheric atomization process of air blast atomizers

The study had the objective of predicting the spray formation of an air-blast atomizer representative of those used in aeronautical combustors. The fuel-sheet behaviour was characterized by measurement of the corrugation wave length and the location of fuel-sheet breakdown.

The results represent the basis for the development and assessment of adequate models. Detailed measurement methods were used on a limited number (four) of configurations.

Task 4.2. Demonstration of limiting capabilities of laser diagnostics

Emissions research which demands testing at or near to engine-operating conditions, is extremely expensive and data acquisition from the combustor interior at these conditions is technically difficult and usually impossible. The traditional 'cut and try' development method is unlikely to yield an optimum solution to the pollution problem within an acceptable cost and timescale. The application of measurement methods and computational tools, developed within the last five years, has contributed significantly to a solution.

In this task, the range of conditions over which non-intrusive diagnostic techniques can be applied had to be explored and, if possible, extended. Data generated were intended to be used by all interested partners for computer model development and validation to enable the optimum concept of future experiments in this field. Subsequently, the refined models had to be used to predict the performance of systems at engine conditions and to support the optimization of the design.

Measurements of both the gas and liquid phase had to be carried out, cold and with combustion, in a pressure range of 1–15 bar.

The laboratories of Rolls-Royce Derby, SNECMA, DLR, Cranfield, Karlsruhe and ONERA were asked to participate in the cross-calibration of optical instruments used within the partnership, to enable a realistic assessment of reliability and accuracy of the data and the selection of the best instruments for use in future experiments.

Task 4.3. Influence of primary zone architecture and stoichiometry on pollution formation in a tubular combustor

This task was dedicated to the study of the influence of primary zone geometry and the air-flow distribution on the CO and NO_x emission level. The objective was to obtain design tools to define future low-pollution combustion chambers.

The work had to be performed on an existing tubular swirl stabilized combustor where SNECMA-defined modifications of the flame tube and the injection system were integrated.

The combustor was intended to be tested at atmospheric and at subatmospheric pressure conditions.

Task 4.4. Influence of fuel distribution and droplet size on homogeneity and pollution formation

Under high combustor inlet temperature and pressure conditions, premixing and prevaporizing is difficult to verify due to auto-ignition and flashback in the premix duct. In this case, mixing of fuel and air has to be performed in the primary zone. The efficiency of pollution reduction in a lean combustor environment is highly dependent on the initial distribution of air and fuel and on the droplet size. It was agreed that the influence of fuel distribution and droplet size on pollutant emission be investigated in Task 4.4 in a rectangular duct, simulating an annular sector of a combustor. The fuel system to be studied in this task consisted of a large number of air and fuel injectors, providing the option of varying the fuel distribution by activating different sets of fuel injectors. Droplet size had to be varied by changing the pressure drop.

Task 4.5. Influence of air distribution on homogeneity in a tubular primary zone combustor

The idea of this task was to focus the investigation activities on the influence of homogeneity, i.e. the mixing of fuel and air in the primary zone.

DRA has considerable experience in the practical application of multiple jet primary zones, where the air is discharged through a large number of orifices into the head of the combustor. Fuel introduced into the shear layers surrounding these jets of air mixes rapidly and, in addition, enables turbulent flame processes to stabilize locally and to burn efficiently. Moreover, the higher the number of jet systems, the more uniform the primary zone combustion process becomes as a whole.

A combustor, such as the one described, provided the basis for a more detailed study of the influence of homogeneity on pollution reduction by varying the quantity and the distribution of the primary zone air flow.

Task 4.6. Smoke production in rich burning primary zones

The basic mechanisms of smoke formation in gas turbine combustor primary zones are still poorly understood. Combustors, employing a fuel-rich reaction zone to inhibit NO_x formation, are likely to promote soot formation and therefore increase the risk of smoke emissions unless adequate burnout is subsequently achieved.

To distinguish between the smoke formation and the consumption process, further diagnostic information from within the combustor reaction zone was required. The objective of this study was to develop reliable means of measurement and to obtain spatially resolved soot concentration data in the primary zone of the model combustor used in Task 4.3.

Task 4.7. Modelling and optimization of an existing premix duct, computational and experimental research

Premixing is a powerful means of NO_x reduction, but only practical at lower combustor inlet temperatures and pressures. An existing combustion chamber geometry with premix ducts tested in an earlier test series was chosen as a model for a computer program.

Calculations and tests had to be carried out for fuel preparation in the premixing duct first and afterwards for combustion in an annular flame tube. Investigations of the influence of evaporation and premixing on NO_x generation and the emission rate by varying the premixing tube geometry were required. Comparison of calculation and test results were conceived to lead to a general assessment of premix design parameters.

Task 5. Selection of pollutant reduction methods

On the basis of the results obtained from Task 4, the pollutant reduction methods summarized in Task 3 had to be critically reviewed and the most promising methods had to be selected. Criteria for selection were the potential of pollution reduction, reliability, safety and weight and volume requirements.

3 Research activities

Task 1. Definition of combustor conditions of future aero engines

The combustor operating conditions expected for engines certified by the end of the century were summarized in Task 1. A survey of data from seventeen engines provided by the industrial partners addressed small and large engines separately. The six engine categories of propfans, turbofans, turboprops, turboshafts, APUs and supersonic turbofans were selected with nine representative engines to cover a wide range of engine applications. These data provided the basis for prediction of future engine emissions levels in subsequent tasks.

Task 2. Assessment of pollution levels using present technology

Correlations were applied in order to estimate the expected emissions for future aircraft gas-turbine engines (defined in Task 1). These correlations are based on existing engines as well as on SNECMA and Turboméca experimental

Table 2.1 Responsible and cooperating partners

Task	Resp. Partner	Cooperating Partners	Subtask 1 No. Resp.	Subtask 2 No. Resp.	Subtask 3 No. Resp.	Subtask 4 No. Resp.
1	RR	MTU, BRR, SN, TM	—	—	—	—
2	SN, TM	BRR, RR	—	—	—	—
3	MTU, BRR	RR, SN, TM, Volvo	—	—	—	—
4.1	SN	TM, Lisbon Rouen, Cert	4.11 Cert	4.12 Rouen	4.13 TM	—
4.2	RR	Cranfield, CERT, ONERA Karlsruhe DLR, Rouen	4.21 Cranfield	4.22 RR	4.23 RR	—
4.3	SN	Volvo, Onera	4.31 SN	4.32 Volvo	4.33 SN	4.34 Volvo
4.4	MTU	DLR	4.41 MTU	4.42 DLR	—	—
4.5	DRA	—	4.51 DRA	4.52 DRA	—	—
4.6	Cranfield	RR	4.61 Cranfield	4.62 Cranfield	4.63 Cranfield	—
4.7	BRR	Karlsruhe	4.71 BRR	4.72 Karlsruhe	4.73 Karlsruhe	—
5	MTU	RR, SN, TM, BRR, Volvo	—	—	—	—

combustors. For each pollutant and for each existing engine a comparison between the ICAO data bank and the correlations mentioned was employed [1].

SNECMA correlations for future large engines were chosen to evaluate the emission index of NO_x, CO and unburnt hydrocarbons. These correlations are related to the engines: CFM56-5al, V2500, CF6-80c2/al, PW2037, RB211-524G, RB211-535E4, Tay MK620-15 as well as a SNECMA experimental combustor.

Turboméca correlations for small engines were based on the generally admitted background on pollutants formation and on Rolls-Royce and Turboméca results with their new turboshaft engines and with several research combustors. EPA L.T.O. cycles were used for turboprop engines, but because no information for turboshaft engines and APUs was available, Turboméca defined 'power/time cycles' for these two types of engines.

Task 3. Summary of pollution reduction methods

Concerning the combustor entry conditions defined in Task 1, each of the partners involved in Task 3 provided a review of the numerous methods of reducing pollutant emissions, giving details of the expected degree of reduction, the field of application and critical aspects of each method in comparison to conventional combustors.

Task 4. Fundamental investigation

Task 4.1. Study on atmospheric atomization process of air blast atomizers

Three identical air-blast (annular sheet) injectors were built by CORIA (University of Rouen): one was used by CORIA itself, another one was installed by IST (University of Lisbon) and the third one was shared between Turboméca and CERT.

Four configurations of the model air-blast atomizer were defined:

Geometry 1:   liquid sheet (400 μm) + inner air
Geometry 2:   liquid sheet + inner air + outer air
Geometry 3:   liquid sheet + inner air + outer air with swirl
Geometry 4:   same as 3 but with different sheet diameters

Subtask 4.1.1. Spray diagnostics The model air-blast atomizer was used for

- Flow visualization and laser Doppler anemometry (LDA) measurements at Lisbon with the first three geometry configurations [2],

- LDA and PDA measurements, high-speed video monitoring and pressure-drop measurements by CERT [3],

- Flow visualization at CORIA [4],

- Malvern measurements of droplet diameters carried out by Turboméca on three different injector geometries with water and kerosene applying several values of air–fuel ratios.

Subtask 4.1.2. Atomization by an airblast atomizer, elementary models A linear analysis of the instability of a liquid annular sheet was derived and validated by CORIA on the basis of flow visualization experiments [4]. This theoretical part is related to the first phase of the atomization process which takes place in a real atomizer. Liquid sheet instability, frequencies and amplitudes and liquid sheet break up length were studied by CERT.

Subtask 4.1.3. Flow field calculations With measured initial conditions, provided by IST and CERT (Subtask 3.1.1), SNECMA and Turboméca completed the calculations, using two different numerical models, and compared the theoretical and experimental results.

The calculations performed by SNECMA with a two-dimensional finite element single-phase code correspond with the three first geometries (inner air, inner air + outer air, inner air + outer air with swirl). The calculations performed by Turboméca with a three-dimensional finite volume code correspond with the geometry configuration 3 (inner air + outer air with swirl) and different values of air velocity.

Task 4.2. Demonstration of limiting capabilities of laser diagnostics

Recently developed laser-based, non-intrusive diagnostic techniques offer the potential for measurement of detailed processes occurring within combustor reaction zones. However, operation at near engine operating conditions is impractical both from test costs and obtaining optical access. Task 4.2, therefore, explored the operating envelopes of various techniques applied to cold fuel sprays and to burning model combustors up to 6 bar [5].

In order to generate confidence in the instrumentation itself and to ensure compatibility between laboratories of various partners within the collaboration, a cross-correlation exercise was undertaken on spray diagnostics. An airspray fuel injector mounted in a perspex airbox was circulated between Rouen University, Rolls-Royce, Cranfield (CIT), DLR and Karlsruhe University. Measurement of drop size and velocity were made at specified conditions and locations within the atmospheric spray. Instruments employed included several phase Doppler anemometers (PDAs), laser amplitude Dopplers (LADs) and Malvern diffraction and holography. The extension of the PDA technique to higher pressures was also explored by Cranfield along with the assessment of the quality of sheet lighting needed to obtain sufficiently high-resolution photographs for image analysis.

Within model combustors the feasibility of PDA, LAD, and CARS was investigated, their limitations identified and solutions proposed. Again, in order to increase confidence in data transfer between laboratories, the CARS system of ONERA and Rolls-Royce were applied to the same tuboannular combustor mounted on the combustion rig at SNECMA and Rolls-Royce, respectively. Full details of rigs, operating conditions and results are included in [5] to [14].

Task 4.3. Influence of primary zone architecture and stoichiometry on pollution formation in a tubular combustor

Subtask 4.3.1. Modifications and tests of the tubular combustor An existing SNECMA tubular combustor was modified to four new configurations. The tests aimed at comparing these various combustors in terms of emissions and lean blow-out limits.

The configurations mentioned split into two groups: a lean primary zone family (the base-line configuration and two modified configurations) and the rich primary zone family (two configurations).

The base-line combustor was tested at ONERA (gas analysis) up to 40 bar. It was tested along with the two other configurations by Volvo at 530 K and 3.5 bar and at 660 K and 7 bar during gas analysis measurements and under atmospheric pressure conditions during the lean blow-out tests [15].

SNECMA designed, built and tested (gas analysis and smoke measurement) two 'rich quench lean' tubular combustors up to 7 bar and 549 K. Volvo tested the second version of the RQL combustor during gas analysis measurements (up to 7 bar and 660 K) and lean blow-out tests [16].

Subtask 4.3.2. Lean blow-out and relight tests at subatmospheric conditions The five tubular combustors defined by SNECMA in the previous subtask were tested by Volvo [16], in subatmospheric environment.

The lean blow-out tests were carried out at 1 bar at 373 K and 0.8 bar at 373 K on the reference liner.

The second version of the rich–lean combustor was submitted to LBO tests at 1 bar at 370 K and at 1.0, 0.8 and 0.6 bar at 290 K. During these test series, the liners ran on a wide flow function range while the LBO was judged visually by comparing the flame with the reading from a thermocouple at atmospheric pressure, whereas at subatmospheric conditions only the thermocouple reading was used.

Subtask 4.3.3. Pollutant calculation Navier–Stokes three-dimensional, reactive, two-phase calculations were performed by SNECMA in order to design the rich–lean configurations [17]. These calculations enabled SNECMA to optimize the air-flow distribution and the jets, penetrating the quenching zone and thus minimizing the near stoichiometric volumes (NO_x generation) and the pressure loss through the combustor.

Subtask 4.3.4. Feasibility study of lean blow out modelling The goal of this subtask was to study the feasibility of lean blow-out-modelling [18], seen in the light of the current state of chemical kinetic modelling and flow modelling.

The process of LBO was investigated, the necessary simplifications required for reasonable blow-out calculations were discussed and a simplification of the very complex schemes for hydrocarbon combustion was proposed.

Using a simplified chemical kinetic theory, LBO calculations were carried out and compared with experiments.

Task 4.4. Influence of fuel distribution and droplet size on homogeneity and pollution formation

Four different fuel injection systems were investigated. The drop size and the flow pattern of the sprays were characterized by light diffraction analysis and by laser light sheet visualization, respectively. The influence of ambient pressure on drop size was measured for pressures up to 6 bar.

The influence of several fuel injectors on emissions was investigated in preceding tests with a 5 nozzle annular multi-injector combustor. A MTU design and a DLR design, both with variations in the swirler configuration and fuel distribution, were compared with a single nozzle design at atmospheric pressure tests without air preheating at different pressure-drop settings and air-to-fuel ratios. Gas analysis was performed with a movable single probe at the combustor exit and profiles of the gas compostion were measured with two traverses.

A multi-injector test rig with a rectangular cross-section was designed and manufactured. The fuel was injected through 14 nozzles instead of three, typical for the conventional design of this sector. Gas probes could be traversed in several planes inside the combustor and at the combustor outlet.

The most promising air-blast atomizer design was tested in the rectangular combustor rig at several air-flow rates and equivalence ratios to study the influence of drop size, homogeneity and dwell time on pollution formation. The pollution was measured in two planes inside the combustor and at the combustor outlet using a movable single probe. Additionally, the lean blow-out limit was measured.

Task 4.5. Influence of air distribution on homogeneity and pollution in a tubular primary zone combustor

Combustor primary zones with a wide range of primary zone entry hole geometries were designed, built and tested. This geometry range includes variations of the spatial distribution of holes, the number and the size of the holes. The experiment was designed to explore the effects on emissions of parameters such as local and global air–fuel ratios and the air-mixing rate and scale.

Isolated primary zones were selected for this task so that secondary effects such as the continued combustion of emissions like carbon monoxide and smoke or the generation of nitrogen oxides (NO_x) in intermediate and dilution zones could be eliminated. The relative simplicity of the primary zones allowed parametric variations over a range of important combustor operating variables. These variations would not have been possible within the complexity of a complete combustor. Parameters such as the air–fuel ratio, the residence time, the air mass flow and the pressure loss were varied systematically and the effects of the individual parameters on emissions were isolated by analysis.

All the combustors were designed to have the same shape, volume, pressure loss and wall cooling air flow so that the results of the tests on different combustors are comparable.

The effects of varying operating conditions, such as the pressure loss, would have changed the atomization quality produced by an airspray atomizer and this would have masked the effects of other variables on the emissions. To avoid this problem, an air assist atomizer with an external air supply was used throughout the programme. This ensured that atomization quality and fuel placement was the same for all combustors and that valid comparisons were possible between different combustors and tests.

Gas samples were obtained from emissions measurements by means of a temperature-conditioned five-hole gas-sampling probe. This probe was traversed across the combustor exit in two-degree increments. The gas samples were analysed for unburnt fuel, carbon monoxide, oxygen, carbon dioxide, nitric oxide, nitrogen dioxide and smoke. Combustion efficiencies, emission indices and gas temperatures were determined analytically by the data input from the gas analysis measurements. Because of the high temperatures and temperature differences measured, it was necessary to apply density weighting corrections to these calculated gas analysis data [19]. After the data processing procedure described, the agreement between gas-analysis air–fuel ratios and rig-metered air–fuel ratios were generally better than ± 1 AFR unit.

Combustors were tested at idling conditions typical of a modern high pressure-ratio engine. At idle conditions, the combustors ran at air–fuel ratios of 60, 80 and 100 AFR to determine the efficiency vs. air–fuel ratio characteristics together with an estimate of the combustion weak stability.

The combustors were also tested at simulated full power conditions with a high air loading and a typical advanced engine cycle inlet temperature. A reduced inlet air pressure was employed due to compressor plan limitations. At the full power conditions, the combustors were tested at overall air–fuel ratios of 24, 28, and 32. About 30% of the combustor air flow was used for wall cooling of which only a very small fraction became involved in the combustion process. So, the actual combustion air–fuel ratios measured in the exit plane reached values of about 16, 19 and 22. Combustor wall temperatures were measured by means of temperature-sensitive paints.

In a preceding phase of this task, three pairs of combustors covering a wide range of air entry hole geometries were evaluated by means of a short, simplified test programme. On the basis of these test results one combustor out of each pair was selected for further and more extensive evaluation in the second phase of the task. For each of these combustors two copies were manufactured with the same number and location of air entry holes as the original sample. However, the free flow area of the holes in one copy was reduced by 20% to represent a low-porosity/high-pressure-loss combustor, while the holes in the others were enlarged by 20% to create a high-porosity/low-pressure-loss combustor.

The combustors used in the second phase of the tests were designed to permit the selection of range of atomizer air–fuel ratios by means of simple mechanical adjustments. In total, the combination of nine combustors, each with three atomizer air–fuel ratios, makes 27 different combustor set-ups for the second phase of the test series, with the same test conditions as in the first phase. However, additional tests were regarded to be necessary in the high-power range, where the standard air mass flow was varied by ± 10%. While the increased air flow resulted in a higher pressure loss and a lower residence time, the reduced air flow led to a lower pressure loss with an increased residence time.

Task 4.6. Smoke generation in rich burn primary zones

The research programme incorporated three distinct elements:

- lay out, design and commissioning of a test rig for *in situ* soot measurements;
- development of suitable laser absorption techniques for the measurement of soot concentration;
- soot volume fraction measurements in rich-burning zones of a model tubular combustor.

A pressure casing was designed to accommodate the SNECMA tubular combustor, providing line-of-sight optical access for laser absorption measurements at stations between the primary and dilution hole locations.

The restriction in optical access provided by the liner prevents spatially resolved measurements applying more conventional integral absorption techniques and tomographic inversion. Therefore, spatial resolution was achieved by direct differential laser absorption, incorporating robustly designed traversable optical probes. Two schemes permitting transverse and axial access to the liner were developed for this study.

Within the combustor, measurements of the local soot volume fraction were made at operating pressures up to 4 bar, inlet temperatures in the order of 200 °C and over widely varying air–fuel ratios ($12 < AFR < 45$). The sight-tube probe proved to be particularly robust in this hostile combustor environment, but reliable in both operation and interpretation.

Task 4.7. Modelling and optimization of an existing premix duct, computational and experimental research

The existing BMW/Rolls-Royce (BRR) combustion chamber [20], which is part of this investigation, consists of 16 premix ducts with air-blast nozzles [21]. These premix ducts are installed at the front of the annular combustion liner, as shown in Fig. 2.1.

For the experiments an existing test rig with burner segments was used. In this set-up, three premixing ducts which ended in an annular segment were exposed to air and fuel flow. The measurement device, installed behind this annular segment, was used to measure the temperatures, pressures end exhaust emissions. The segment burner was designed for a pressure up to 6 bar.

To describe the three-dimensional mixing and evaporation of fuel droplets [22] in the premixing zone, calculations were carried out at the 'Institut für Strömungsmaschinen' at Karlsruhe (ITS).

With the computer code employed, the transport equations of a turbulent, chemically reacting and three-dimensional flow in general curvilinear coordinates can be solved numerically by use of a finite volume discretization [23]. In cases where a strong mutual coupling caused by recirculation had to be considered, the convergence rate of the solver mentioned was rather low. So, the adaptation of the existing computer code at ITS failed.

In view of the lack of the computer calculations originally planned, a strongly simplified computer model was composed, based on an original mixing tube followed by a can-type combustor.

BRR calculations were made for the original premix duct and for an advanced configuration with flat air-blast atomizers and a counter-flow air swirl shown in Fig. 2.1, using the computer code 'COMB-3D' [24] for gaseous fuel.

Burner parts for the segment burner and the improved premixing duct were designed and manufactured by BRR. A very intensive air–fuel mixing inside the mixing tube with large turbulence is the objective of this design. So, the improved mixing tube consists of six flat air-blast atomizers located between six outer diameter swirl holes and six inner diameter swirl holes, arranged in counter swirl (Fig. 2.1). It proved to be necessary to locate a deflector behind the flat airblast nozzles to tilt the fuel/air mixture jet to the center line of the tube.

To get a feeling of the spray quality both a swirl spray nozzle DELAVAN 46817-13 and the BRR air blast atomizer D128-X.4 were measured by PDPA at ITS, Karlsruhe.

The existing segment burner was assembled for cold tests at ITS-Karlsruhe. Looking at the exit of the centre mixing tube through the quartz glass window, PDPA measurements along and across the optical axis were accomplished.

For the same segment burner arrangement, but located in the hot test facility at ITS, spray and flow measurements with burning fuel were carried out up to 3.5 bar. In addition, emission measurements (NO_x, CO, CO_2, O_2) were carried out.

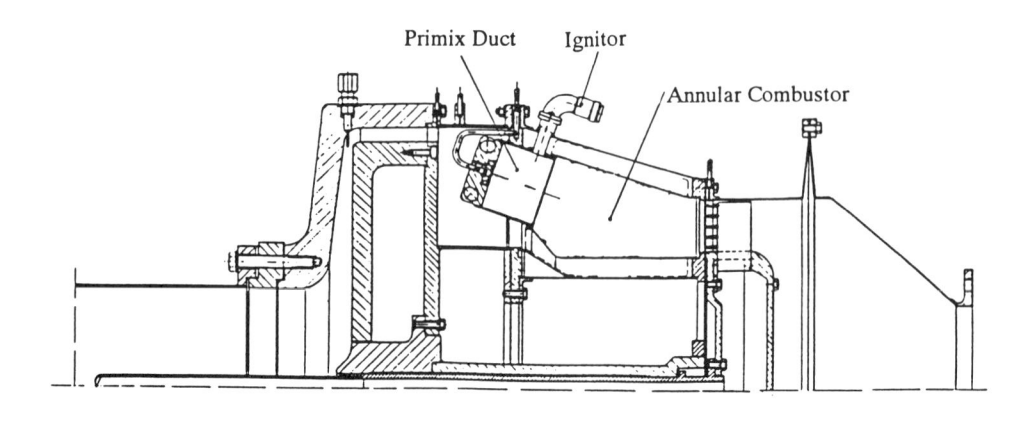

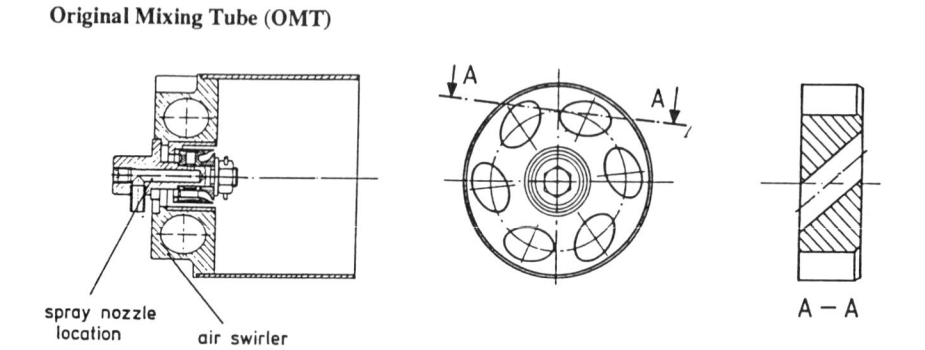

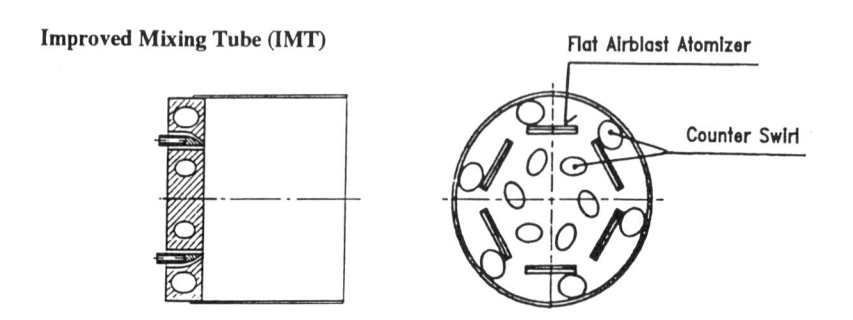

Figure 2.1 Test combustor arrangement.

The hot test comprised atmospheric tests as well as high pressure experiments. The maximum operating point, ran at ITS Karlsruhe corresponds to 80% load. For all experiments preheated clean air was used. Due to the modular set-up of the preheater the air inlet temperature was about 550 K for the conditions under investigation, the fuel mass flow rate was reduced respectively.

For the new improved mixing tube with flat air-blast nozzles and counter-flow air swirlers BRR investigated:

- the flow field at the mixing tube exit by measurements;
- the air flow characteristic of the mixing tube by measurement;
- the ignition behaviour and the ascertainment of the lean blow-out limitations.

Task 5. Selection of pollutant reduction methods

Based on the results of the investigations from Task 4 as well as on a literature study, all industrial partners involved critically reviewed the preliminary selected NO_x reduction methods. Further investigations were recommended.

4 Research results

Task 1. Definition of combustor conditions of future aero engines

The full set of data provided by all partners and a complete description of the task is given in [25]. Table 2.2 summarizes the combustor operating conditions for the six engine categories at take-off, idle and cruise operating points. Defining these engine conditions, the EPA landing and take-off cycles (LTO) were used for the majority of engines, but not for turboshaft helicopter engines, for which an AIA (Aerospace Industry Association) defined cycle was selected.

Task 2. Assessment of pollution levels using present technology

Calculation for pollutants mass flow during the 'pollution cycle' was executed for the six engine categories (supersonic turbofans, propfans, turbofans, turboprops, turboshafts and APUs). This constitutes a good basis to estimate the impact of aeronautical emission on world-wide pollution without evolution of combustion technologies.

Task 3. Summary of pollution reduction methods

In conventional combustion chambers, the pollutant emissions are influenced by the fuel preparation system, the residence times of reactants in the combustion chamber and the cycle as such. Optimization of the cycle over the past 25 years has resulted in a reduction of fuel consumption of almost 40%. The full set of information provided by all partners is given in [26].

Table 2.2 Summary of combustor operating conditions

Engine category	Engine designation	Take-off T (K)	Take-off P (kPa)	Take-off AFR	Idle T (K)	Idle P (kPa)	Idle AFR	Mn	Alt (m)	Cruise T (K)	Cruise P (kPa)	Cruise AFR
Propfan	B	860	3550	44.4	500	820	144.0	0.80	10500	820	1500	39.1
Turbofan	E	917	4760	36.9	551	701	114.8	0.83	10668	846	1758	39.9
	G	522	520	38.5	328	140	72.9	0.60	0	542	620	40.1
Turboprop	I	658	1310	39.7	515	545	67.6	0.40	6000	616	735	39.3
	J	722	1828	35.2	555	700	70.3	0.40	6000	667	985	37.4
Turboshaft	K	658	1310	39.7	515	545	67.6	0.20	1500	640	1110	41.2
	M	720	1830	35.2	555	700	75.4	0.20	1500	689	1459	38.8
APU	N	514	556	46.1	505	496	106.6			514	556	46.1
Supersonic Turbofan	Q	735	1622	36.2	453	343	99.7	2.00	18288	917	783	41.4

Nomenclature:
T Combustor inlet temperature
P Combustor inlet pressure
AFR Combustor air fuel ratio
Mn Mach number.

In contrast, nitrogen oxide emissions, represented on the basis of the ICAO parameter, which describe the emissions for a given flight cycle close to airports, remained more or less unchanged. The reduction in the emissions of unburnt hydrocarbons and carbon monoxide resulted from more efficient combustion of the fuel.

To reduce the NO_x-emission, the combustion chamber must be improved.

A 30% reduction in the emission of nitrogen oxides can be achieved by staged combustion, in which combustion takes place in one or both stages of the combustor, depending on the engine operating conditions. The stages may be radial [27, 28, 29] or axial [30, 31]. The NO_x reduction attainable depends on the uniformity of the air–fuel mixture in the lean primary zone [31]. The emission of carbon monoxide and unburnt hydrocarbons remains low. The advantage of this technology lies in the possibility of optimization of the emissions at both high and low power levels without risking critical approaches for other operating parameters such as stability. This technology is listed in the category of moderate NO_x reduction at lower technical risk. The disadvantage of this combustor concept lies in the mixing of both zones and in the large surface area to be cooled compared to conventional combustion chambers which implies an increase in weight [27]. Despite the fact that there still exist some problems concerning combustor control without variable geometry, the fuel preparation and temperature distribution at the combustor exit, practical application by 1995 seems a real possibility. In the USA, the necessary field of experience was derived from NASA programmes over the past 20 years, but by and large, this experience is not available within Europe.

As a result of premixing and prevaporization of the fuel, the nitrogen oxide emissions can be reduced by 70–80%, but this technology currently is only applied in low-pressure-ratio engines. The greater the degree of vaporization and the greater the homogeneity of the fuel–air mixture, the lower will be the emission level [32, 33]. The major problems in the realization of this concept are concerned with auto-ignition and flashback. For this reason, fundamental auto-ignition and flashback investigations of liquid fuel were carried out [34–38].

Because of related problems, the practical application of combustion chambers in which the fuel is prevaporized and premixed with air is not expected to become reality before the end of the century. Again, the technology will be developed in the USA in a multimillion dollar NASA research programme, aimed at supersonic transport application beyond the year 2005.

Another method to reduce nitrogen oxide emissions by the injection of additives can be split into two steps: whilst the reaction temperatures are reduced by the injection of water (at full load), the nitrogen oxides are converted into molecular nitrogen by the addition of ammonia [42, 43]. Both methods are already applied in power stations. However, the aero engine application is associated with the problem of an additional tank and since water injection requires nearly as much water as fuel, the load capacity and the range of the aircraft decrease as a result of the gain in weight. From today's point of view, practical application is not realistic for gas turbines in aero applications.

Since methane or natural gas can be mixed without any problems concerning auto-ignition, a reduction of 80% is achievable using a lean primary zone with premixing [44–47]. The advantage of these fuels lies in the minimal formation of carbon monoxide and unburnt hydrocarbons. Methane or natural gas is used as the main fuel in industrial gas-turbine installations but as far as aero engines are concerned, it was only used experimentally [48, 49].

Existing experimental data of burning hydrogen in a gas-turbine combustor were apparently obtained from conventional combustion chambers and, therefore, the different properties of hydrogen could not be considered sufficiently. Thus, only a slight reduction in nitrogen oxides is to be expected by the use of hydrogen as a fuel [50].

A summary of the NO_x reduction potential of these methods is shown in Fig. 2.2 [26].

```
XX=5% of Present Combustor Technology
xx=5% Variation
```

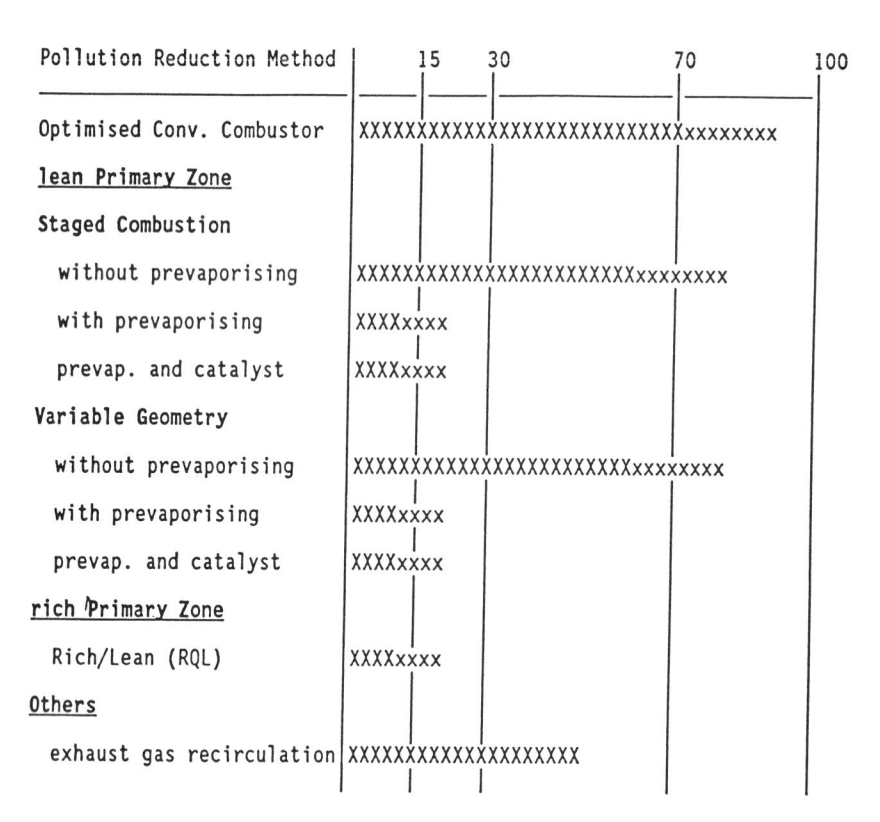

Pollution Reduction Method	15	30	70	100
Optimised Conv. Combustor	XXXXXXXXXXXXXXXXXXXXXXXXXXXxxxxxxxx			
lean Primary Zone				
Staged Combustion				
without prevaporising	XXXXXXXXXXXXXXXXXXXXXXXXXXxxxxxxx			
with prevaporising	XXXXxxxx			
prevap. and catalyst	XXXXxxxx			
Variable Geometry				
without prevaporising	XXXXXXXXXXXXXXXXXXXXXXXXXXxxxxxxx			
with prevaporising	XXXXxxxx			
prevap. and catalyst	XXXXxxxx			
rich Primary Zone				
Rich/Lean (RQL)	XXXXxxxx			
Others				
exhaust gas recirculation	XXXXXXXXXXXXXXXXXX			

Figure 2.2 Summary of NO_x reduction potential [26].

Task 4. Fundamental investigations

Task 4.1. Study on atmospheric atomization with air-blast atomizers

Subtask 4.1.1. Spray diagnostics The mean and turbulent velocity measurements performed at IST at low air velocities showed the effect of swirling flow on the mean shear and mixing performances. Some instability and asymmetry of the flow, shown by IST measurements and visualization, could be confirmed by CERT results at a higher velocity level.

The effects of the air–fuel (liquid) ratio, inner and outer velocities and pressure of swirl on the mean SMD measured by Malvern at Turboméca were studied extensively. The local SMD distribution, histograms, axial, tangential and radial droplet velocities (enclosed at various downstream distances at CERT) confirmed and detailed the Turboméca results. These results were used to provide boundary and reference conditions for computations.

Subtask 4.1.2. Atomization by an air-blast atomizer, elementary models The first phase of the pulverization process, namely the initial sheet instability, was analysed by CORIA. This was achieved for an annular liquid sheet, typical of aircraft air-blast atomizers. An initial linear analysis was developed including one of the reducing assumptions previously found in the literature.

The calculations were validated by flow visualization with the experimental air-blast atomizer.

For the lowest liquid flow velocities (less than 1 m/s) and a wide range of air-flow velocities (up to 50 m/s) an excellent agreement was found between theory and the experiment.

For higher liquid velocities (greater than 2 m/s) a large scale of new instability was encountered, which is rarely described in the literature. An explanation for this phenomenon was considered to lie in pressure effects. For highest liquid velocities (greater than 10 m/s), some longitudinal filaments appeared (rasta effect). A complete explanation for this phenomenon has not yet been found. One direction for further research should be in the investigation of Rayleigh–Taylor or Görtler vortices.

The effect of air velocity (increase) on the break-up length (decrease), instability amplitude (maximum near 40 m/s) and frequency (increase) was studied at CERT.

Subtask 4.1.3. Flow field calculations With the calculations, simulating the two first geometries (air flow velocity: 20 m/s) SNECMA obtained the air-flow velocity distribution downstream of the airblast injector, comparable to the measured one (see Subtask 4.1.1) [51]. Simulating the third geometry (air flow velocities: 20 m/s and 100 m/s) proved to be difficult with this computer code because of the influence of the swirl which had to be described correctly. Further investigations are planned.

Turboméca calculations on one-phase flow at 20 m/s show very good agreement with the three mean velocity components and the values of the kinetic energy, measured by IST [51]. Because of some differences found between CERT measurements at 100 m/s and IST measurements at 20 m/s, a new set of measurements as well as another calculation on a two-phase flow with new boundary conditions was carried out by CERT.

Task 4.2. Demonstration of limiting capabilities of laser diagnostics

Drop size and axial velocity distributions across a fuel spray were measured successfully using PDA at pressures up to 12 bar and a fuel flow of 0.025 kg/s in a plane 70 mm downstream the injector. Secondary scattering in denser sprays provided the limits where the signal–noise ratio was reduced to unacceptable levels. Employing a more powerful laser than the 10 MW equipment used here would probably extend the range to higher pressures and/or to planes closer to the injector [6].

It could be shown that a sheet thickness of 1 mm from a NdYAG laser was suitable in order to obtain sufficient image clarity for spray photographs to be used for image analysis [6, 7]. For denser sprays, future work will aim at laser sheets of 0.25–0.5 mm thickness.

Applying a two-component PDA to model combustors, a pressure limit of 5 bar was identified beyond which the signal–noise ratio was unacceptable. Absorption of input and signal beams by soot increased with pressure and was accompanied by a widening of the fuel injector cone angle which increased the rate of window fouling, again reducing signal strength. The flame luminosity joining the soot generation also led to problems in manually focusing the receiver upon the probe volume against a very bright background. A more powerful laser is not thought to extend this pressure limit any further due to the soot luminiscence problem [9, 10].

In order to measure the air velocity, micron-sized alumina particles were injected into the air mainstream, whereas the PDA instrument was used without the sizing option as a simple LDA system. This proved to be feasible in a simple isothermal flow, but in the presence of fuel drops or cold combustion, the lower intensity signals from the seeding particles were swamped by those of the larger fuel drops.

The alternative drop-sizing technique of LDA was also applied to a model combustor, burning at atmospheric pressure [12]. Data quality was limited by the optical access to the rig and by the data rate, except for two points within the spray, where drop size and axial velocity were obtained at fuel flows up to 80 g/s. In setting the probe beam volume to a diameter sufficiently large to measure the largest drops in the spray, a dilemma may occur in maintaining a single drop in the volume of measurement in denser spray regimes. Identified frequency filter discrepancies were addressed along with the improvement of the data collection rate.

Application of CARS to model combustors was shown to be possible at pressures up to 6 bar [8, 11]. Limitations were imposed by signal breakdown as drops or soot particles passed through the measurement volume leading to signal rejection. High turbulence levels also led to problems at very strong fluctuations of the signal about the detector position. This was reduced by minimizing the path length through the turbulent medium, but in a tuboannular combustor opposing air jets were blocked to reduce turbulence levels. To maintain the spatial and temporal resolution, a practical pressure limit of 2.5 bar was recommended [8].

Cross-correlation of two CARS systems showed good agreement in cooler regions of the combustor but severe discrepancies in areas of peak temperatures. CARS signals are inversely proportional to gas temperature and in highly turbulent gas mixtures, where pockets of hot and cooler gas can exist simultaneously within the CARS measurement volume, there may be a biasing towards the stronger cold signals. This biasing effect led to discrepancies up to 600 K in peak temperatures in the CARS system, which had a significantly larger probe volume. A new laser with improved beam quality and hence smaller probe volumes should resolve this problem [14].

The cross-correlation of spray diagnostic instruments showed surprisingly good agreement considering the geographical and chronological spread of the testing and the inclusion of both horizontal and vertical sprays [3]. Phase Doppler anemometry (PDA) proved to be the most robust technique with fast data collection rates and the ability to measure the drop size and up to two velocity components simultaneously. However, the way in which the instrument is set up, i.e. the selection of frequency filters for velocity ranges and the selection of the drop-size range, can have significant influence on the results and should be tailored explicitly to suit the data investigated. The statistical expression of drop size will also influence the magnitude or the effect of these instrument settings, e.g. D90—the 90% undersize diamenter value is extremely sensitive to the presence of just a few large drops. 'Mid-range' parameters such as D30 or D50 are less sensitive and thus preferable values for general spray comparisons.

Task 4.3. Influence of primary zone architecture and stoichiometry on pollution formation in a tubular combustor

Subtask 4.3.1. Modifications and tests of the tubular combustor In gas analysis measurements at Volvo, ONERA and SNECMA:

- On the *lean tubular combustors*, the influence of the hardware modifications between the three configurations is studied:
  - the lowest NO_x emission is obtained with the 'modification 1' combustor (combustor with an additional primary zone internal liner);
  - the lowest CO emission is obtained with the base line configuration;
  - the lowest UHC emission is not obtained with any particular configuration, it depended on the running conditions.

- On the *rich–lean combustors*, it was shown that rich zone wall cooling and

Figure 2.3 Rich quench lean combustor.

coking are key issues. An NO_x emission index reduction of 50% compared with the baseline lean combustor was achieved.

Subtask 4.3.2. Lean blow-out and relight tests at subatmospheric conditions During the test series on the three 'lean' configurations the pressure effect was studied (the LBO performances at 0.6 bar are not as good as at 1 bar) and the influence of hardware modifications was investigated ('modification II' and the reference combustors stayed at the same LBO level, while the 'modification I' showed a lower LBO limit).

Compared with the 'lean' liners, the rich–lean combustors provided the best LBO performance. The pressure effect of this kind of combustor was not as easy to study as of 'lean' configurations (the evolution of the LBO limit with the pressure is not important).

Subtask 4.3.3. Pollution calculations A comparison between the two configurations of the RQL combustor was performed with a three-dimensional SNECMA computer code. The latter configuration showed a better compromise between the 'RQL effect' and the quenching zone pressure loss than the first one tested. Good agreement could be stated between the measured and the calculated NO_x emission index.

Subtask 4.3.4. Feasibility study of lean blow-out modelling In order to attain reasonable computer times both the chemical kinetic and the fluid dynamical calculation had to be simplified. A chemical kinetic scheme, including the most important features involved in blow-out mechanism was demonstrated. Only very fundamental comparisons with experimental results were performed but the results of these comparisons were encouraging enough to draw the conclusion that lean blow-out modelling is feasible.

Task 4.4. Influence of fuel distribution and droplet size on homogeneity and pollution formation

Subtask 4.4.1. Design and manufacture of the rectangular combustor and investigations of fuel nozzles Lean low NO_x combustion can be obtained only if the fuel is rapidly mixed with air before burning in the primary zone of the combustor. Under these conditions, mixing of the fuel and air has to be performed in the primary zone and the efficiency of pollution reduction under lean conditions is highly dependent on the initial distribution of air and fuel and on the droplet size.

For lean combustion, about 60–70% of the air must be injected through the combustor head. With conventional combustors the amount of air through the nozzles is at about 5–15%. The problem to cope with in this case is to design the flame tube head in such a manner that it allows 60–70% of the air to be injected.

The air-blast nozzles and the additional three fuel-air mixing tube geometries were tested in a can combustor at a pressure of 6 bar (tube B at 4 bar) and at the inlet temperature of 670 K. Tube A was characterized by a cylindrical geometry, tube B exhausted into a small diffusor while tube C was swirled. The can combustor had only cooling films and no additional mixing air. The measured lean blow-out limits (LBO) stated at these conditions were as shown in Table 2.3.

The air-blast atomizer had a combustion efficiency of nearly 100% over the entire stability range. The efficiency of all other tubes was poor. Based on the results of these investigations, the air-blast atomizer was selected to be tested in the rectangular combustor.

Subtask 4.4.2. Testing of the influence of fuel distribution and droplet size The test results of the rectangular combustor investigations are fully documented in [52]. The rectangular combustor system is shown in Fig. 2.4.

The measured profile of the air–fuel ratio inside of the combustor and at the combustor outlet indicates the homogeneity of the mixture. At a distance of one channel height from the combustor head the profile became plain.

At atmospheric pressure conditions the inlet temperature was increased from 573 K to 700 K at an AFR of 32. The NO_x emission index increased from 2.30 g/kg to 3.04 g/kg. The efficiency remained unchanged at 99.6% as well as the CO emission index at 15 g/kg. The unburnt hydrocarbons decreased from 0.11 g/kg to 0.08 g/kg.

If the overall air–fuel ratio was increased from 30.9 to 39.9 and proceeded to

Table 2.3 Blow-out lines

Nozzle	Air blast	Tube A	Tube B	Tube C
LBO	34.0	28.8	24.5	18.5
ϕ	0.43	0.5	0.59	0.79

Figure 2.4 Lean-burning rectangular combustor.

42.4, the NO_x emission index measured was in the range of 3.01 g/kg, 1.33 g/kg or 1.16 g/kg in the combustor outlet plane.

To investigate the influence of atomization and homogeneity the pressure drop was varied from 3.95% to 6.52%. The NO_x emission index decreased in all planes of measurement because of the smaller fuel droplets, the wider fuel spray angle and the higher air turbulence level which led, all in all, to a more homogeneous mixture. In planes located more than one channel length behind the injectors, the efficiency stated was high.

The lean blow-out limit was tested at the two pressure loss levels 6.4% and 3.8%, respectively:

Pressure drop	3.8%	6.4%
LBO (air–fuel ratio)	70.9	68.3

These lean blow-out limits were measured at atmospheric pressure conditions but they are too low for engine use. At higher pressure levels the limit will increase.

The values found indicate that a lean combustor can not be realized without an additional combustor control system. In this case, one or two of the three fuel injector rows can be switched off, which increases the blow-out limit to:

Pressure drop	3.8%	6.4%
LBO (one row of nozzles)	248.9	239.4
LBO (two rows of nozzles)	99.2	95.5

The measurements in three planes inside of the combustor allowed determination of the influence of residence time on NO_x formation and combustion

Research results

efficiency. The NO_x emission index of the traverses 3, 4 and 5 in Fig. 2.5 with lower outlet temperatures are nearly independent of the residence time. The higher the primary zone temperature, the more intensive the dependency.

The test results indicate that NO_x emissions in a homogeneous primary zone without additional mixing jets are independent of the residence time inside the combustor.

However, a sufficient residence time is required to realize a high combustion efficiency. Depending on the conditions of combustion, the residence time for a measured combustion efficiency greater than 99% was found to be between 3 ms and 5 ms (Fig. 2.6).

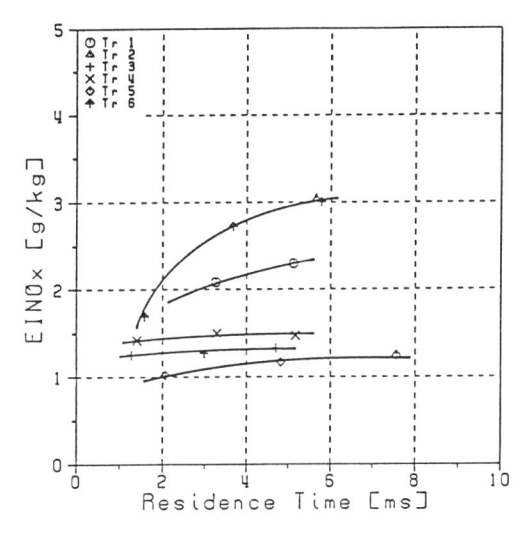

Figure 2.5 Influence of residence time on measured NO_x emission of the rectangular combustor.

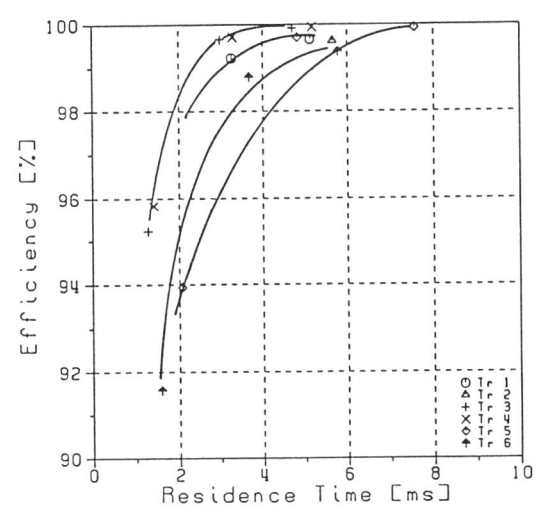

Figure 2.6 Influence of residence time on combustion efficiency of the rectangular combustor.

To compare the measured NO_x emission indices with the estimates for future engines, the pressure influence must be taken into consideration. The comparison of the pressure-corrected data with the values of the correlation [1] showed an NO_x level of the lean combustor lower than 50% of the correlation value. In Fig. 2.7, the NO_x emission indices measured are plotted in comparison with the correlation.

Task 4.5. Influence of air distribution on homogeneity and pollution in a tubular primary zone combustor

Essentially, the research work generated four types of results:

- The first type consists of the processed and integrated emission data measured for each of the test points for each combustor set-up. These data were plotted as a function of the overall air–fuel ratio [53, 54, 55].

- The second type of results was obtained by detailed statistical analyses of the emissions [55, 56].

- The third type of result was obtained from water analogy testing of combustor models. This work showed the fluid dynamic flow mechanisms in

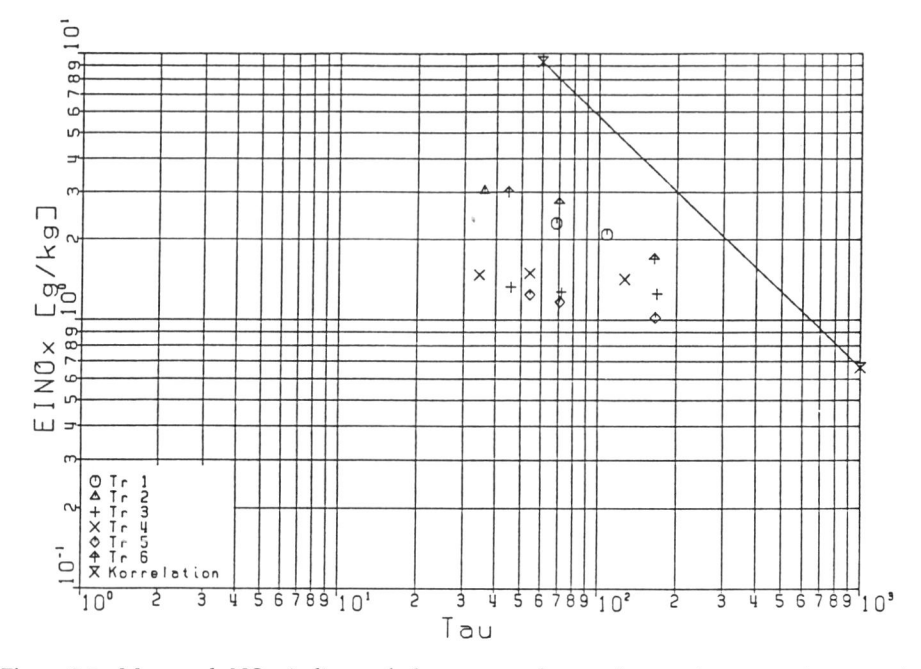

Figure 2.7 Measured NO_x indices of the rectangular combustor in comparison to the correlation.

operation of each of the combustors investigated. Time-resolved measurements of dye tracer flows inside the combustor model were used to provide indications of the mixing degree and the residence time [57].

- In the fourth type of results, mathematically lumped-parameter models of the combustors were arranged to fit the measured NO_x emissions results [58]. This modelling process in combination with the water analogy results provided an excellent understanding of the combustion mechanisms taking place inside of the combustors.

The extensive and detailed results from the work undertaken in Task 4.5 were fully reported in [19, 53 to 58]. Some of the most significant findings will be summarized.
One combustor type was designed with a large number of small and well distributed air entry holes to support very fast and uniform mixing of the fuel and the air. The specific objective was the attempt to achieve a weak, low temperature combustion to reduce NO_x formation. Compared to the more conventional combustor designs, the well-mixed design showed better smoke and combustion efficiency performance, especially at the richer conditions. This improvement in combustion performance and higher flame temperatures resulted in higher NO_x generation under rich conditions. In general, for all of the combustors tested, any change in geometry or an operating parameter (such as increasing pressure loss) that improved smoke and efficiency also produced an increase in NO_x.
Reductions in NO_x emissions, relative to the conventional standard [59], could be made in any of the combustors by running them at weaker overall air–fuel ratios. Provided that combustion efficiencies were comparable, the reductions in NO_x emissions from all the combustors were similar at the weakest operating conditions. Nevertheless, amongst all the combustor builds tested, the overall performance of a relatively small number could be seen superior to the others. Although NO_x emissions were reduced by weak operation, the analysis of detailed gas analysis data showed that the reduction achieved was a result of weak, cool combustion but rather a result of reduced residence time.

Task 4.6. Smoke production in rich burning primary zones

Differential line-of-sight absorption measurements were processed to yield radial profiles of soot volume fraction across the liner. The absorption from the primary zone appear to be heavily contaminated with fuel droplets. Stronger laser absorption or scattering was observed here, often in the absence of significant luminosity. In order to promote soot generation further downstream in the combustor, providing possibly more discrimination relative to fuel droplets, the combustor was operated relatively rich (AFR = 12) on occasions.

In view of these fuel droplet effects at the primary hole station, measurements at higher pressures were confined to the dilution zone station. Peak and integral soot measurements suggest an approximately linear variation with pressure. Radial profiles reveal an annular development of the sooting region with a clearly defined minimum on the combustor centre-line, characteristic of the non-combusting fuel injector studies which also form part of this programme.

BRR Calculations

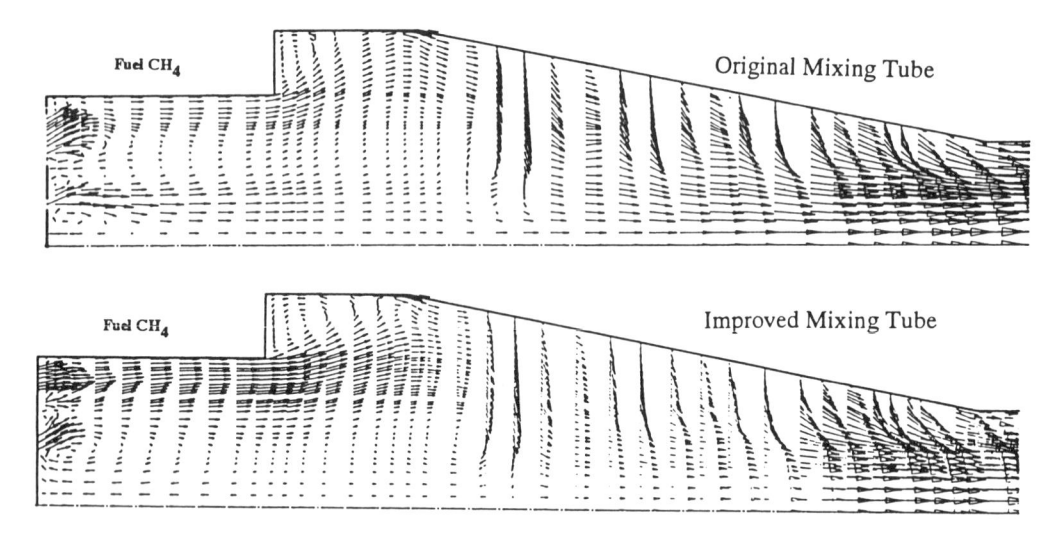

ITS Calculations

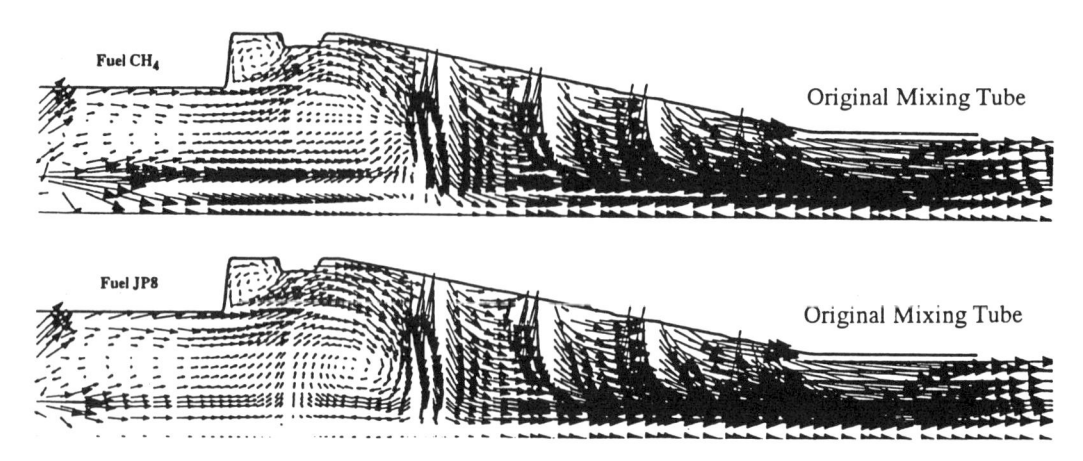

Figure 2.8   CFD-calculations of the BRR-combustor.

Task 4.7. Modelling and optimization of an existing premix duct, computational and experimental programme

The computation results of the $x-y$ flow field for the two above-mentioned configurations are depicted in Fig. 2.8. This figure shows a remarkable recirculation zone in the centre of the improved mixing tube burner at the mixing tube outlet and high axial velocities at the outer diameter of the mixing tube outlet.

The calculations carried out by the ITS Karlsruhe with their own computer code [60–63] indicate a similar flow behaviour for gaseous fuels and for the liquid fuel JP8 as shown for the first version in Fig. 2.8. A pronounced recirculation area was established, located immediately before the first mixing airflow holes which improved mixing and flame stabilization in case of using JP8.

Air mass flow measurements at improved mixing tube (BRR) showed a pressure loss caused by the installation of the deflectors at the nozzle outlet. In the first test series no ignition occurred by using the flat air blast nozzle without deflectors. The inertia of the blowing air of the nozzle apparently was too large. Insertion of deflectors permitted ignition to be reached depending on the air–fuel ratio and the air-flow parameter. A narrow gap was found between the minimum fuel ignition and the flame extinction limit. As mentioned before, the fuel atomization becomes worse by introduction of the deflectors. Large burning fuel droplets, caused by the deflectors, could be clearly observed moving through the flame. This behaviour is not desired and can be influenced by introducing inclined flat air-blast nozzle slots.

The cold measurements of the segment burner at ITS showed a small droplet diameter at the large z-position. It is supposed that large droplets are separated close to the wall by the large tangential velocity.

Atmospheric preheat tests were carried out at ITS. In Fig. 2.9 typical temperature profiles are plotted versus an annular coordinate. Inserted lines are to indicate the position of the segment combustor side walls and the position of the three mixing tubes [64, 65]. The three mixing tubes could not be identified as three peaks in the temperature profiles. Here, only two maxima were observed as a result of of the highly swirling flow field in the annular flame tube. Applying burning and non-burning conditions to the fuel spray behaviour clearly indicated the effect of the heat release due to combustion and the resulting interaction with the flow field. Fig. 2.9 shows a comparison of the characteristic diameters of the volume distribution.

At elevated pressures with reheat (ITS), the temperature profiles (Fig. 2.10) were quite similar to those obtained at atmospheric conditions. Again, only two peaks were detectable. These two peaks could also be stated as minima of the lambda values (see Fig. 2.11). A similar profile was obtained for the oxygen and the CO-concentrations, shown in the same figure.

In Fig. 2.11, the CO- and NO_x emissions are given as emission index values (ECO, ENO). The ECO also corresponds to the lambda values. The determination of the absolute ECO level was critical regarding the application of an uncooled probe. The ENO values were quite low, confirming earlier measurements with this type of combustor. Compared to the ECO values, the ENO values did not show a clear profile. Due to the low level of ENO, the fluctuations were below the sensitivity of the measuring device used.

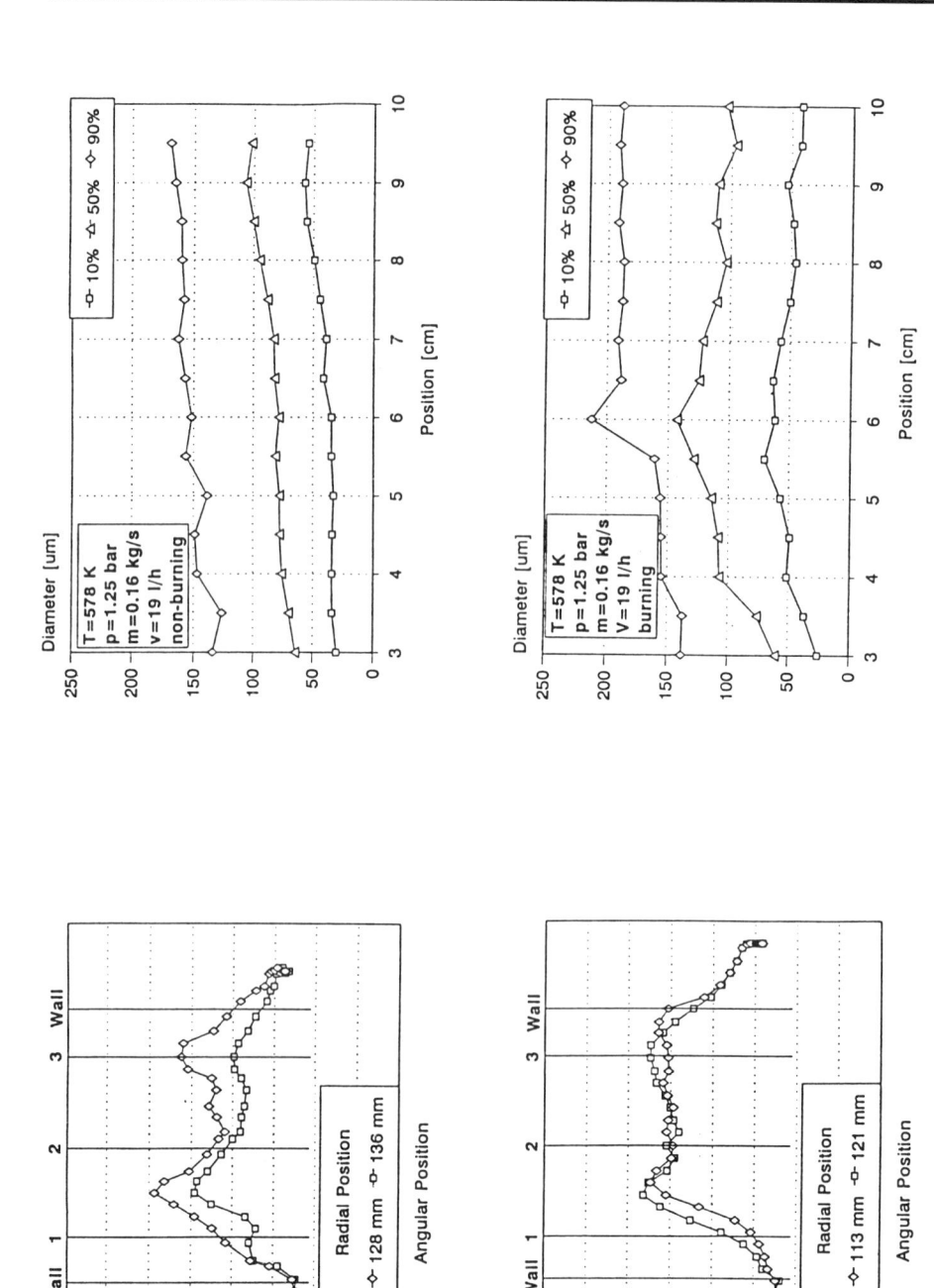

140   Low emission combustor technology (LOWNOX I)

Figure 2.9 Results of the atmospheric measurements of the BRR-combustor.

Research results

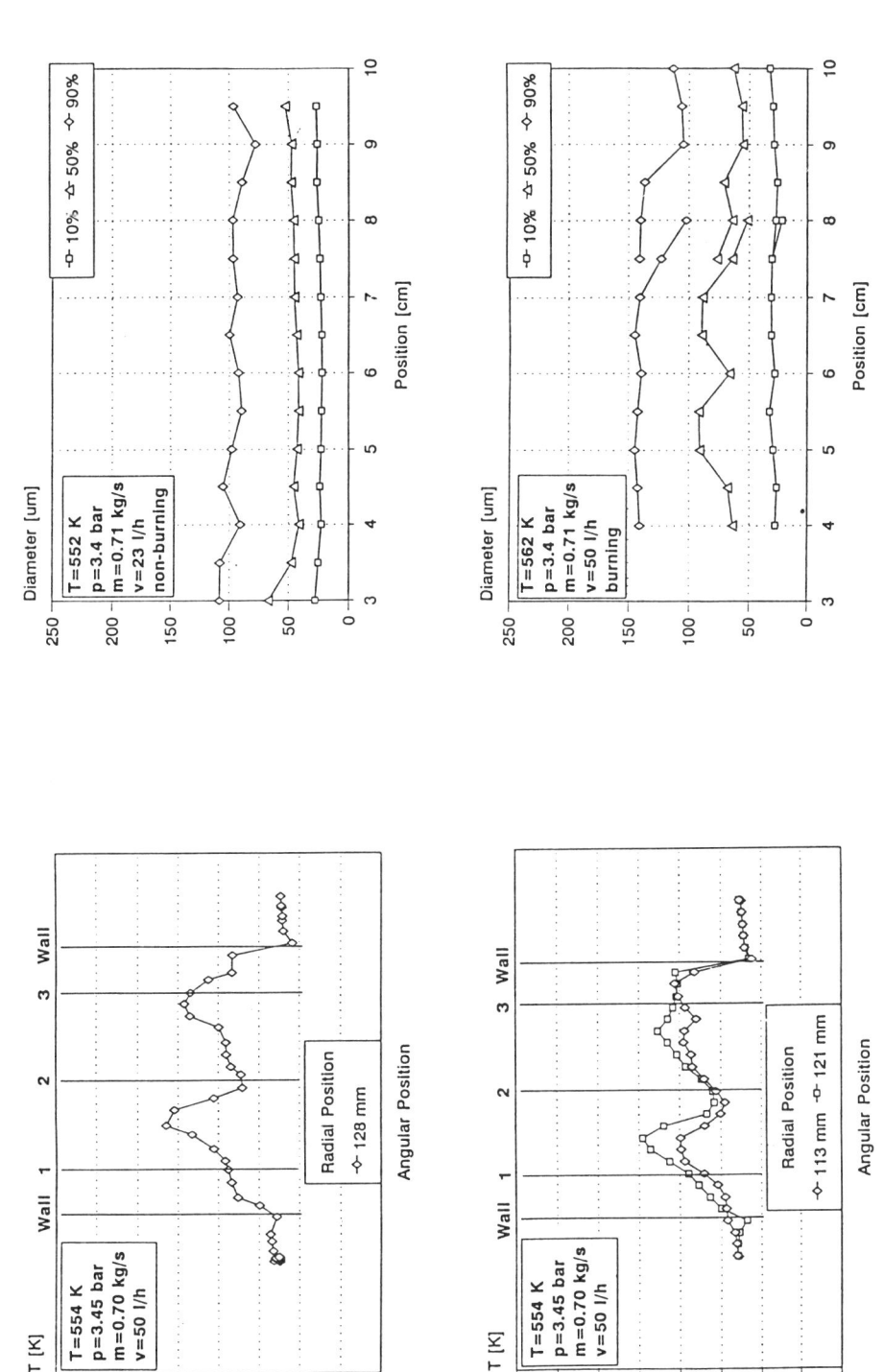

Figure 2.10 Temperature profiles of the BRR-comustor at elevated pressure.

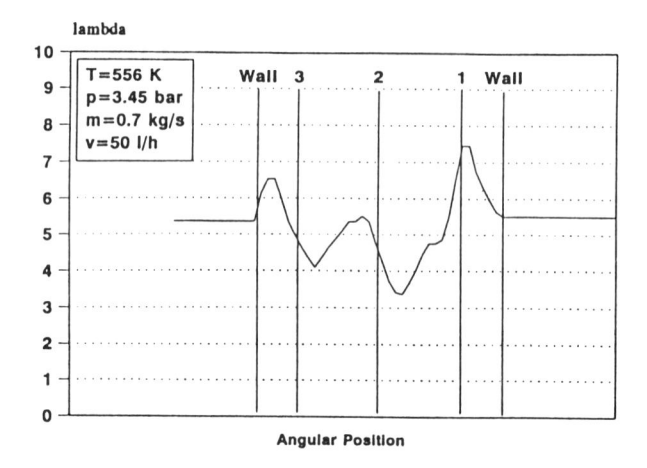

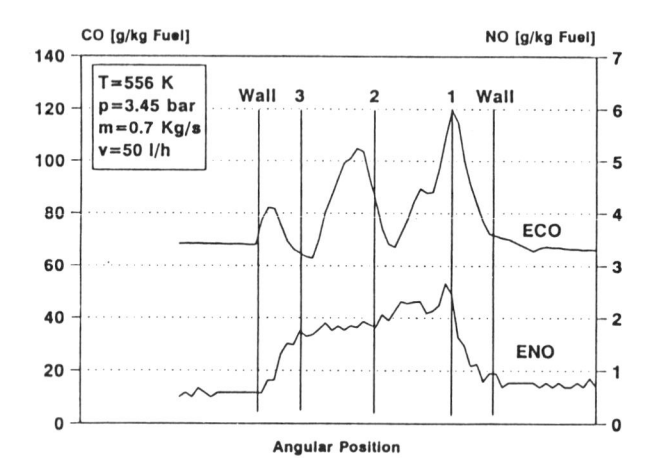

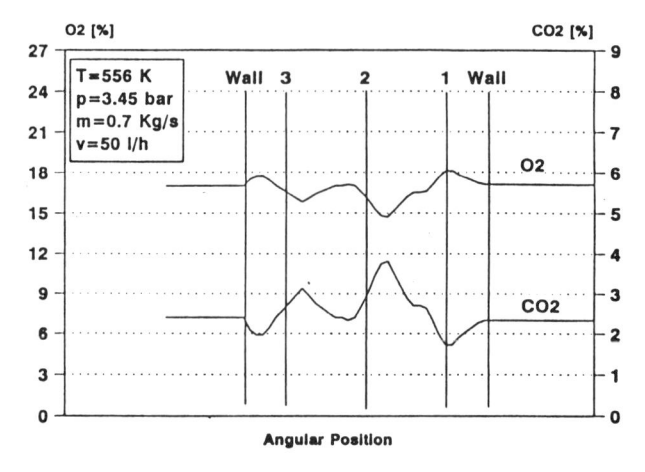

Figure 2.11 Emission measurements of the BRR-combustor.

Task 5. Selection of pollution reduction methods

In view of the results of this project [66], three concepts were selected in terms of emission reduction potential, reliability, safety and weight that appear viable for use in the next generation of aero engines.

In the near future, aero engines fitted with advanced combustors are expected to reduce the nitrogen oxide emission at about 30%, compared to today's aero engine standard. Any further reduction would require more concerted development efforts which are unlikely to take place before the end of the century.

A 20–40% reduction in nitrogen oxide emission can be achieved with a lean burning double annular configuration. Depending on the operating point, combustion occurs in one or both stages. The NO_x reduction potential varies largely with the homogeneity of the air–fuel mixture in the primary zone.

Lean combustion, using premixed and prevaporized fuel can reduce nitrogen oxide emissions by 80–90%, although in this case the potential of application of this method seems to be restricted to gas turbines of low pressure ratios. At high pressures, auto-ignition of the air–fuel mixture and flashback in the vaporizer may occur. An additional problem which needs solving is how to achieve a homogeneous mixture of air and fuel to ensure lean, low NO_x combustion. Inhomogeneous distribution aggravates the NO_x formation in a sporadically stoichiometric combustible mixture. Also, conventional combustor control will hardly be able to cope with the narrow stability range associated with lean combustion.

Another method of reducing nitrogen oxide emissions lies in the use of a rich primary zone. The formation of NO_x is minimized by the lack of oxygen and the low reaction temperatures. Following further optimization, reduction in NO_x emissions are expected to each 80–90%. A lot of technical problems must be solved before applying this method in an engine, i.e. wall cooling and rapid mixing.

5 Conclusions

Task 1. Definition of combustor conditions of future aero engines

Depending on the type of engine, the pressure ratios as well as the combustor inlet temperatures of future aero engines will increase noticeably. However, the combustor outlet temperature will increase only moderately.

Task 2. Assessment of pollution levels using present technology

The SNECMA correlations used show that all the selected future large engines fit the current ICAO recommendations (D_p/F_{oo} denotes the concentration D of the component p at take off thrust under ISA conditions, F_{oo}):

$$NO_x: D_p/F_{oo} < 40 + 2\pi_{oo} \text{ g/kg}$$
$$CO : D_p/F_{oo} < 118 \text{ g/kg}$$
$$HC : D_p/F_{oo} < 19.8 \text{ g/kg}$$

No overall characteristics exist for turboshafts, turboprops and APUs but the comparison with the large engines shows that the contribution of small engines to NO_x pollution is clearly lower but not negligible and that NO_x reduction targets clearly should be functions of the engine size.

Task 3. Summary of pollution reduction methods

A first step in the direction of a low emission combustion is the radially staged combustor without premixing and prevaporization. A 30% reduction of the nitrogen oxide emission is considered to be possible with relatively low technical risk. Further reduction requires more intensive research work.

To achieve a 30% reduction by the mid-nineties and a further reduction to 80% at the end of the century, the fundamental combustion research must be carried out urgently which includes the development of combustor control methods.

Task 4. Fundamental investigations

Task 4.1. Study on atmospheric atomization with air-blast atomizers

The comprehensive understanding obtained between calculated and measured phenomena (velocities, kinetic energy) during the study of the atomization process of an air-blast atomizer proves the applicability of two computer codes to predict the flow field downstream of an air-blast injector. The model of sheet instability was found to be satisfactory when physically relevant. Two additional phenomena were demonstrated and could be explained partially: high-scale dilatational waves and the rasta effect. From this point of view, visualization was determinative.

Task 4.2. Influence of primary zone architecture and stoichiometry on pollution formation in a tubular combustor

The application of non-intrusive, laser-based diagnostic techniques was demonstrated successfully on both isothermal sprays and in burning model combustors. Experimental problems were identified and solutions proposed where appropriate. A pressure limit of approximately 5 bar was generally

Conclusions

Table 2.4 Feasibility of laser diagnostics: summary

Technique	Experimental demonstration	Limitations
Phase Doppler anemometry → Drop size and velocity distributions	• 12 bar isothermal spray • 5 bar combustion rig • Good cross-correlation agreement • LDA in air flow only	• Secondary scattering • Dirty windows, high soot levels → high background luminosity • Seeding signals swamped by larger drops
Amplitude Doppler anemometry → Drop size and velocity distributions	• 1 bar isothermal spray • 1 bar combustion rig • Cross-correlation velocity OK, sizes low	• Size limit imposed by probe beam diameter • Further investigation of filter effects needed • Improvements needed in data collection rate
Holography → Drop size	• 1 bar isothermal spray	• Poor image quality, non-focused drops • Few spherical drops to be sized • Complementary information on drop shapes • Double pulsed holography needed for velocity
Photographic image processing → Drop size	• 1 bar isothermal spray	• Density limit for individual drops • Thin sheet lighting (<1 mm) & sharp images • Time-consuming procedure • Complementary spary visualization data • Double exposure photography needed for velocity
Cars → Gas temperature	• 1 bar single & 3-sector combustors • 2.5 bar for spatial & temporal resolution • 6.8 bar data collection still possible	• High turbulence levels → beam fluctuations • Refractive index gradients → steering + phase disturbance • Fuel drops/soot reduce efficiency • Large probe volume causes cold biassing

imposed on the optical techniques applied, caused by the high luminosity of sooting flames. Practical aspects such as window fouling may be addressed in the basic rig design but such difficulties inevitably increase with the operating pressure. If these detailed measurements within the combustion zone are used to validate CFD predictions on the experimental range, the models may be used to extend predictions up to engine operating conditions. The experimental range investigated and the limits identified are summarized in Table II.

Following the definition of the diagnostics requirements, further detailed measurements within the model combustor will be carried out in subsequent programmes both to provide CFD validation data and to enlarge the understanding of low-power combustion phenomena.

Task 4.3. Influence of primary zone architecture and stoichiometry on pollution formation in a tubular combustor

Following the investigation activities of the primary zone architecture, the 'rich quench lean' concept seems to be a promising technology. The 50% NO_x emission index reduction obtained (Fig. 2.12) can be improved in the succeeding phase (interime phase) by using calculations with a three-dimensional computer code (which provided acceptable evaluation of pollutant levels) to optimize the geometry and the air-flow split of the RQL combustor.

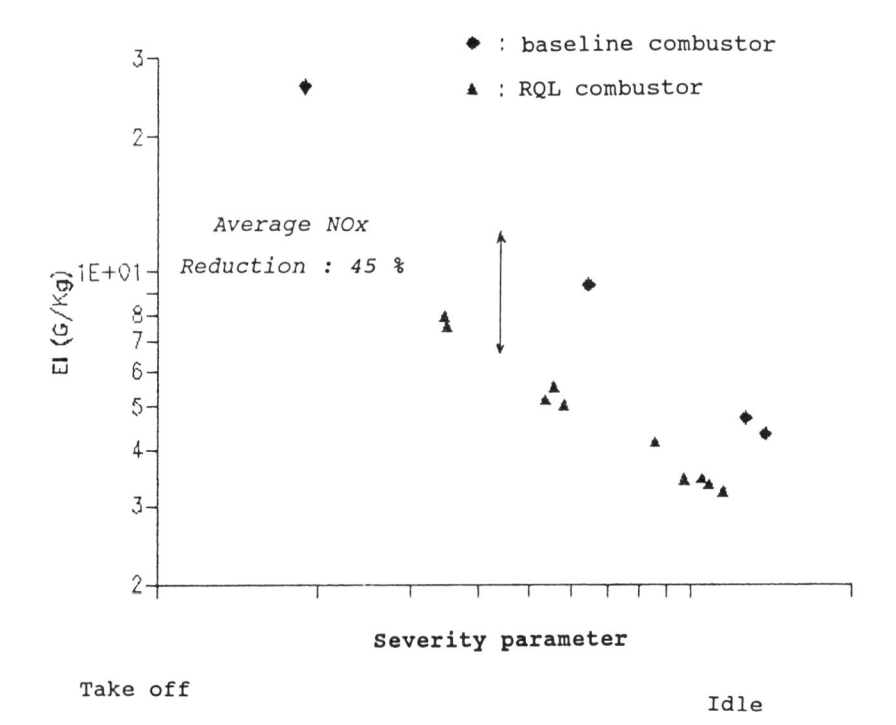

Figure 2.12 Rich quench lean technology performance in NO_x emission.

Conclusions

It is important to remark the interesting lean blow-out performance of this type of a combustor—the LBO limits obtained during measurements as well as the modelling.

Task 4.4. Influence of fuel distribution and droplet size on homogeneity and pollution formation

Lean combustors fitted with three rows of fuel injectors and tested without additional mixing jets suggest the following conclusions:

- the strongest influence on NO_x formation was found at primary zone temperatures greater than 1900 K. Due to the homogeneous primary zone mixture of the tested configuration low NO_x values were measured.

- at primary zone temperatures below 1900 K, the residence time inside the combustor has no strong influence on NO_x formation. The residence time can be selected to obtain a higher combustor efficiency.

- the lean blow-out limit of the lean combustor was as poor as expected. An additional combustor control system is required to operate such a type of combustor in an engine. To increase the lean blow-out limit, two of the three fuel nozzle rows of this design had to be disconnected at lean conditions which led to an excellent lean blow-out behaviour.

The estimated NO_x reduction potential of the tested multi-jet combustor without premixing was about 50%, compared with the results of a correlation for present combustor technology.

Task 4.5. Influence of air distribution on homogeneity in a tubular primary zone combustor

The overall conclusions from the investigations on a lean primary zone of a tubular combustor can be summarized:

- It appears to be possible to achieve excellent idling efficiencies, smoke emissions and satisfactory stability along with a 30–40% reduction of the NO_x level with simple, unstaged combustors.

- The reductions in NO_x were a result of reductions in residence time because combustion reaction rates were much faster than any fluid dynamic mixing rates that were achieved.

- Further substantial improvements of simple unstaged combustors seem to be unlikely, although they result in useful emission reductions—particularly in NO_x. Further research should be aimed at significantly higher NO_x

reductions by application of more advanced combustion techniques such as fuel or air staging, rich burn/quick quench combustors or lean premixed prevaporized combustors.

Task 4.6. Smoke production in rich burning primary zones

A technique for the *in situ* measurement of local soot concentration within a tubular combustor was successfully demonstrated. Integral laser absorption was combined with a traversable water-cooled probe to enable differential measurements. Defining the difference between laser absorption in parts of the combustor due to particulate soot and to fuel droplets proved to be difficult. Laser beam steering due to refractive index gradients made the operation of the axially mounted reflector probe significantly more complicated than its transversely mounted sight-tube alternative. The spatially resolved radial profiles of soot concentration across the liner showed a distinctive annular structure, characteristic of the fuel injector.

Task 4.7. Modelling and optimization of an existing premix duct, computational and experimental research

The following conclusions can be drawn from the modelling and the optimization of a premix duct of a lean annular combustor:

- No literature-documented solution was found for problems of multidomain technique for the very complex BRR combustor geometry. Further time-consuming work has to be spent before a first reliable calculation of the mixing and burning of the liquid fuel is justified.

- The existing computer code at ITS represents a preferable instrument for air-flow calculations, the mixing process and the droplet vaporization.

- Simple combustor configurations should be examined numerically and experimentally to consider carefully the influence of the mixing and evaporation of the fuel in the mixing tube.

- First tests with the improved mixing tube indicate that deflectors are not suitable because of large losses. To stabilize the flame, a higher flow recirculation rate in the mixing tube has to be aspired.

- Air-blast nozzles with an immediate mixing of air and fuel to enable a lean combustion represents a very promising method to minimize the flame temperature and thus to improve the NO_x-emission rate.

- ITS in Karlsruhe was successful in obtaining data from the fuel spray droplets under complicated circumstances, caused by the optical measurements through the quartz glass window.

Task 5. Solution of pollution reduction methods

In a first step towards a low emission aero engine, a radially or axially staged combustor can be used to reduce nitrogen oxide emissions by 20–40%. The second step is characterized by the use of an RQL combustor and a lean premixed combustor in lower-pressure engines ($p_\text{combustor} > 20$ bar) to realize a potential reduction of 80–90%. A substantially greater development effort backed up by in-depth basic research is required for the LPP approach for large engines. Compared with the RQL concept, the problems of the LPP concept seem to be more of a fundamental nature rather than on the engineering side.

Continuing this task in Phase II, further investigations of all three concepts will be carried out in an attempt to measure and approve the specific NO_x reduction potential.

6 References

[1] Bardey, X., Bouchie, Y. and Grienche, G. Final Report for Task 2, SNECMA/Turboméca, Note 5155/YKC/90, October 1990.

[2] Heitor, M.V. *Final Report, Instituto Superior Technico on Subtask 4.1.1: Spray Diagnostics*, Brite/Euram: Aero 1019, February 1992.

[3] 24 Month Progress Report, CERT/ONERA/DERMES on Task 4.1, *Study on Atmospheric Atomization Process of an Airblast Atomizer*, February 1992.

[4] Camatte, P. and Ledoux, M. *Modelization of Annular Sheet Instability* Task 4.1.2: Final Report 1992, Brite/Euram AERO 1019.

[5] Hawkins, H.L. *Demonstration of Limiting Capabilities of Laser Diagnostics*, Task 4.2: CRR89748 Final Summary Report, May 1992.

[6] Tam, I.C.K. and Jasuja, A.K. *Measurement of Dense Fuel Sprays*, Task 4.2.1: Final Summary Report, Cranfield, 1992.

[7] Trichet, P. *Participation in Brite/Euram Low Emissions Combustor Technology Programme*, Interim Report 3/2420.00/CERT/DERMES (CERT), March 1992.

[8] *Brite/Euram Project 1019 Low Emission Combustor Technology*, Final Report ONERA: Participation as SNECMA Sub-Contractor on Task 4.2 5903/YKC/92, 1992.

[9] Cormack, G., *Phase Doppler Anemometry in a Quartz-Sided combustor Rig at Atmospheric Pressure*, Brite/Euram Project 1019 Low Emission Combustor Technology. Task 4.2.2: CRR89679 (Rolls-Royce), May 1992.

[10] Cormack, G., *Phase Doppler Anemometry in a Twin-Sector Rectangular Combustor at 1 to 5 bar*, Brite/Euram Aero/ 1019 Low Emissions Combustor Technology, Task 4.2.2: CRR89742, May 1992.

[11] Black, J.D. *CARS Measurement on a Two-Sector Combustor Rig*, Brite/Euram Aero/1019 Low Emissions Combustor Technology. Task 4.2.2: (RR/OH) 1228, April 1992.

[12] Lisiecki, D., *Demonstration of Ability to Measure Sprays and Gas Velocities in a Model Combustor*, Brite/Euram, Task 4.2.2 Rouen University, February 1992.

[13] Cormack, G., *Cross-Correlation of Spray Diagnostics Techniques*, Brite/Euram Aero/1019 Low Emissions Combustor Technology, Task 4.2.3: CRR897424 (Rolls-Royce), May 1992.

[14] Black, J.D., *CARS Temperature Measurements in SNECMA Tubular Combustor*, Brite/Euram Aero/1019 Low Emissions Combustor Technology, Task 4.2.3: RR/OH 1220 (Rolls-Royce), April 1992.

[15] Bouchié, Y., *Giving the Description of the Five Configurations of the Burner of the Test Programme*, Progress Report: SNECMA Note No. 9370-573, May 1992.

[16] Kaaling, H., *Lean Blow-Out Tests at Subatmospheric Conditions*, Final Report on Subtask 4.3.2: Volvo-Note No. 9370-573, May 1992.
[17] Bouchié, J., *Giving the Description of the Modelling of the Combustor, Combustor Calculation, Calculation Results and Comparison between Theoretical and Experimental Results*, Progress Report: SNECMA-Note No. 6089/YKC/92, June 1992.
[18] Berglind, T., *Lean Blow-Out Modelling, a Feasibility Study*, Final Report: Volvo, June 1992.
[19] Tilston, J.R., *The Averaging of Gas Analysis Data Allowing for Density and Velocity Weighting*, RAE Techn. Memo P1213, Defence Research Agency, August 1991.
[20] Collin, K.H., *A Small Annular Combustor of High Power Density, Wide Operating Range and Low Manufacturing Cost* AGARD Conference Proceedings No. 422/42, October 1987.
[21] Wittig, S., Leuckel, W., Sakbani, Kh., Aigner, M., Horvay, H. and Sattelmayer, Th., *Experimentelle und theoretische Untersucheung der Strömung und Tropfenbildungsmechanismen in Zerstäuberdüsen für Gasturbinenbrenner*, Abschlußbericht FVV-Vorhaben Nr. 267, Heft 347, 1984.
[22] Wittig, S., Klausmann, W., Noll, B. and Himmelsbach, J., *Evaporation of Fuel Droplets in Turbulent Combustor Flow*, ASME 88-GT-107.
[23] Bauer, H.J., *Überprüfung numerischer Ansätze zur Beschreibung turbulenter ellipitischer Strömungen in komplexen Geometrien mit Hilfe konturangepasster Koordinaten*, Dissertation, Lehrstuhl und Institut für Thermische Strömungsmaschinen, Universität Karlsruhe, o. Prof. Dr.-Ing. S. Wittig, 1989.
[24] Eickhoff, H., Koopmann, J., Neuberger, W. and Rachner, M., *Entwicklung eines Berechnungsverfahrens zur Bestimmung der dreidimensionalen Strömungs- und Verbrennungsvorgänge in Gasturbinenbrennkammern*, FVV-Vorhaben Nr. 300, 1985.
[25] Matcalfe, M., *Definition of Combustor Conditions of Future Engines*, Task 1: CRR89674, June 1990.
[26] Burbank, J., *Summary of Pollution Reduction Methods* Brite/Euram Aero 1019, Task 3: Report 1991 MTUM-B91ET-0084.
[27] Bayle Labouré, G., *Development of New Combustor Technologies at SNECMA*, ISABE 85-7008.
[28] Gastebois, Ph. and Carnel, J., Chambre de Combustion anti-NO_x. A écoulement aérodynamique variable pour Turboréacteur, AGARD CP-205, 1976.
[29] Ciepluck, C.C., Davis, D.Y. and Gray, D.E., Results of NASA's energy efficient engine program, *Jet Propulsion*, 3(6), 1987.
[30] Roberts, R., Diehl, L. and Fiorentino, A.J., *The Pollution Reduction Technology Programme for Can-Annular Combustor Engines*, Description and Results, AIAA No. 76-761.
[31] Gardner, W.B. and Gray, D.E., *The Energy-Efficient Engine (E^3) Advancing the State of the Art*, Asme 80-GT-142.
[32] Lyons, V.J., *Fuel/Air Nonuniformity Effects on Nitric Oxide Emission*, AIAA 81-0327.
[33] Roffe, G. and Ferri, A., Prevaporization and Premixing to Obtain Low Oxides of Nitrogen in Gas Turbine Combustors, NASA CR-2495, 1975.
[34] Spadaccini, L.J. and TeVelde, J.A., Autoignition characteristics of Aircraft-type fuels, *Combustion and Flame*, 46, 1982.
[35] Marek, C.J., Papathakos, L. and Verbulecz, P.W., *Preliminary Studies of Autoignition and Flashback in a Premixing Prevaporizing Flame Tube Using Jet A at Lean Equivalence Ratios*, NASA TM X-3526, 1977.
[36] Roffe, G. and Venkataramani, K.S., *Experimental Study of the Effects of Flameholder Geometry on Emissions and Performance of Lean Premixed Combustors*, NASA CR-135424, 1978.

[37] Roffe, G. and Venkataramani, K.S., *Experimental Study of the Effects of Cycle Pressure on Lean Combustion Emissions*, NASA CR-3022, 1978.
[38] Duerr, R.A. and Lyons, V.J., *Effect of Flame Holder Pressure Drop on Emissions and Performance of Premixed Prevaporized Combustors*, NASA Technical Paper 2131, 1983.
[39] Sjöblom, B.G.A. and Zetterström, K.-A., *A Double Recirculation Zone Two-stage Combustor*, AIAA 79-7019.
[40] Rizk, N.K. and Mongia, H.C., *Ultra Low-NO_x, Rich–Lean Combustion*, ASME 90-GT-87.
[41] Lew, H.G., Carl, D.R., Vermes, G., DeZubay, E.A., Schwab, J.A. and Prothroc, D., *Low NO_x Heavy Fuel Combustor Concept Program Phase I*, NASA CR-165482, 1981.
[42] Kimball-Linne, M.A. and Hanson R.K., Combustion-driven flow reactor studies of thermal $DeNO_x$ reaction kinetics, *Combustion and Flame*, **64**, 1986.
[43] Just, T. and Kelm, S., Mechanismen der NO_x-Entstehung und Minderung bei technischer Verbrennung, 5. DVV-Kolloqium, Clausthal-Zellerfeld, 1986 *Die Industriefeuerung*, Heft 38.
[44] Smith, K.O., Angello, L.C. and Kurzynske, F.R., *Design and Testing of an Ultra-Low NO_x Gas Turbine Combustor*, ASME 86-GT-263.
[45] Becker, B., Berenbrink, P. and Brandner, H., *Premixing Gas and Air to Reduce NO_x Emissions with Existing Proven Gas Tubine Combustion Chambers*, ASME 86-GT-157.
[46] Grobmann, J. and Norgren, C.T., *Turbojet Emission, Hydrogen Versus JP*, NASA TM X-68258, 1973.
[47] Norgren, C.T. and Ingebo, R.D., *Emissions of Nitrogen Oxides from an Experimental Hydrogen Fueled Gas Turbine Combustor*, NASA TM X-2997, 1973.
[48] Smith, K.O., *Ultra-Low NO_x Gas Turbine Combustion System*, Gas Research Institute, 1986.
[49] Sosounow, V.A. and Orlov, V., *Experimental Turbofan Using Liquid Hydrogen Liquid and Natural Gas as a Fuel*, AIAA 90-2421.
[50] Suttrop, F. and Dorneiski, R., Low NO_x-potential of hydrogen-fueled gas turbine engines, *First International Conference on Combustion Technologies for a Clean Environment*, Portugal FH-AC-TB 06-84-91-05, September 1991.
[51] *Minutes of expert meeting*, Brite/Euram: Aero 1019, April 1992.
[52] Joos, F. and Hassa, H., *Influence of Fuel Distribution and Droplet Size on Pollution Formation*, Brite/Euram Aero 1019, Task 4.4 Report MTUM-B92ET-0051, 1992.
[53] Tilston, J.R., Wedlock, M.I. and Marchment, A.D., *Low Emissions Combustor Technology — The Influence of Air Distribution on Homogeneity on Pollutant Formation in the Primary Zone of a Tubular Combustor*, RAE Contract Report RAE/CR/PROP/002, Defence Research Agency, April 1992.
[54] Wedlock, M.I., Tilston, J.R. and Marchment, A.D., *Low Emissions Combustor Technology — The Influence of Air Distribution on Homogeneity and Pollutant Formation in the Primary Zone of a Turbulent Combustor*, DRA Contract Report DRA/CR/AP5/92/01, Defence Research Agency. August 1992.
[55] Tilston, J.R., Marchment, A. and Wedlock, M.I., *The Influence of Air Distribution on Homogeneity and Pollutant Formation in the Primary Zone of a Tubular Combustor. Part 3. Design of the Experiment and Summary Report*, DRA Contract Report DRA/CR/AP5/92/02, Defence Research Agency, August 1992.
[56] Britton, R., *Application of Taguchi Statistical Experiment Design to DRA Combustor, Experimental Data*, Rolls-Royce CRR89731, February 1992.
[57] Bateson, J.W., *Flow Studies on a Perspex Primary Zone Research Model*. Report No. B49 907, Burnley, October 1991.
[58] Ludäscher, M., *No_x Prediction Model for an Aero Engine Combustor*, RAE Tech.

Memo P 1211, Defence Research Agency, July 1991.
[59] Lipfert, F.W., *Correlation of Gas Turbine Emissions Data*, ASME Paper No. 72-GT-60., March 1972.
[60] Wittig, S., Noll, B., Klausmann, W., Leuckel, W. and Krämer, M., *Einfluß der Zerstäubung und Luftführung auf die Gemischaufbereitung in Gasturbinen-Brennkammern*, FVV-Vorhaben Nr. 362, 1986.
[61] Noll, B., *Evaluation of a Bounded High Resolution Scheme for Combustor Flow Computations*, AIAA-J-, pp. 64–69, 1992.
[62] Noll, B., A generalized conjugate gradient method for the efficient solution of three-dimensional fluid flow problems, *Numerical Heat Transfer*, **20**(2), 207–221, 1992.
[63] Noll, B., *Möglichkeiten und Grenzen der numerischen Beschreibung von Strömungen in hochbelasteten Brennräumen Habilitation*, Lehrstuhl ind Institut für Thermische Strömungsmaschinen, Universität Karlsruhe, o. Prof, Dr.-Ing. S. Wittig.
[64] Schneider, M. and Hirleman D., *Effects of Refractive Index Gradients on Phase Doppler Particle Sizing*. Submitted for Presentation at the International Syposium on Special Topics in Chemical Propulsion: Non-Intrusive Combustion Diagnostics, Scheveningen, The Netherlands, May 10–13, 1993.
[65] Kneer, R., Schneider, M., Noll, B. and Wittig, S., *Effects of Variable Liquid Properties on Multicomponent Droplet Evaporation*, ASME-902-GT-131, 1992.
[66] Burbank, J., *Selection of Pollution Reduction Methods*, Brite/Euram Aero 1019, Task 5, Report: MTUM-B92ET-038, 1992.

3 Investigation of the wake mixing process behind transonic turbine inlet guide vanes with trailing edge coolant flow ejection

C. H. Sieverding*, T. Arts, R. Dénos
von Karman Institute

J. Amecke, C. Kapteyn
Deutsche Forschungsanstalt für Luft- und Raumfahrt

F. Martelli, V. Michelassi
Universita di Firenze

S. Colantuoni, G. Santoriello
Alfa Romeo Avio

T. Schröder
MTU Motoren- und Turbinen-Union

J. Bernard, Y. Lapidus
Societé National d'Etudes et de Construction de Moteurs d'Aviation

This report, for the period January 1990 to June 1993, covers the activities carried out under the BRITE/EURAM Area 5: 'Aeronautics' Research Contract No. AERO-CT89-0001 (Project: AERO-P1046) between the Commission of the European Communities and the following:

*Von Karman Institute, VKI (coordinator)
Deutsche Forschungsanstalt für Luft- und Raumfahrt, DLR
University of Florence
Alfa Romeo Avio
MTU Motoren- und Turbinen-Union München GmbH
Société Nationale d'Etude et de Construction de Moteurs d'Aviation, Snecma
Contact: Professor C. H. Sieverding, Von Karman Institute, Turbomachines Dept., Chaussée de Waterloo 72, B-1640 Rhode-St-Genèse, Belgium (Tel: +32/2/358 19 01, Fax: +32/2/358 28 85)

Advances in Engine Technology. Edited by R. Dunker
© 1995 John Wiley & Sons Ltd.

List of symbols

a	sonic velocity
A	area
b	slot width
B	blowing rate $(\rho_c/V_c)/\rho_g \cdot V_g$
c	chord
c_f	friction coefficient
C_m, c_m	mass flow coefficient $\dfrac{\dot{m}_c}{\dot{m}_g}$
c_p	specific heat at constant pressure
c_v	specific heat at constant volume
C_p	pressure coefficient (defined in equation (3.2))
C_b	pressure coefficient (defined in equation (3.1))
d	diameter
e	specific internal energy
E	total specific energy
f	unknown vector
F	flux vector
g	pitch
h	blade height
H	shape factor
i	incidence angle
I	momentum ratio $(\rho_c V_c^2)/(\rho_g V_g^2)$
k	thermal conductivity coefficient
K	turbulence kinetic energy
$\dot{m}$	mass flow
M	Mach number or molecular mass
Ma	Mach number
o	throat
P	pressure
Pr	Prandtl number
q	heat flux vector
R	gas constant or low Reynolds term
r	radius
R_c	Reynolds number based on outlet velocity and chord length
s	curvilinear coordinate
S	surface distance
te	trailing-edge thickness
T	temperature
Tu	degree of turbulence
t	time
u,v	velocity components
V	absolute velocity
W	relative velocity
x,y	coordinates

List of symbols

α	absolute flow angle referred to axial direction
β	relative flow angle referred to axial direction
γ	stagger angle
δ	wedge angle
δ^*	displacement thickness
ε	turning angle or kinetic energy dissipation rate
ζ	kinetic energy loss $\left(\zeta = \dfrac{V_{2,is}^2 - V_2^2}{V_{2,is}^2}\right)$
Θ	momentum thickness except if otherwise stated
ϑ	CO_2 concentration
κ	specific heat ratio
μ	viscosity coefficient
ρ	density
τ	stress tensor

Subscripts

ax	axial
b	base
bl	blade
c	coolant
g	gas
h	hub
i	inviscid
is	isentropic
l	laminar or local
LE	leading edge
PS	pressure side
R	relative
SS	suction side
TE	trailing edge
t	turbulent or tip
te	trailing edge
v	viscous
0	total
1	inlet
2	outlet-measurement plane
3	homogeneous flow downstream

Abbreviations

PS	pressure side
SS	suction side
TE	trailing edge

BL   Baldwin–Lomax
NS   Navier–Stokes
IR   infra-red
PLL   phase-locked loop
AVDR axial velocity density ratio
CT   compression tube

0 Abstract

The aim of the project was to study the axial evolution of the outlet flow field behind a transonic turbine guide vane with trailing edge cooling. The research was undertaken, (a) in view of the lack of detailed information on the decay process of the pressure and temperature gradient fields in the inter-blade row space of the turbine stages, in particular as regards the wake mixing process, which plays a predominant role for the amplitude of the time-varying aerodynamic and heat-transfer characteristics of the turbine rotor, and (b) to provide a valuable test case for the validation of two-dimensional and three-dimensional Navier–Stokes (NS) codes, which are supposed to play an increasingly important role in the turbine design process.

The project comprised four important tasks: (a) the aero design of an advanced transonic trailing edge cooled turbine blade with two different trailing edge configurations, ejection trough a trailing edge slot and ejection from a slot on the pressure side near the trailing edge, denoted hereafter as the TE and PS ejection blades; (b) the detailed testing of both blade configurations in a straight cold-flow suction-type cascade facility; (c) the testing of the PS ejection blade in a compression tube annular cascade facility; and (d) the calculation of the cascade flow fields with two-dimensional and three-dimensional NS codes adapted specifically for handling blades with coolant flow ejection.

The blades were designed for an outlet Mach number $M_{2,is} = 1.05$, an outlet flow angle of $\alpha_2 = 73°$ and a Reynolds number of $Re = 10^6$. The design process made use of a semi-inverse design method, two-dimensional and three-dimensional direct Euler solvers and boundary-layer integral methods. The blade profiles for both blades are identical except for the cut back of the rear pressure side in case of ejection from the pressure side. This solution appeared to minimize the differences between the blade velocity distributions for both blades and therefore allow easier analysis of the test results with regard to the effects of the two different trailing-edge coolant-flow configurations on the blade losses.

The test program in the straight cascade facility encompassed two major tasks: (a) the comparison of the two-dimensional performance characteristics of both blade configurations over a wide range of outlet Mach numbers and coolant flow rates ($M_{2,is} = 0.7$–1.2 and $\dot{m}c/\dot{m}g = 0$–4%) through schlieren photographs, blade-velocity distributions and wake traverses, with air as coolant flow, and (b) systematic wake traverses for both coolant-flow ejection configurations from the near-wake to the far-wake region at design Mach number and coolant flow rate, with carbon dioxide as coolant flow, to investigate the coolant-flow–mainflow mixing via CO_2 concentration measurements.

The blade with the pressure-side ejection was then selected for full flow field measurements in an annular cascade. The use of a compression tube as a gas generator allowed the running of the tests at an air temperature of 440 K and the use of air at ambient temperature as coolant flow. The flow field was measured in three axial planes at nominal outlet Mach number and coolant flow rate. The bulk of the experimental measurements consisted in pressure and flow angle measurements. The development of a low-blockage probe contributed significantly to the reliability of the probe measurements. Problems in the development of a fast response temperature probe did not allow continuous wake traversing for temperature measurements. The excessive testing time of point-by-point thermocouple measurements strongly limited the program for the investigation of the decay of the wake temperature profile. In addition to surveying the flow with stationary measurement systems, an attempt was made to measure the guide vane exit flow field with a high-speed rotating probe, as a first step towards measurements in a full stage turbine. This activity included the successful development of a low-noise wide-bandwidth infra-red opto-electronic data transmission system.

Existing two-dimensional and three-dimensional N–S solvers were modified to incorporate the possibility of coolant flow ejection and implement appropriate turbulence models. The codes required mesh improvement and point clustering in the jet mixing region. While retaining the general H-grid structure, the grid generator is based on a mixed elliptic and algebraic method. The boundary conditions in the jet area required special attention. The solution retained was to fix the mass flow, the flow angle and the total temperature of the coolant jet. The total pressure at the coolant slot exit results from the calculation. Two turbulence models were used: an algebraic turbulence model based on the Baldwin–Lomax formulation (BL) and a low-Re form of a $k–\varepsilon$ model. The BL turbulence model was implemented into an implicit code, the $k–\varepsilon$ model into an explicit code. The three-dimensional NS code used only the implicit formulation together with the BL model. Different methods are used to calculate the blade losses: (a) by averaging the total pressure in the exit plane of the computational domain and (b) by applying the conservation equations for mass, momentum and energy to a control volume between the trailing edge plane and a far downstream plane with uniform flow conditions, the latter providing results which are in better agreement with the experimental data.

Among the wealth of information from the two-dimensional tests, the information on the effect of Mach number and coolant flow rate on the trailing edge base pressure is most interesting because of the importance of the base pressure on the blade losses in the transonic domain on one side, and the entirely different evolution of the base pressure for the two trailing-edge geometries on the other side. The ability of the two-dimensional NS code to visualize the near wake flow pattern in the trailing-edge separation region contributed significantly to the understanding of the measured data. The inferior performance of the PS ejection blade is the result of the combined effect of the differences in the trailing-edge base pressures, the trailing-edge shock strength and the position of the maximum suction side velocity and the associated point of boundary layer transition. As regards the mixing of the coolant flow with the main flow one observes an extremely fast decrease in the maximum CO_2 concentration in the near-wake, $c_m \simeq 10\%$ at 10% of the axial chord, followed by a rather slow decrease in the far-wake region, $c_m = 4–5\%$ at a downstream distance of 50% of the axial chord. Analogous tendencies are found for the wake temperature profiles in the annular cascade tests. The concentration and total pressure profiles appear to be similar to each other and can be approximated by a Gaussian distribution.

In comparing the straight and annular cascade tests it appears that the losses in the annular cascade are lower than what one might expect on the basis of the straight cascade tests. In fact the strong increase of the two-dimensional blade losses at $M_{2,is} \approx 1.0$ from $\zeta = 3.5\%$ to 7% at the nominal coolant flow rate $c_m = 3\%$ is not observed in the annular cascade, in which the profile losses remain at a level of $\zeta = 3–3.5\%$ over most of the blade span (i.e. outside the secondary flow region) in spite of Mach numbers as high as $M_{2,is} = 1.09$. This different behaviour is explained by three-dimensional flow effects which are particularly stong in the trailing-edge region, as witnessed by three-dimensional NS visualizations.

The measurements with the fast rotating probe can be viewed as a success. However, the blockage effect of the probe appears to play a dramatic role for the relative total pressure profile. This finding indicates enormous difficulties in the interpretation of the relative total pressure when measured with probes mounted on a rotor blade in a real turbine stage.

1 Introduction

Although the project should mainly be thought of as a contribution to the investigation of such fundamental aerodynamic problems as wake mixing, coolant jet-wake interference and decay of pressure and temperature gradients downstream of the blade rows, certain aspects are of direct interest to the turbine designer because they have to do with stage efficiency and rotor blade life.

Blade trailing edge losses count among the most important penalties for the overall turbine-stage efficiency. They are proportional to the trailing-edge thickness and the trailing-edge base pressure and may amount to 30% of the profile losses for blades operating at transonic outlet Mach numbers. It is also known that the ejection of coolant flow from the trailing edge considerably affects the base pressure and thereby the trailing-edge losses. The mechanism by which this occurs is the modification of the dynamics of the near-wake region, which is very sensitive to any mass injection into the trailing-edge base region. Most presently available information refers to coolant flow ejection through the trailing edge, but the actual tendency seems to favour the ejection of the coolant from the pressure side just ahead of the trailing edge. The reason for this trend seems to be the apparent reduction of the trailing edge thickness. However, this may be a false argument because there exists necessarily a metal lip between the pressure-side flow and the coolant jet, which is to be considered as a second trailing edge, known as the pressure-side (PS) trailing edge as opposed to the suction-side (SS) trailing edge, which is situated somewhat further downstream. Base-pressure data for this type of trailing-edge configuration are needed for an in-depth analysis of the aerodynamics of the two trailing-edge cooling methods.

For each stage there exists an optimum axial distance between guide vane and rotor which is a function of (a) the inter-blade row mixing losses, (b) the rate of decay of the inter-blade row pressure and temperature gradient field and (c) the endwall friction losses. Both the pressure and temperature gradient fields affect the heat transfer on the rotor blade. An incorrect evaluation of the upstream non-uniform flow field on the rotor boundary layer transition may lead to a blade failure due to unforeseen heating or to an overestimation of the needs of coolant flow. Hence the knowledge of the decay of the gradient fields in the inter-blade row space is of utmost importance for the evaluation of the blade life of the rotor and the overall turbine efficiency.

The research program described hereafter is the result of a combined effort of two research institutes (DLR and VKI), one university (Florence) and three aero-engine companies (ARA, MTU and SNECMA), who joined their forces in a spirit of true international collaboration, free from any corporate interest. Two guide vanes were designed with coolant flow ejection from the pressure side and through the trailing edge, respectively. Both blades were tested in the DLR cold flow straight cascade tunnel, using either air at ambient temperature or carbon dioxide as coolant flow. The latter allowed the study of the coolant flow/main flow mixing through CO_2 concentration measurements. Subsequently the PS ejection blade was tested in the CT3 compression tube annular cascade facility at VKI, which allows proper simulation of the temperature ratio between main gas and coolant flow. Parallel to the straight and annular cascade testing, the University of Florence adapted existing two-dimensional and three-dimensional NS solvers for handling blades with coolant-flow ejection. The fully integrated experimental–theoretical approach proved to be a valuable undertaking which benefited both sides.

2 Aerodynamic design of the BRITE turbine inlet guide vane

The overall design characteristics of the IGV were chosen to be representative for advanced aero engines and the mean blade section was optimized following actual design practice.

However, time and cost limitations imposed the use of cylindrical blades for the annular cascade rather than highly three-dimensional shaped blades. Nevertheless, this was not considered to be a real drawback since the main objective is not the optimization of the IGV but the wake-mixing process downstream of the IGV. The choice of the actual blade dimensions was governed by the requirement of their compatibility with the test section dimensions of the DLR and VKI wind tunnels.

2.1 Blade profile design

The blade was designed by ALFA AVIO, but the design has benefited from several revisions by the partners, in particular from MTU and SNECMA.

Two different blade configurations were considered: blade A with coolant ejection through the trailing edge, called hereafter the TE ejection blade, and blade B with coolant ejection from the pressure side near the trailing edge, called hereafter the PS ejection blade. It appeared that, for comparison reasons of the wake mixing behind both blades, the blade velocity distributions as well as the blade shape should be the same over most of the blades. On the other hand it was argued that an optimized blade with coolant ejection through the trailing edge, and an optimized blade with coolant ejection from the pressure side would be necessarily different in blade shape and velocity distribution, a fact which would make the comparison of the wake-mixing process difficult. Hence, the decision was taken to design an optimal mean blade section with provision for coolant flow ejection through the trailing edge and modify this blade for an ejection of the coolant flow from the pressure side.

2.1.1 Design procedure

The flow chart in Figure 3.1 shows the intensive use of advanced design and analysis methods in the design procedure.

The two-dimensional Euler code solves the two-dimensional Euler equation of a transonic, inviscid, adiabatic flow of a perfect gas. The shocks are treated by using a shock-capturing technique. The numerical procedure is based on a time-marching procedure, damping surface technique and finite-volume area discretization.

The three-dimensional Euler code solves the three-dimensional Euler equation written in a system of cylindrical coordinates, and applied to the flow in an axial turbine blade row. The general assumption of the method limits its application to the computation of a three-dimensional, transonic, inviscid, adiabatic and rotational flow of a perfect gas. The shocks are treated by using a shock-capturing technique. The numerical procedure is based on a time-marching procedure, damping surface technique and finite-volume discretization.

The code used for boundary-layer analysis is a special version of a two-dimensional finite difference boundary-layer code based on the Patankar and Spalding scheme. The program solves the momentum equation as a minimum, plus any number of diffusion equations. The flow may be laminar or turbulent. The turbulence model is a Prandtl mixing-length scheme. The effect of curvature on the transition, the free-stream influence and some transition models are also included.

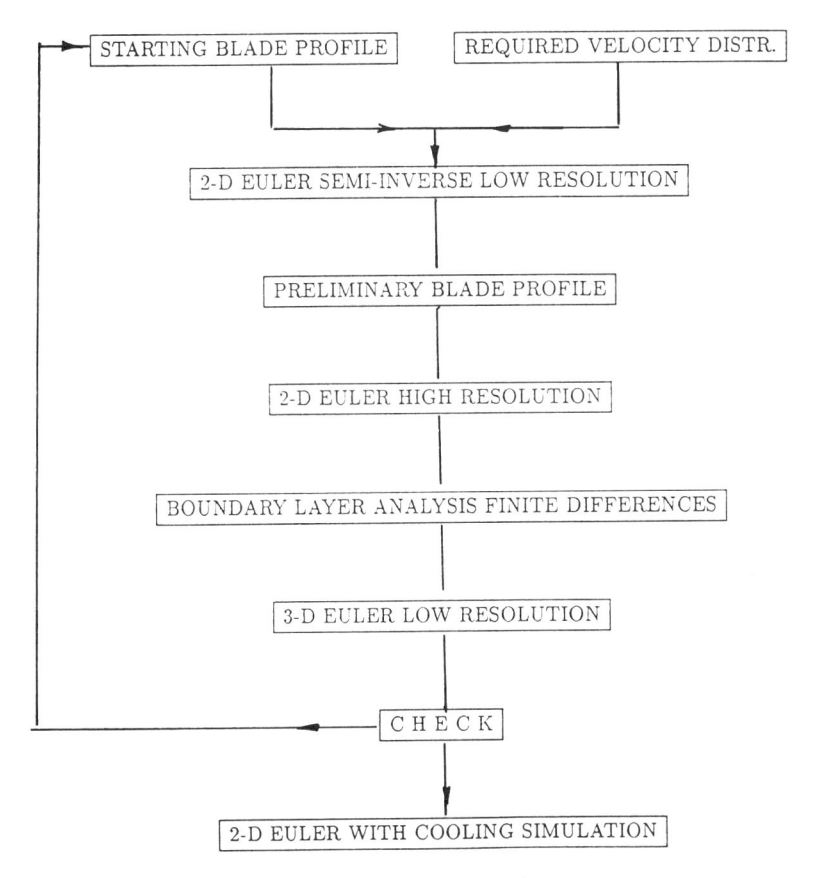

Aerodynamic design of the BRITE turbine inlet guide vane

```
STARTING BLADE PROFILE          REQUIRED VELOCITY DISTR.
             │                           │
             └───────────┬───────────────┘
                         ▼
         2-D EULER SEMI-INVERSE LOW RESOLUTION
                         │
                         ▼
              PRELIMINARY BLADE PROFILE
                         │
                         ▼
                2-D EULER HIGH RESOLUTION
                         │
                         ▼
         BOUNDARY LAYER ANALYSIS FINITE DIFFERENCES
                         │
                         ▼
                 3-D EULER LOW RESOLUTION
                         │
                         ▼
                       CHECK
                         │
                         ▼
              2-D EULER WITH COOLING SIMULATION
```

Figure 3.1 Design procedure.

2.1.2 Design requirements

The design conditions for the mean blade section design were defined as follows:

- isentropic outlet Mach number, $M_{2,is} = 1.05$
- outlet flow angle, $\alpha_2 \simeq 73°$
- pitch to chord ratio, $g/c = \sim 0.75$
- blade surface velocity distribution:
    —*on the suction side*: smooth acceleration up to the shock, minimum shock strength, smooth deceleration behind the shock
    on the pressure side: avoid front pressure side velocity peak
- Reynolds number based on chord and outlet velocity: $Re = 1.0 \times 10^6$ (compromise between typical value for helicopter engines and large jet engines)

- Turbulence level $Tu = 1\text{–}2\%$ (natural Tu-level in wind tunnels).
  Note that the Tu-level is far below the level in real engines, but it was considered that the use of turbulence grids would increase significantly the measurement problems and make the interpretation of the mixing process much more difficult. Also, a high upstream turbulence level was considered to be inappropriate for comparison of the experimental data with NS calculations.

2.1.3 TE-ejection blade

As pointed out before, the IGV is designed with cylindrical blades. The blade profile is optimized for the flow conditions at mid-blade height.

Blade and cascade geometry The solid blade profile, the rear suction side angle and curvature distributions are presented in Figure 3.2. The blade coordinates, the surface angle and the surface curvature are given in Appendix 1. The slight wavy SS contour downstream of the throat is the result of the prescribed velocity distribution. This is by no means a design flaw. What counts is the boundary-layer development, for which surface curvature of course plays an important role. Figure 3.2 shows that the curvature distribution remains smooth everywhere.

The main characteristics of the blade section are summarized below:

- chord length, c 72.0 mm
- axial chord length, c_{ax} 43.1 mm
- pitch to chord ratio, g/c 0.75
- throat width, o 14.98 mm
- gauging angle, $\arccos(o/g)$ 73.9°
- stagger angle, γ 51.90°
- trailing edge thickness, te 1.70 mm
- trailing edge thickness to chord ratio, te/c 2.36%
- trailing edge wedge angle, δ_{te} 6.4°
- rear suction side turning angle, ε_{ss} 8.2°
  (angle between tangent to throat and trailing edge).

The coolant flow will be ejected through a slot in the trailing edge. The ratio of slot width to trailing edge thickness is approximately $\frac{1}{3}$. For details, see Figure 3.5. The rather large trailing edge thickness is a result of a compromise of the trailing edge thickness to chord ratio between helicopter engines and big jet engines.

Blade velocity distribution The two-dimensional Euler calculation was performed on this blade section at the following aerodynamic conditions:

Inlet total temperature, T_{o1} 440.0 K

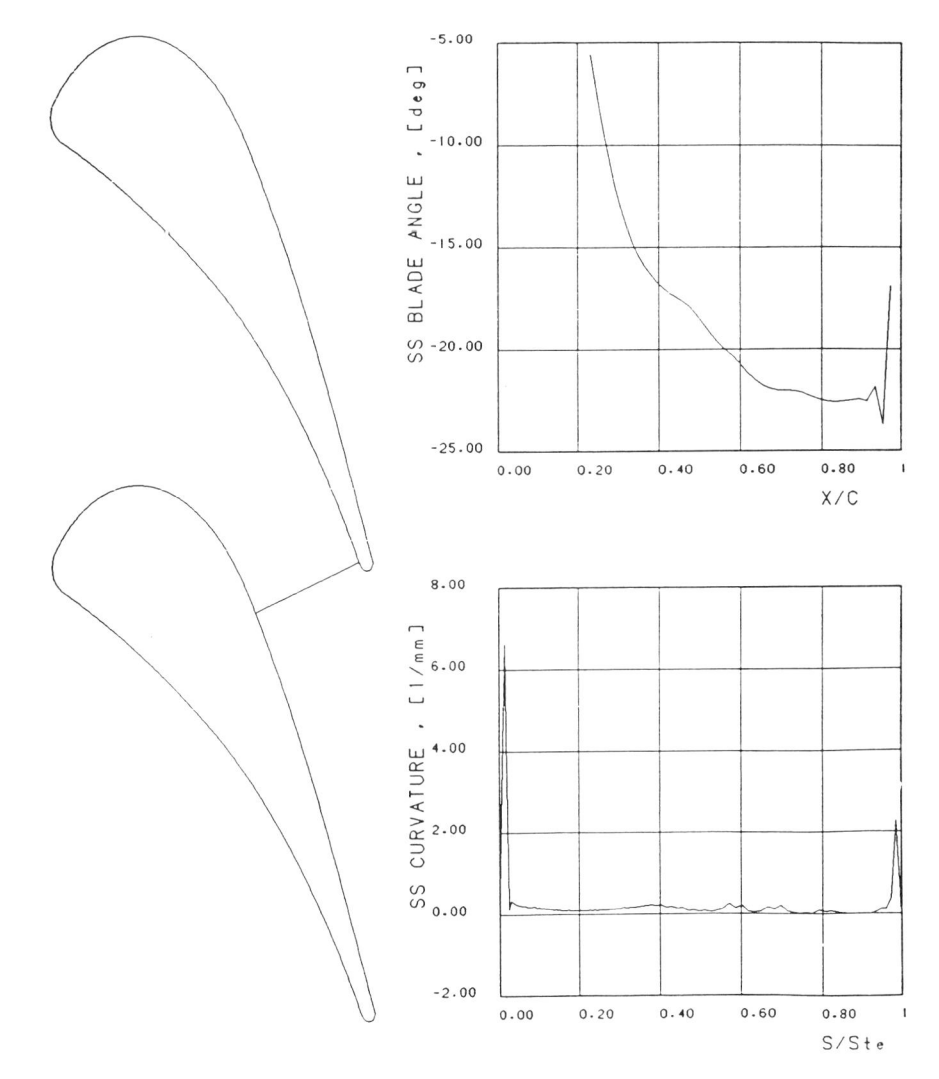

Aerodynamic design of the BRITE turbine inlet guide vane

Figure 3.2 Cascade geometry.

Inlet total pressure, P_{o1} 1.62 bar
Total-to-static pressure ratio, P_{01}/P_2 2.009
Inlet flow angle, α_1 0.0°

The discretization was on an H-mesh size 141 × 41 with 101 points on each side of the blade; the artificial viscosity coefficient is fixed at 0.98. The convergence is checked on the evolution of the exit flow angle. It is judged satisfactory if the residual fluctuations are less than 0.1°. The isentropic Mach number distribution vs. the x/c coordinate is shown in Figure 3.3. The maximum Mach number is 1.2. An oblique shock decelerates the flow from 1.2 to 1.07. The computed exit flow angle is $\alpha_2 = 72.9°$.

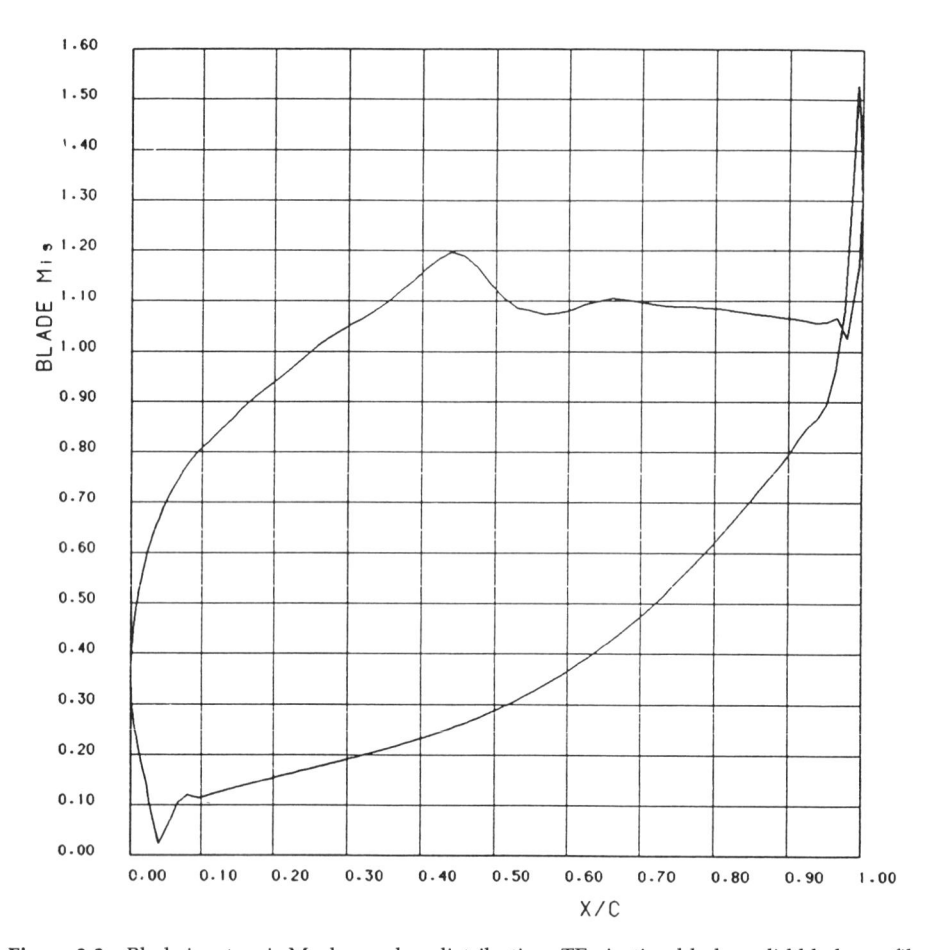

Figure 3.3 Blade isentropic Mach number distribution: TE-ejection blade, solid blade profile.

Blade boundary-layer calculation The program uses a finite difference method and the Prantdl mixing-length turbulence model. On the suction side the Reynolds number for transition is based on Seyb's formulation.

The results of the boundary-layer analysis on the pressure and suction sides are reported in Figure 3.4. They show the following distributions:

- V velocity
- θ momentum thickness
- H shape factor (δ^*/θ)
- cf friction coefficient

For a turbulence level $Tu = 2.0\%$ the transition point on the suction side occurs at about 52% of the total suction side length for a Reynolds number based on the momentum thickness of $Re = 386$. Separation occurs at 56% of the total suction side length for a Reynolds number $Re = 448$, followed by a rapid reattachment.

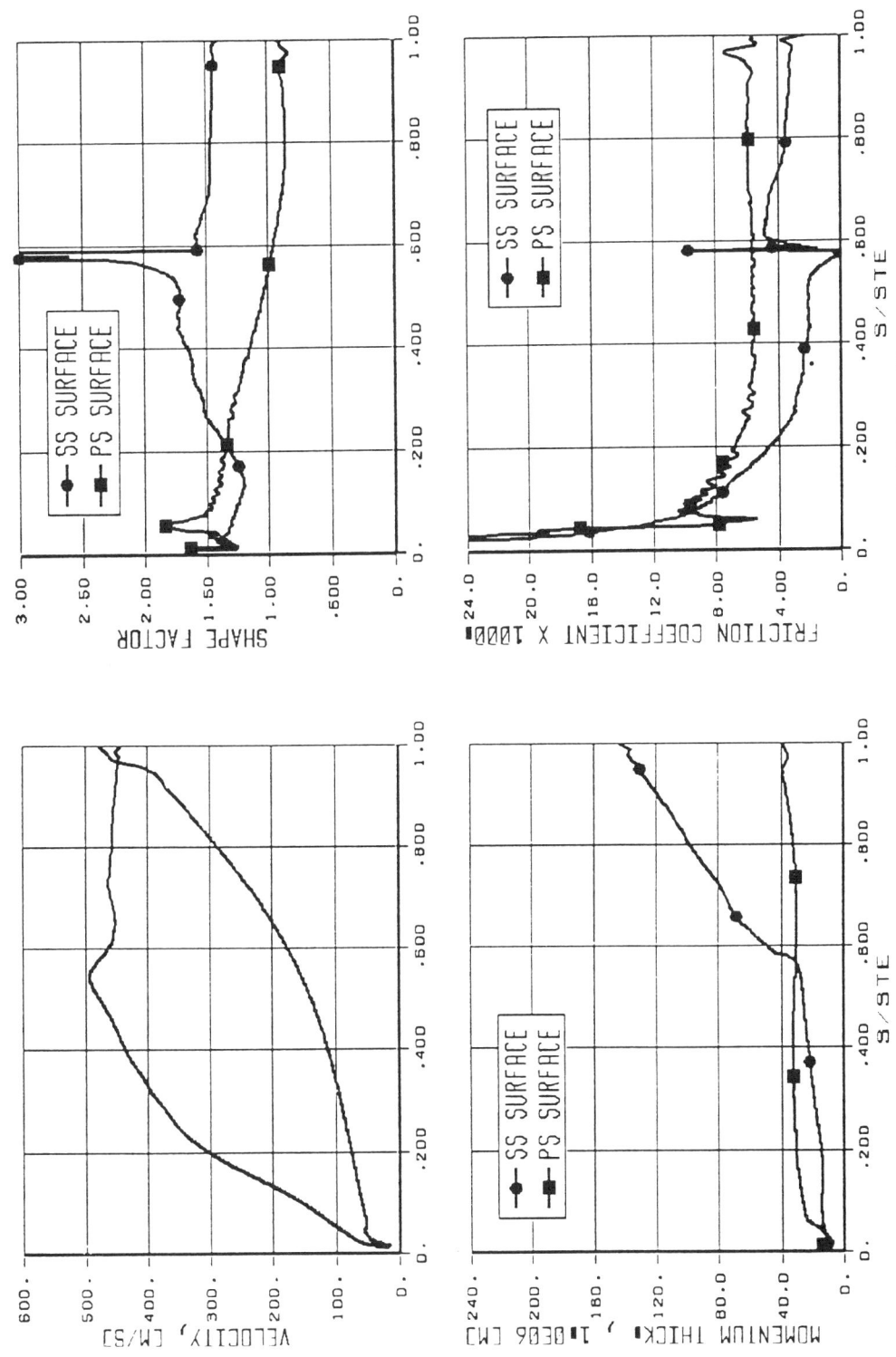

Aerodynamic design of the BRITE turbine inlet guide vane 165

Figure 3.4 Boundary-layer analysis: TE-ejection blade, solid blade profile.

2.1.4 PS ejection blade

Blade geometry As it appeared to be preferable to minimize the differences in the blade velocity distributions for both blades and thereby allow to analyse more easily the test results with respect to the effect of the two coolant flow configurations on the losses, it was decided to generate the pressure side ejection blade by a simple cut back of the rear pressure side. This cut back results in a backwards facing step from which the coolant flow is ejected. Figure 3.5 compares the two trailing edge coolant flow configurations. Compared to the TE-ejection blade, the PS-ejection blade has a slightly increased throat width and the throat position has moved upstream.

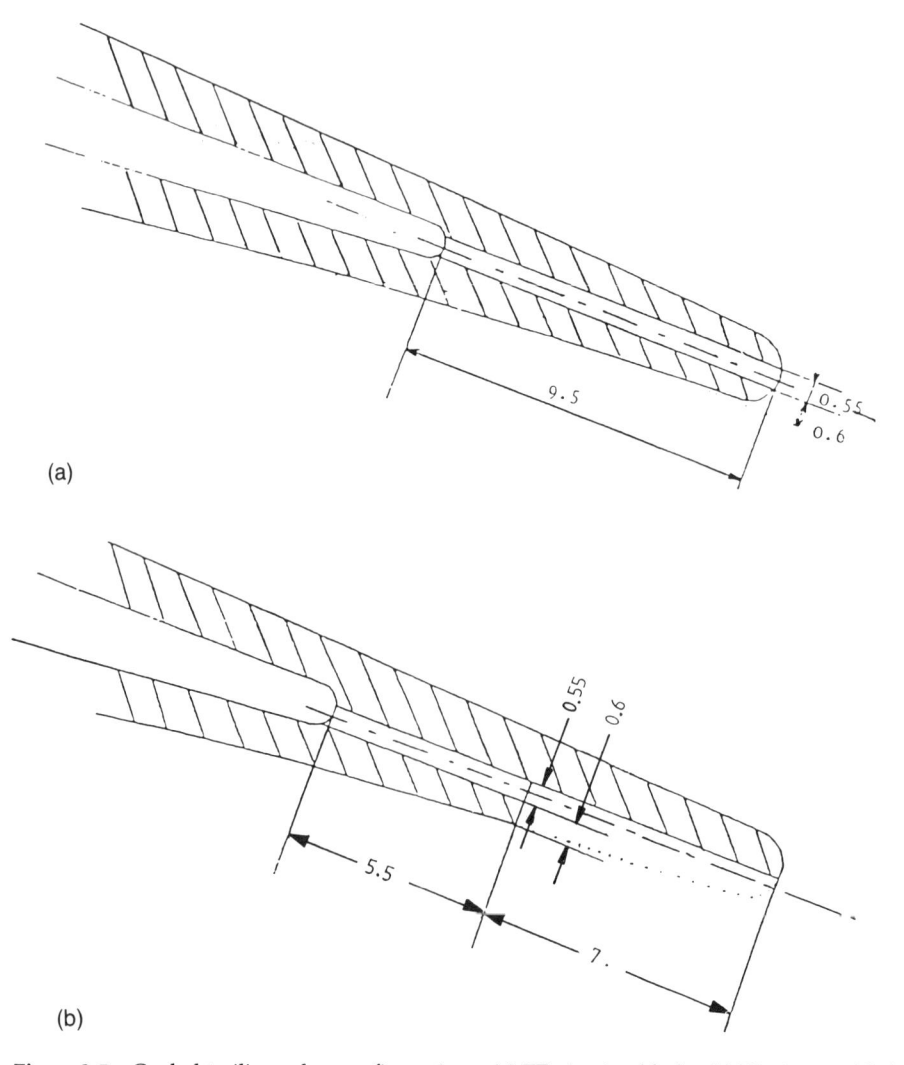

Figure 3.5 Cooled trailing-edge configurations: (a) TE-ejection blade; (b) PS-ejection blade.

Aerodynamic design of the BRITE turbine inlet guide vane

Blade velocity distribution The blade Mach number distributions for the two different configurations, are compared in Figure 3.6. Results of boundary layer analysis of the SS velocity distribution are presented in Figure 3.7.

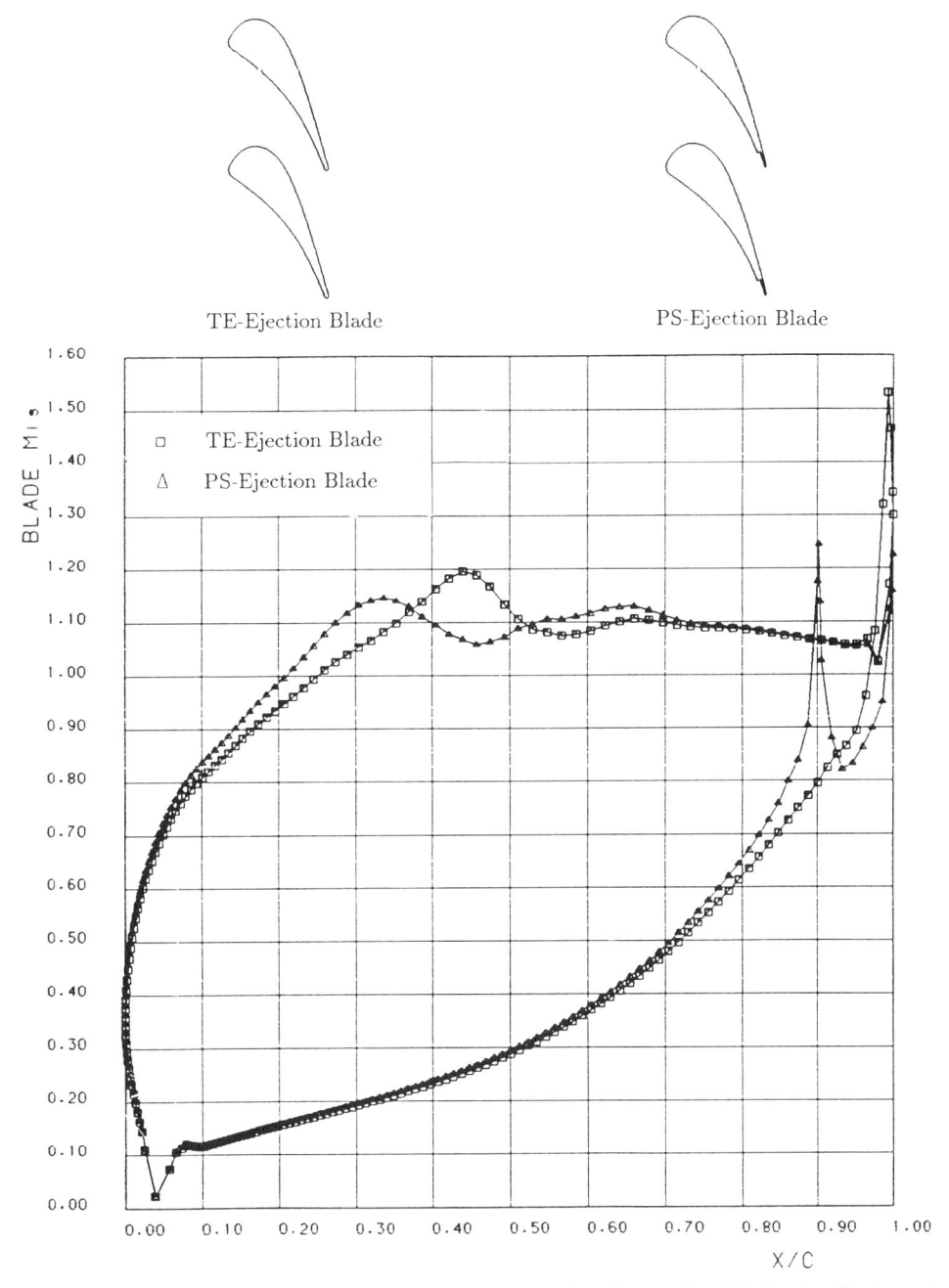

Figure 3.6 Comparison of isentropic Mach number distribution for TE-ejection blade and PS-ejection blade (solid blade profiles).

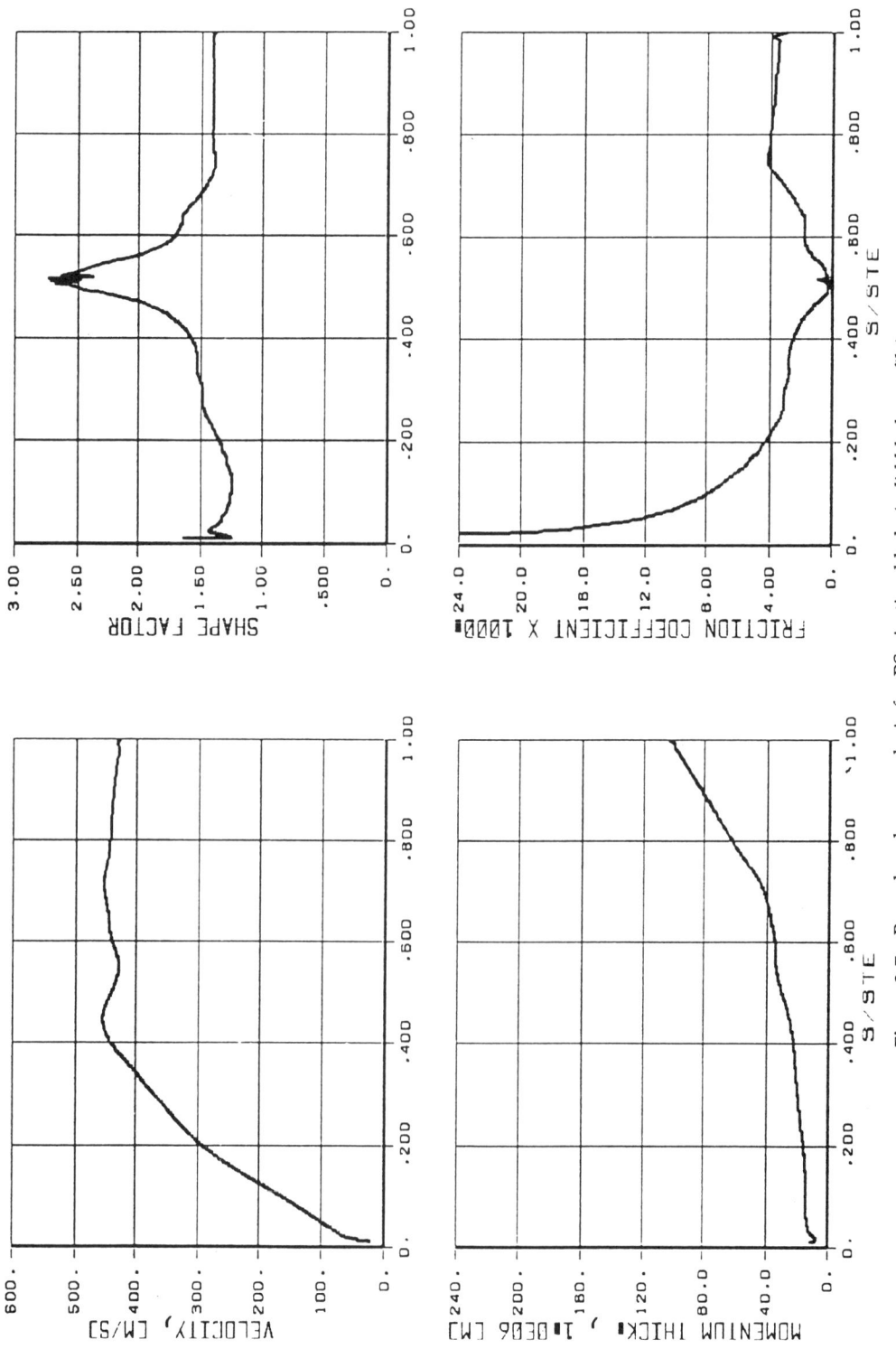

Figure 3.7 Boundary-layer analysis for PS-ejection blade (solid blade profile).

2.2 Effect of coolant flow ejection

2.2.1 Coolant flow conditions

The required nominal values of the coolant parameters are the following:

'Mass flow rate', $C_m = m_c/m_g$	3.0
'Blowing rate', $B = (\rho V)_c/(\rho V)_2$	1.5
'Momentum ratio', $I = (\rho V^2)_c/(\rho V^2)_2$	0.8

2.2.2 Simulation of coolant flow ejection with Euler code

The coolant jet simulation is performed by an adapted version of a two-dimensional Euler code. The following parameters of the coolant jet have to be specified:

m_c coolant mass flow
T_{oc} total temperature of the coolant flow
α_c flow angle of the coolant jet
b width of the slot

However, it turned out that T_{oc} had to be adjusted during the calculation in order to achieve the required values of mass flow rate (C_m), blowing rate (B) and momentum ratio (I).

2.2.3 Effect of coolant flow ejection on velocity distribution at nominal conditions

Figure 3.8 presents the comparison of the isentropic Mach number distributions for the TE ejection blade at zero and 3.2% coolant flow rates.

Figure 3.9 shows the comparison of the isentropic Mach number distributions for the PS ejection blade at zero and 3.6% coolant flow rates.

Figure 3.10 shows the comparison of the isentropic Mach number distributions between the two cooled blade configurations.

2.3 Three-dimensional analysis of the BRITE annular inlet guide vane

2.3.1 Geometry

The three-dimensional annular geometry of the IGV was obtained by radial stacking of the two-dimensional mid section. The stacking line is the vector radius through the throat centers of two-dimensional radial sections (Figure 3.11). The endwalls are cylindrical. Table 3.1 summarizes the geometrical characteristics of the three principal three-dimensional sections.

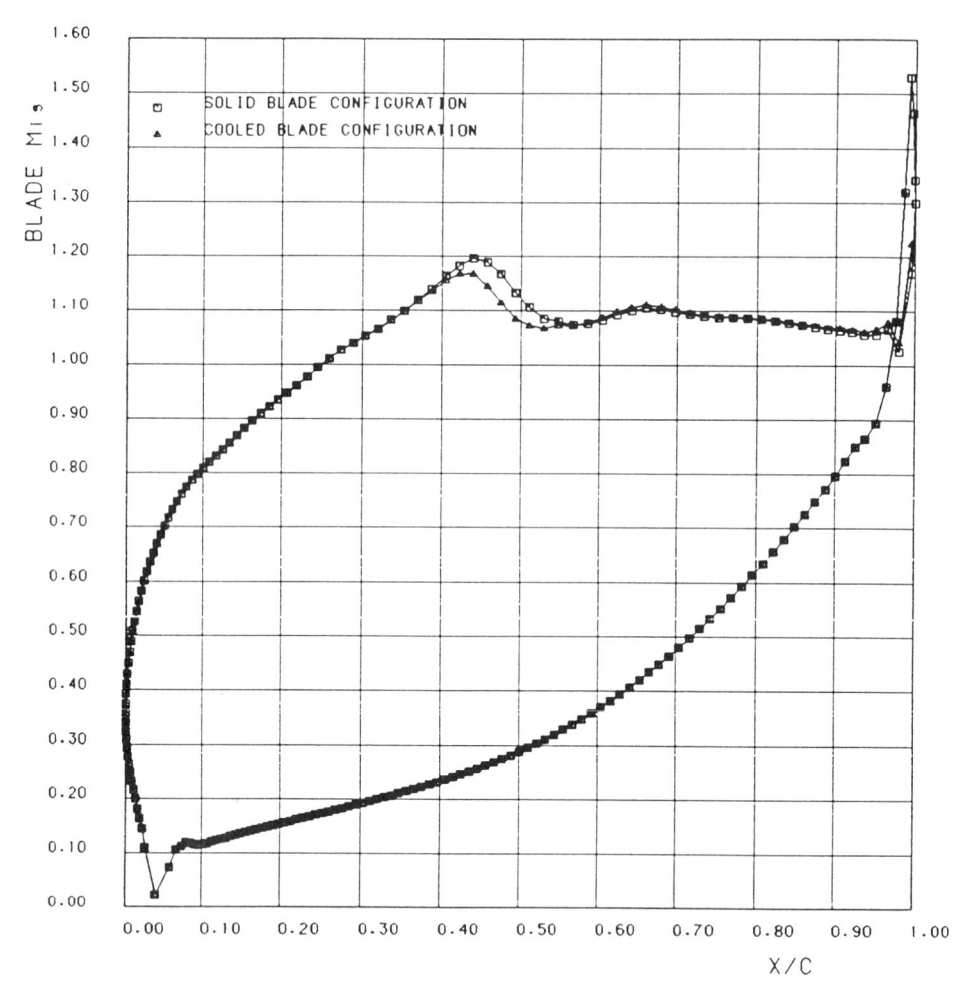

Figure 3.8 Isentropic Mach number distributions of TE-ejection blade at zero and 3.2% coolant flow rate.

2.3.2 Three-dimensional Euler calculation

The three-dimensional Euler calculation was performed on the TE-ejection blade at zero coolant flow rate with a code based on time-marching technique and finite volume discretization procedure. The computational domain (H-mesh) consists of 77 spanwise surfaces (41 of them on the blade), 15 surfaces in the blade-to-blade plane and 21 surfaces from hub to tip endwalls (Figure 3.12).

The inlet conditions are purely axial flow and constant total pressure and total temperature profiles. The static pressure was imposed at the hub downstream of the blade row, such that a mid-span outlet Mach number $M_{2,is}$ close to 1.05 was obtained. An isentropic specific heat ratio equal to 1.4 was used.

Aerodynamic design of the BRITE turbine inlet guide vane

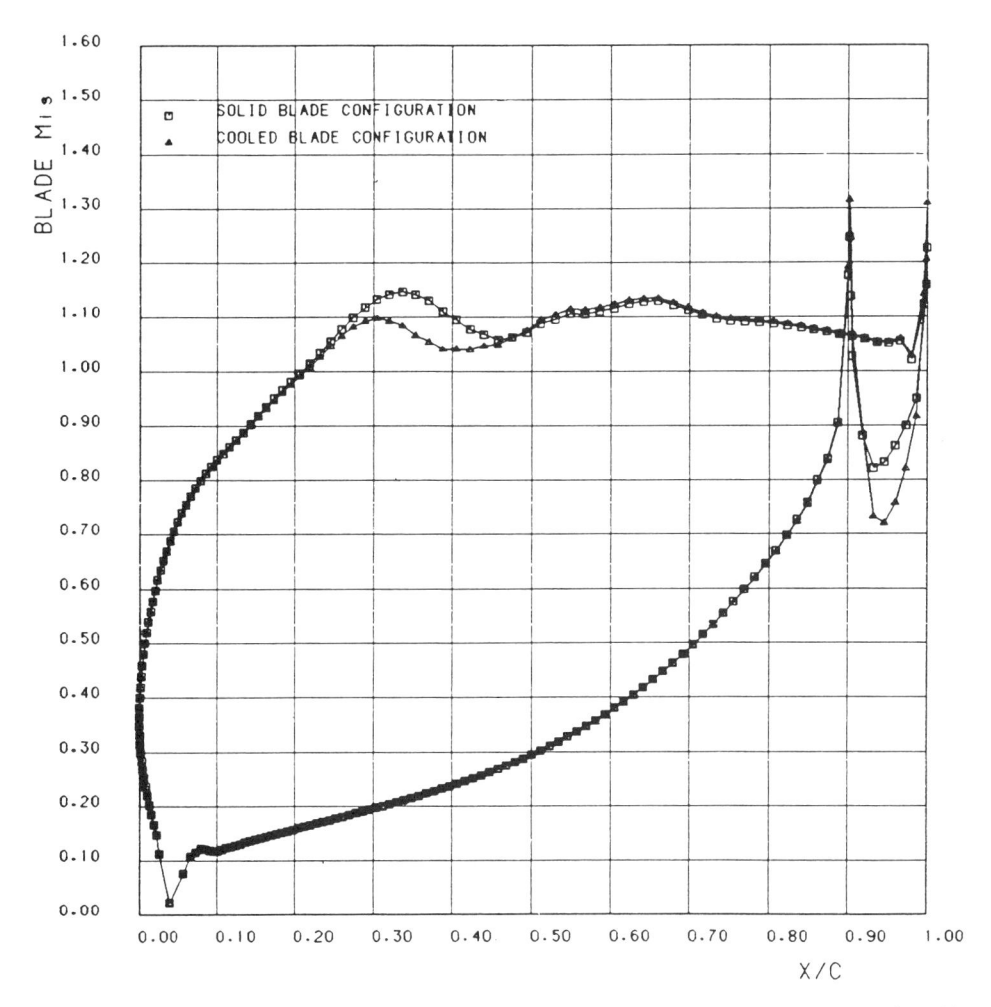

Figure 3.9 Isentropic Mach number distributions of PS-ejection blade at zero and 3.6% coolant flow rate.

In Figure 3.13 the blade Mach number distributions for the hub, mid and tip sections are plotted. The exit Mach number for hub, mid and tip sections are respectively 1.13, 1.05 and 0.987. The exit flow angle presents slight variation around 72.5°.

Figure 3.14 compares the blade Mach number distribution at mid-blade height with the corresponding two-dimensional cascade calculation. The only significant difference between two-dimensional and three-dimensional calculations is the higher SS peak Mach number before the shock impingement. However, this is explained by a slightly higher outlet Mach number for the three-dimensional calculations, i.e. $M_{2,is} = 1.07$ instead of 1.05 for the two-dimensional calculations.

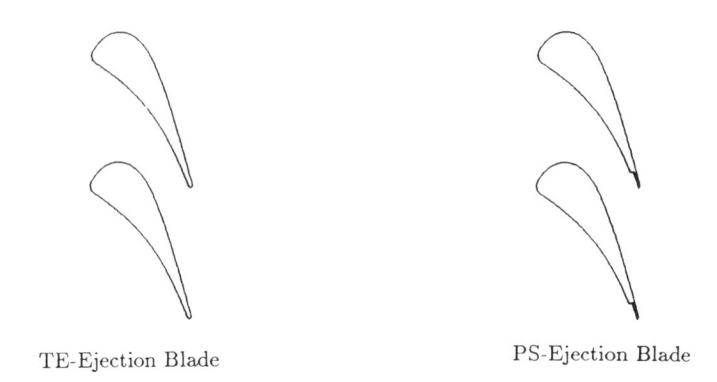

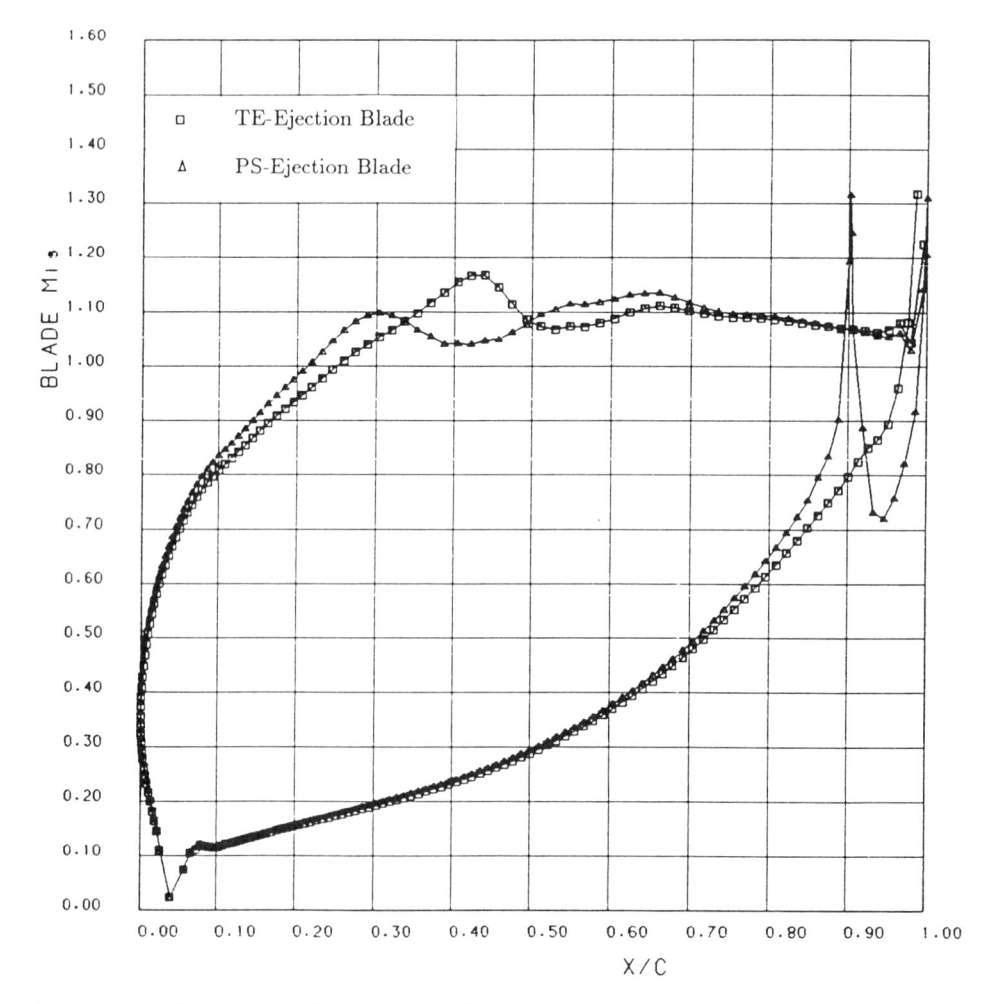

Figure 3.10 Comparison of isentropic Mach number distributions between TE-ejection blade and PS-ejection blade at nominal coolant flow rate ($c_m \simeq 3\%$).

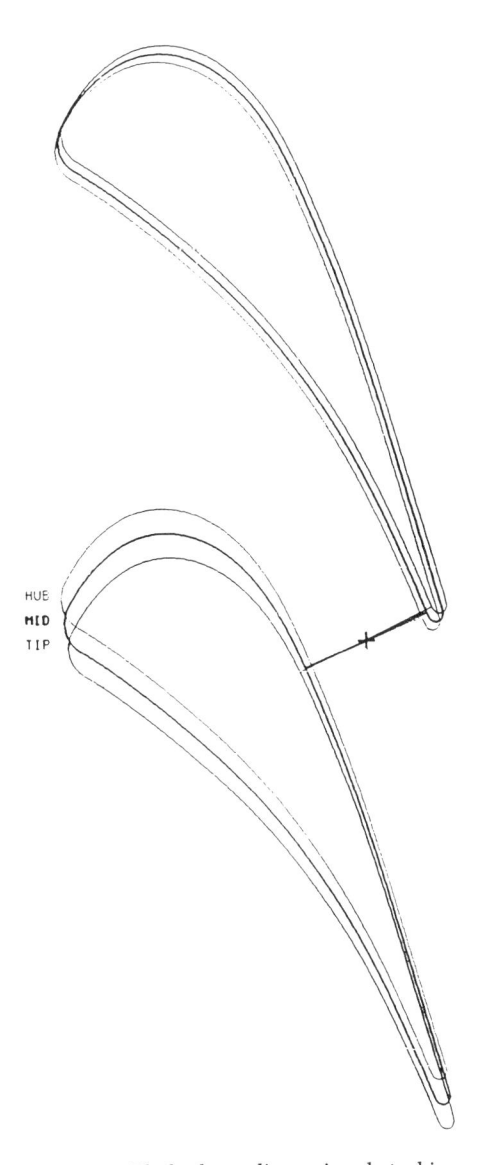

Aerodynamic design of the BRITE turbine inlet guide vane

Figure 3.11 Blade three-dimensional stacking.

Table 3.1 Geometrical characteristics

	Diameter (mm)	Throat (mm)	Pitch (mm)	Gauging angle (degrees)
Hub	689.0	13.56	50.34	74.37
Mid	739.7	14.98	54.04	73.90
Tip	790.4	16.50	57.75	73.40

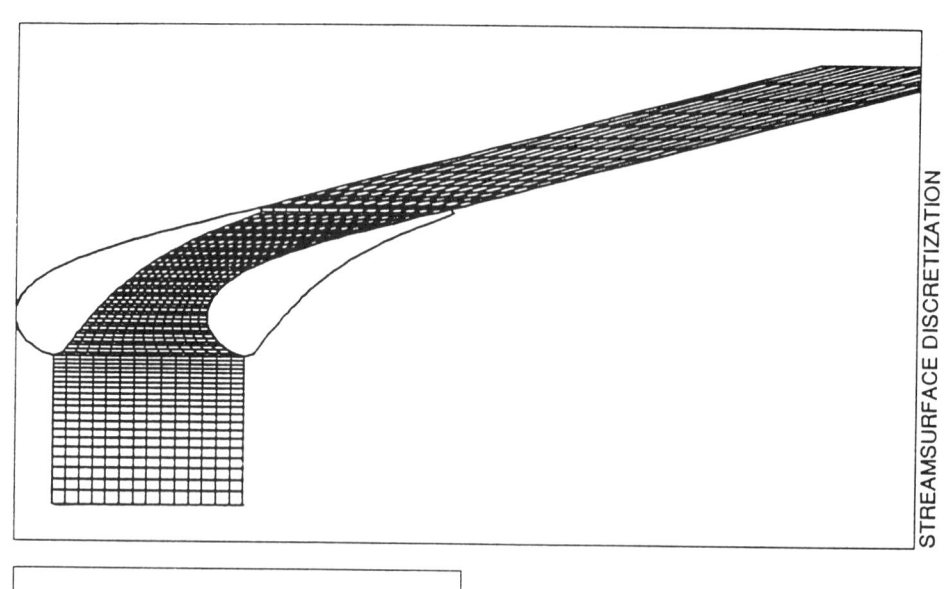

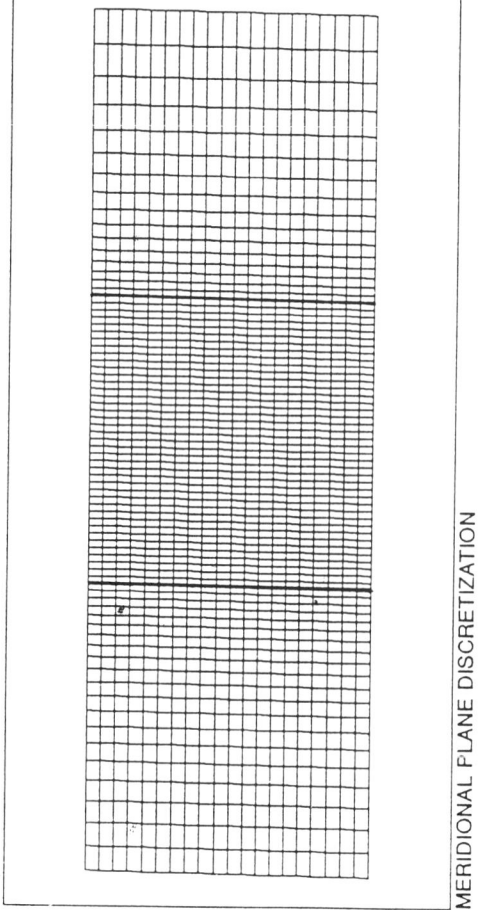

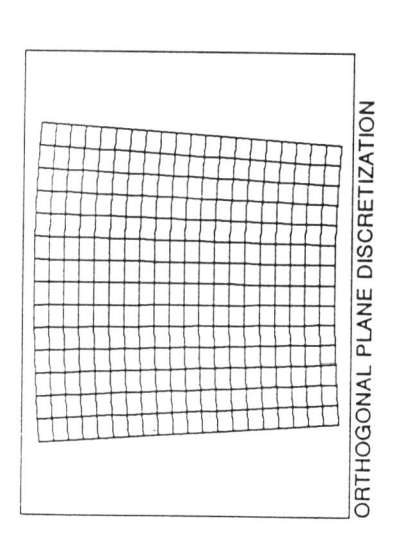

Figure 3.12 Computational grid for three-dimensional Euler analysis.

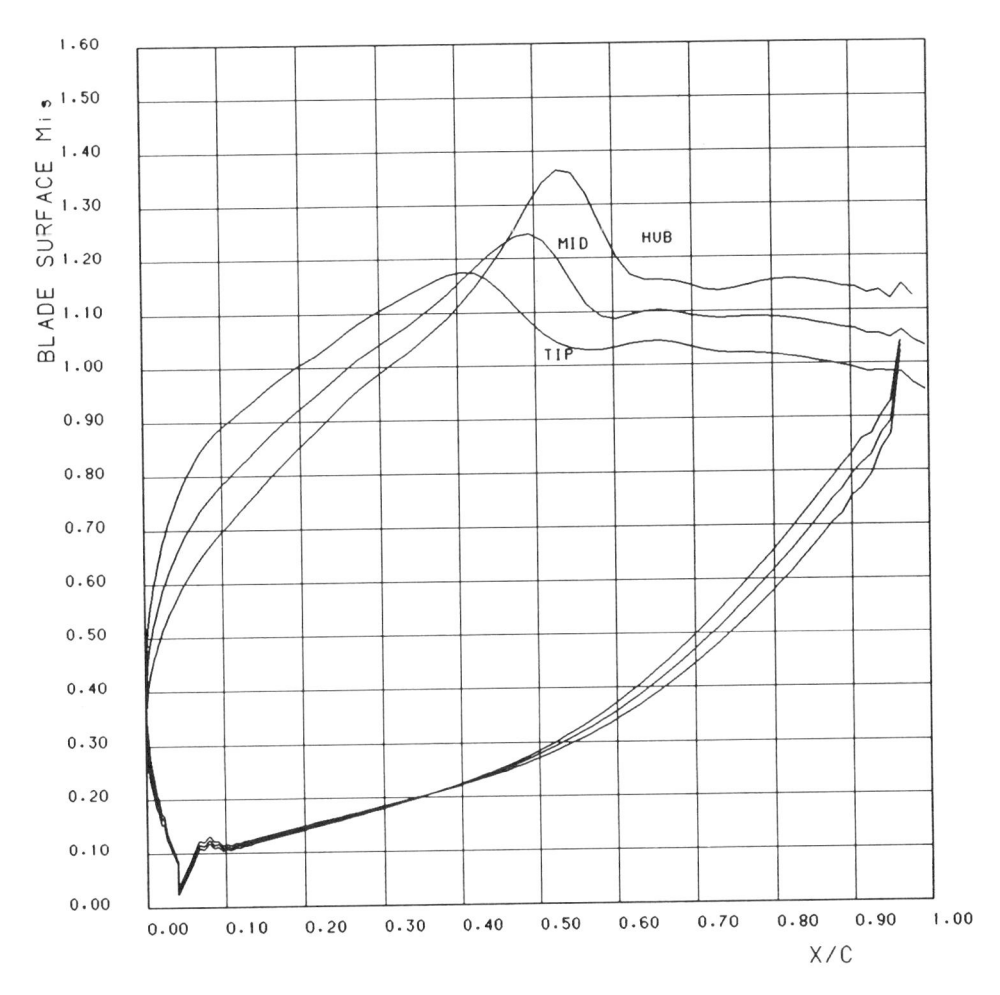

Figure 3.13 Isentropic Mach number distributions at hub, mean and tip of BRITE annular cascade calculated with three-dimensional Euler code.

3 Mechanical design of blades and cascades

SNECMA was in charge of the mechanical design and the manufacturing of blades and cascades for the DLR and VKI wind tunnels:

- TE and PS ejection blades for the cold-flow straight-cascade tunnel at DLR,
- annular cascade with PS ejection blades for the VKI compression tube annular cascade facility.

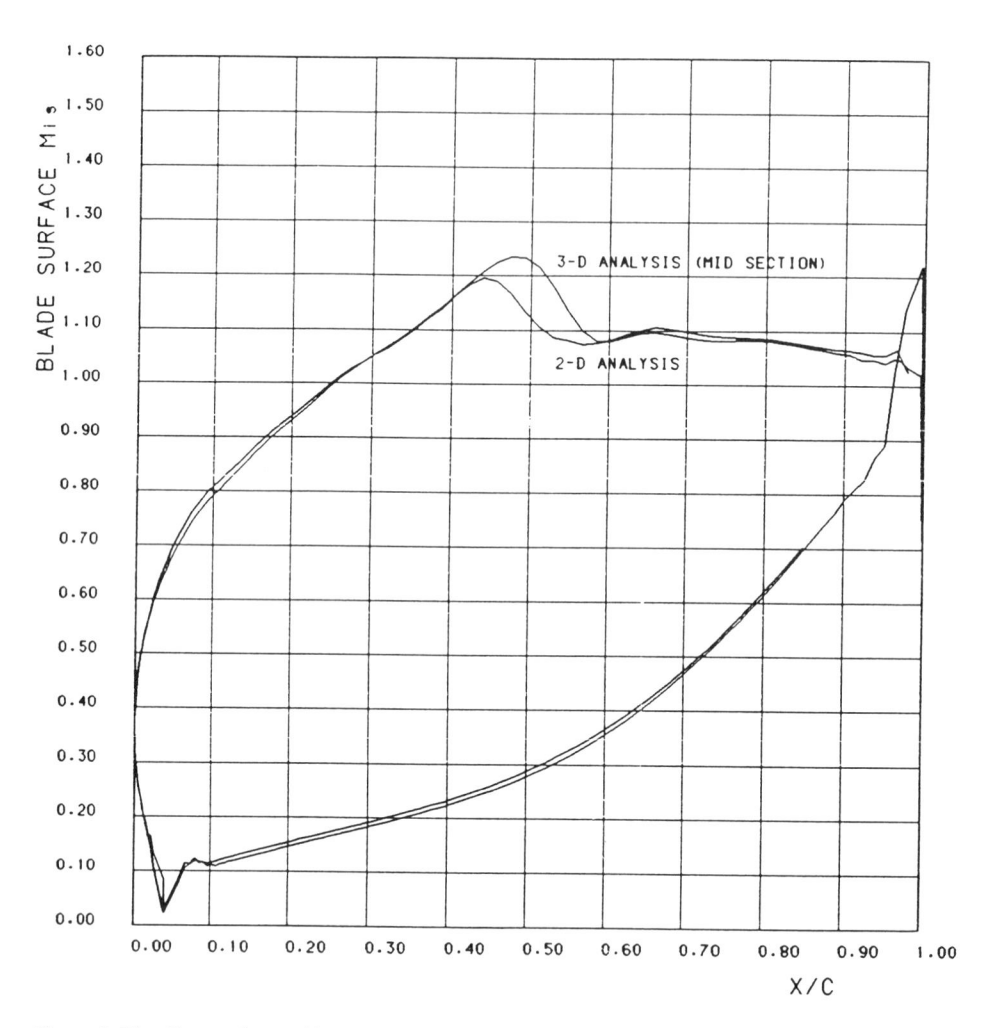

Figure 3.14 Comparison of isentropic blade Mach number distributions for straight cascade and mean section of annular cascade calculated with two- and three-dimensional Euler codes, respectively.

3.1 Straight cascade blades

3.1.1 Mechanical design

Mechanical designs of both cooled blade configurations are shown in Figures 3.15–3.17.

To permit uniform flow ejection along the trailing edge height, coolant air is supplied from both blade ends through lugs supporting the blades in the side walls.

For both coolant ejection configurations the trailing edge slots cover 90% of the 125 mm blade height. A constant slot width for all blowing rates is insured

Mechanical design of blades and cascades

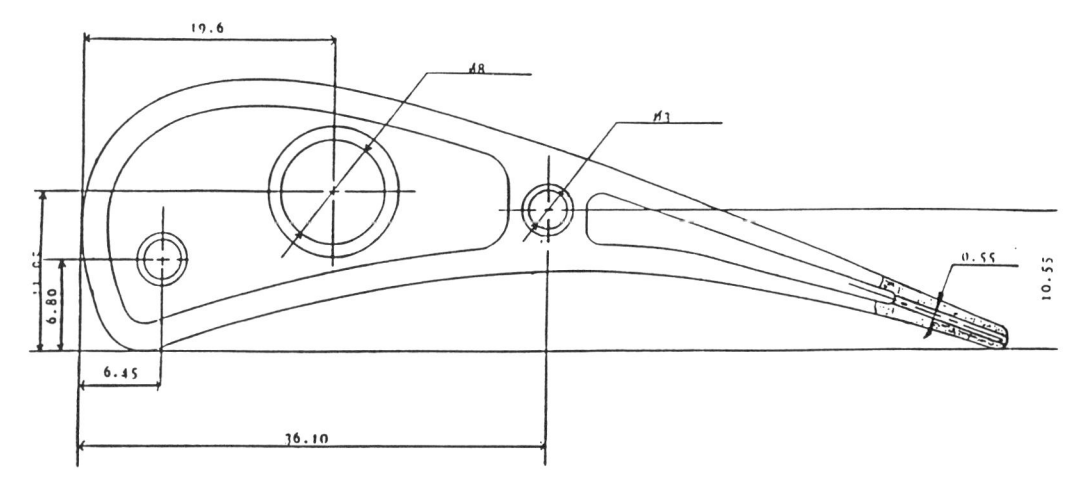

Figure 3.15 TE (trailing edge) ejection blade.

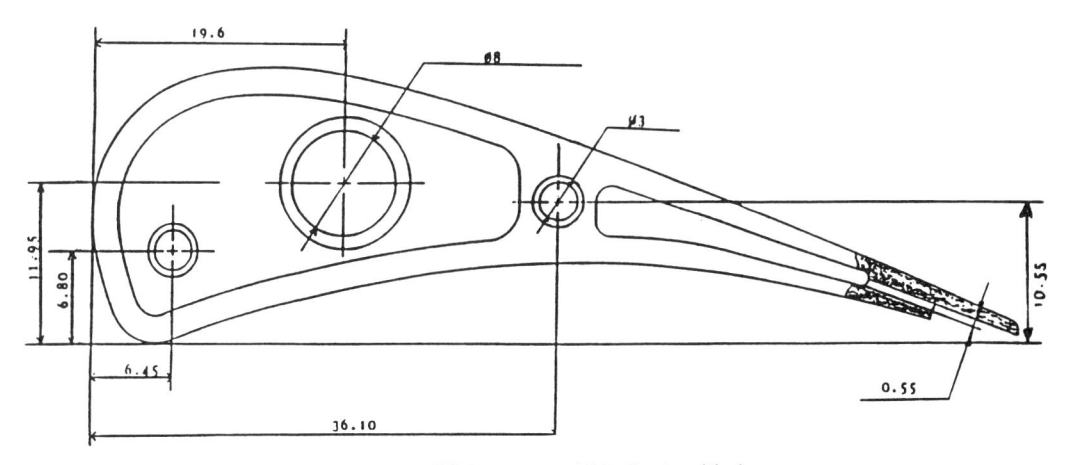

Figure 3.16 PS (pressure side) ejection blade.

by placing two support ribs at 25 mm distance from mid-span. The distance between these two ribs corresponds to the annular cascade height. This has enabled consistency to be maintained in measurements between the straight and annular cascades.

Moreover, inside the blades, far from the trailing edge, two 6 mm × 6 mm pins have been placed symmetrically with respect to mid-span.

TE-ejection blade The TE ejection blade is presented in Figure 3.18(a) The internal coolant passage is designed to decrease continuously up to a sudden contraction at the beginning of the trailing-edge coolant slot, which has a constant width of 0.55 mm and a length of 9.5 mm. The slot width represents about $\frac{1}{3}$ of the trailing-edge thickness.

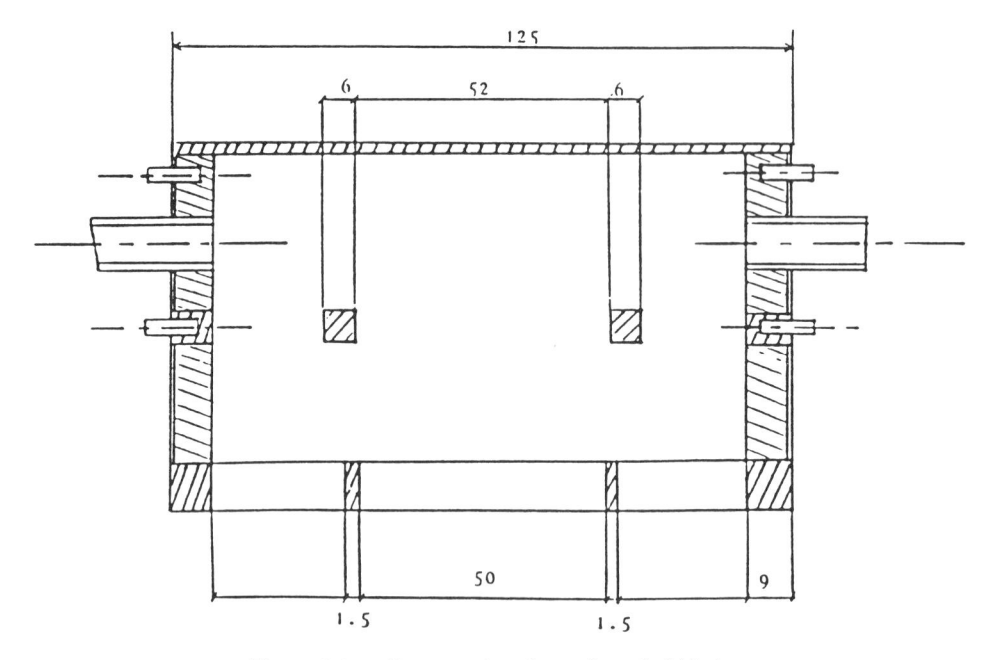

Figure 3.17 Cross-section through cooled blades.

PS-ejection blade This configuration is shown in Figure 3.18(b). The following parameters remain the same as for the TE ejection blade:

- the slot cross-section,
- the ratio of slot height to pressure-side lip thickness,
- the slot angle.

The length of the pressure-side cut back is 7 mm, i.e. 10% of the chord, which is representative of modern aircraft engines with this trailing-edge coolant configuration.

3.1.2 Blade instrumentation

Only one blade of each two-dimensional cascade is equipped with wall static pressure tappings. 39, respectively 38, pressure tappings are distributed along the blade mid-span of the TE ejection blade and the PS ejection blade. There are more tappings on the blade suction side in the shock impact region, and their location is identical for both blades, except in the trailing-edge region.

For the TE ejection blade, the base pressure tapping is placed on the rib, positioned 25.75 mm away from midspan, Figure 3.19. For the PS ejection blade the base pressure tappings are located at midspan in the suction side and pressure side trailing edges, respectively, Figure 3.20.

Mechanical design of blades and cascades

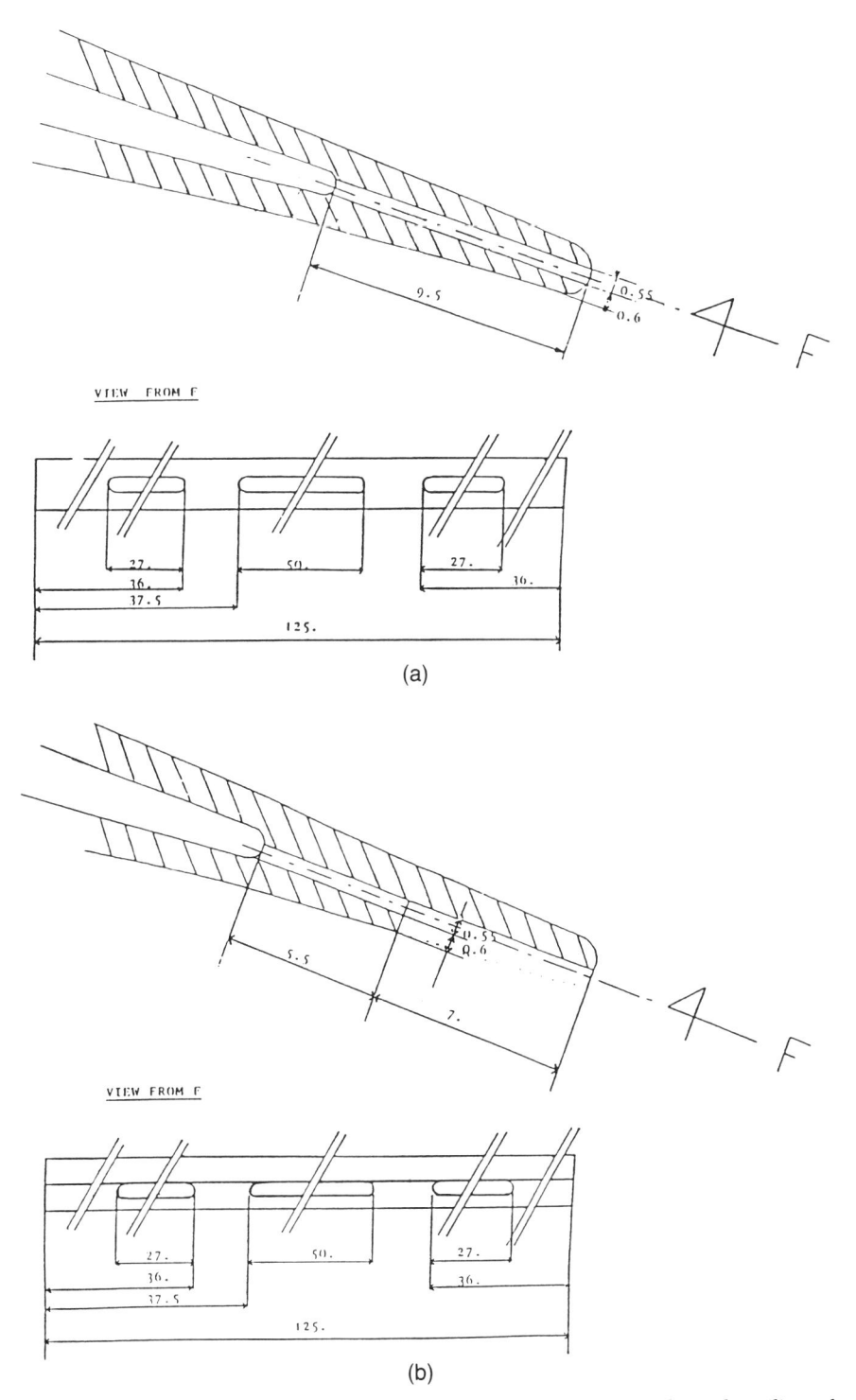

Figure 3.18 Details of trailing edge coolant configurations: ejection (a) through trailing edge; (b) from pressure side.

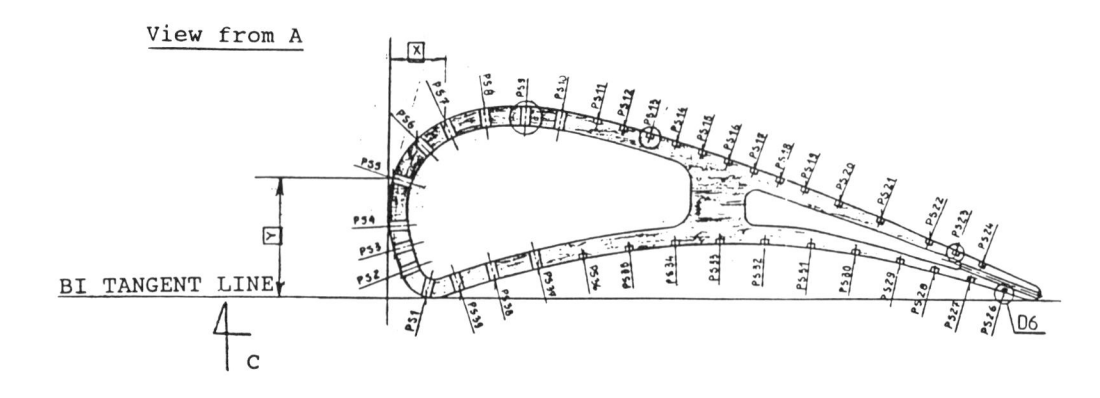

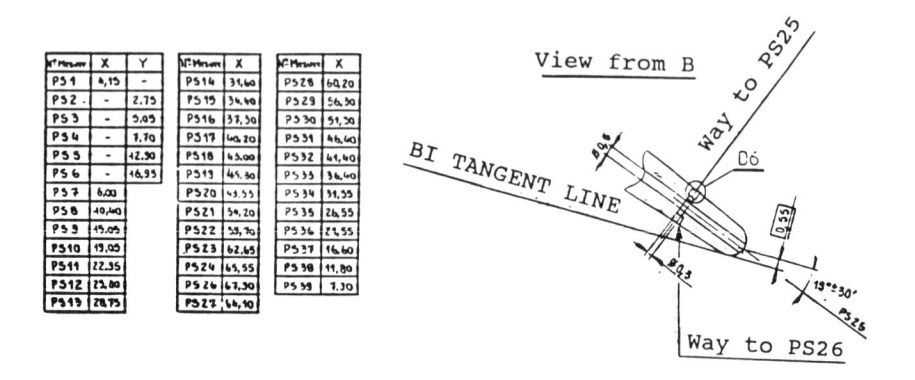

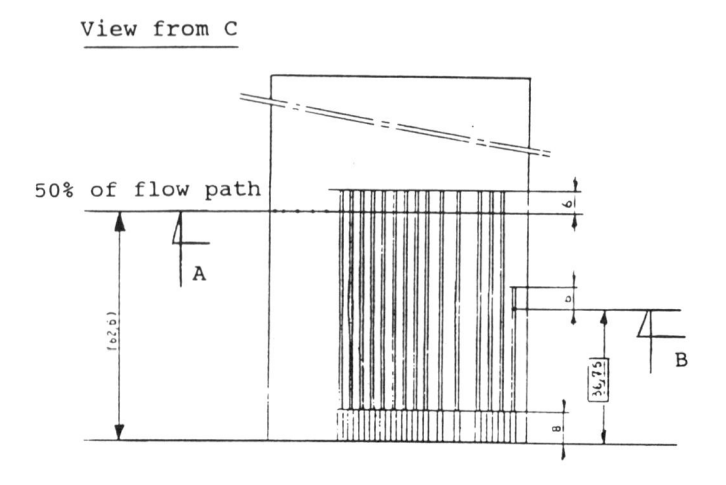

Figure 3.19 TE-ejection blade: position of pressure tappings.

Mechanical design of blades and cascades

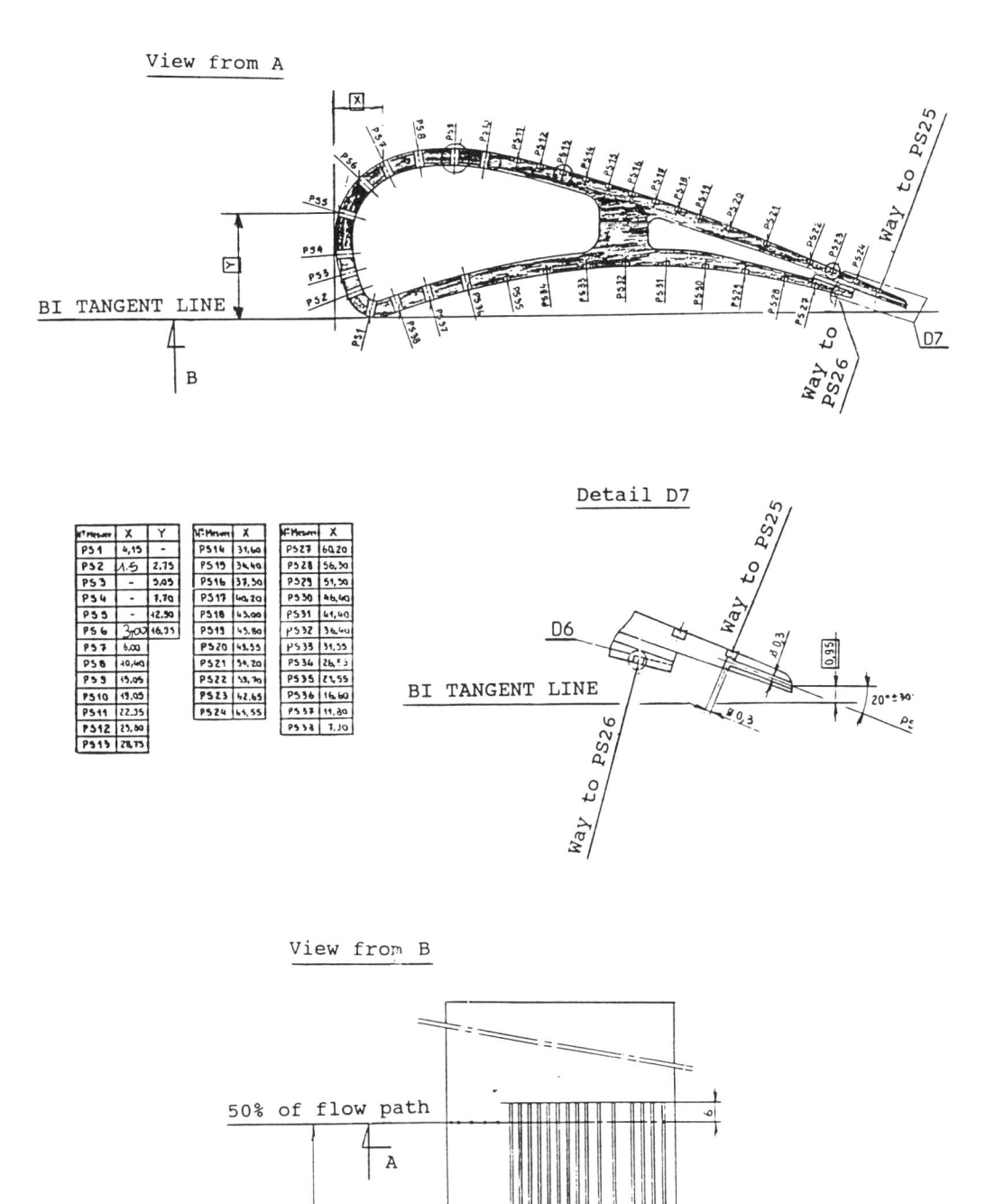

Figure 3.20 PS-ejection blades: position of pressure tappings.

3.1.3 Manufacturing of blades

The blades were manufactured using electro-erosion techniques. The pins inside the blade cavity complicated considerably the manufacturing process. In a first step the cavities on both sides of the pins were cut out by the wire-cutting technique. Then the bridge between the cavities was partially eroded away, leaving only the two pins.

For the non-instrumented blades, closure caps were brazed on both ends and pins were attached to these. For the instrumented blades, only one end was closed so as to leave the way free for instrumenting the blade by DLR. The final end closure was done at SNECMA.

3.2 Annular cascade design and manufacturing

The annular cascade design was carried out in close collaboration with VKI to re-use as much as possible existing hardware of the test section of the annular cascade facility.

The annular cascade consists of 43 cooled blades with ejection from the rear pressure side (this configuration was chosen for its lower cost and easier manufacturing).

SNECMA manufactured five elements of the annular cascade (see Figure 3.21):

- 43 cooled blades,
- hub end wall (inner band),
- tip end wall (outer band),

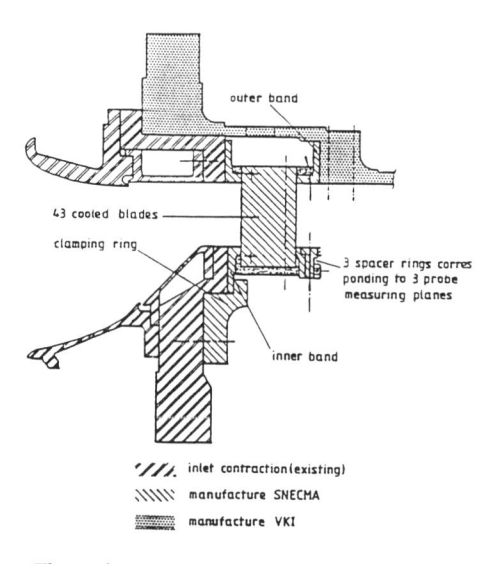

Figure 3.21 Three-dimensional cascade manufactured by SNECMA.

Mechanical design of blades and cascades

- clamping ring,
- three spacer rings for three different measuring planes.

A photograph of the front view of the annular cascade at the VKI test rig is presented in Figure 3.22.

The blade geometry is shown Figure 3.23. It corresponds to the straight cascade adapted to the annular cascade. The blade is pure cylindrical. The stacking axis is radial and corresponds to the center of the throat at the mid-span section (see Figure 3.24). The blades are fixed by a special compount in the endwalls.

In the annular cascade, coolant air enters the blades from the top. The inner ends of all blades are closed by a cap. The trailing-edge slot is slightly longer

Figure 3.22 Front view of annular cascade.

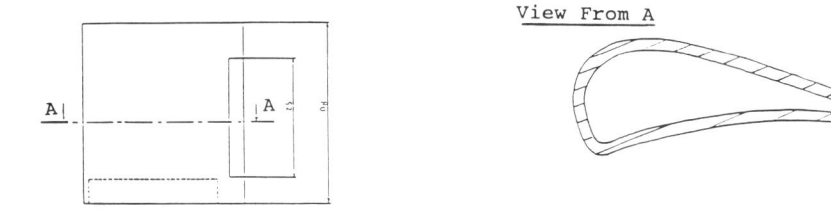

Figure 3.23 Three-dimensional blade geometry.

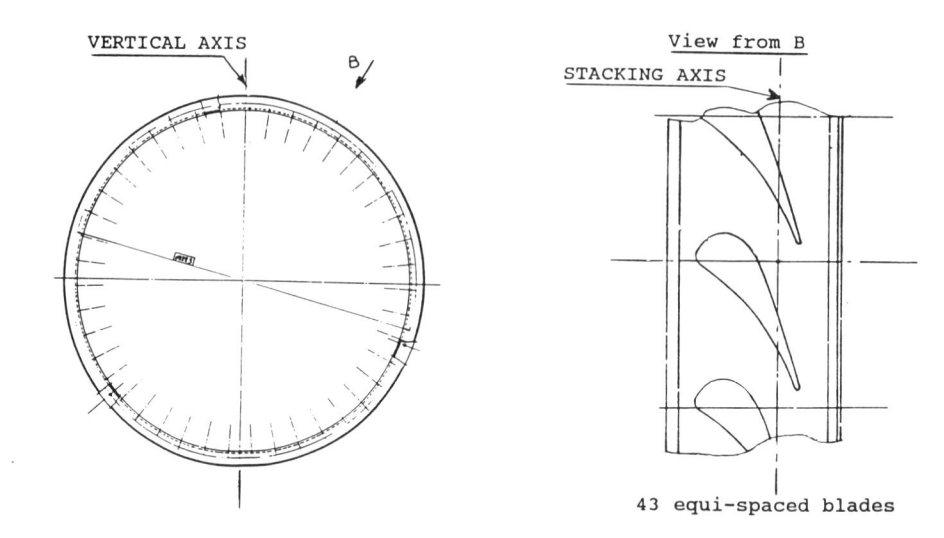

Figure 3.24 Three-dimensional blade cascade.

than the blade height, 52 mm compared to 50.7 mm, to avoid any backwards facing step at the coolant slot exit at the endwalls.

Three measurement planes were defined, corresponding to a distance of 25, 54 and 100% of the axial chord downstream of the cascade exit plane. The axial position of the probe is changed according to the measurement planes by a special spacer ring for each plane.

33 pressure tappings covering three adjacent pitches are located on the hub and outer tip end walls in each measuring planes. The blades are not instrumented.

4 Description of test facilities and measuring systems

4.1 DLR straight cascade facility

4.1.1 Overall description

The wind tunnel for straight cascades at DLR Göttingen [3.1] is of the blow-down type with atmospheric inlet. The general arrangement is shown in Figure 3.25.

The ambient air enters first a silicagel dryer, passes subsequently the inlet line equipped with a butterfly valve, two screens and a honeycomb flow straightener and enters the cascade downstream of the contraction. The test section is installed in a spacious walkable plenum chamber.

Downstream of the cascade the flow passes the adjustable diffuser and the main butterfly valve and enters at last the large vacuum vessel (10 000 m^3). This vessel is evacuated by two sets of sliding-vane vacuum pumps.

Description of test facilities and measuring systems

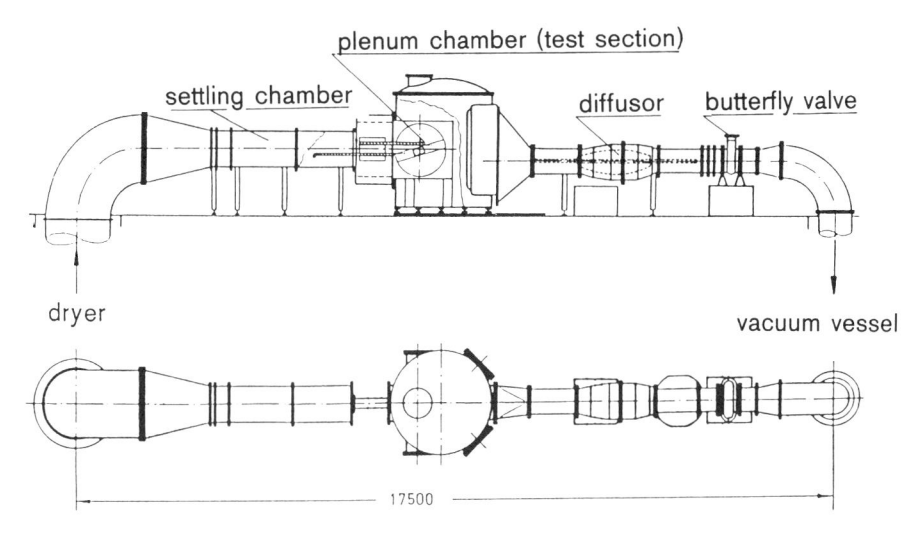

Figure 3.25 Wind tunnel for straight cascades at DLR Göttingen.

The inlet total pressure of the cascade is, disregarding the small pressure loss in the dryer and the inlet line, always equal to the ambient pressure. This assures a high stability of the pressure without any control equipment. But on the other hand the Reynolds number cannot be varied independently. In a limited range it is a function of the Mach number.

The diffuser is composed of an axisymmetrical housing and a centred body movable in axial direction. The main purpose is to provide an adjustable throat in order to maintain the pressure downstream of the cascade constant independent of the pressure in the vacuum vessel. Additionally a certain pressure recovery is expected depending on the cascade outlet angle and Mach number.

The vacuum pumps can provide a continuous operation of the tunnel at low Mach numbers downstream of the cascade. At higher Mach numbers (typically $Ma_2 > 0.8$) only an intermediate operation is possible, but with sufficient blowing time for the application of the standard measuring techniques.

The valve in the line upstream of the wind tunnel is closed only during the regeneration (dewatering) of the silica-gel dryer. The valve downstream of the cascade is the main operating valve of the wind tunnel.

The wind tunnel is equipped with an industrial electronic control system (Simatic) and all electric components except the data-acquisition system are part of this. The locking of all safety-relevant doors and openings is controlled by this system.

4.1.2 Test section

Every cascade is mounted with fixed geometrical parameters (stagger angle, pitch–chord ratio) in a support frame, which is installed in the test section

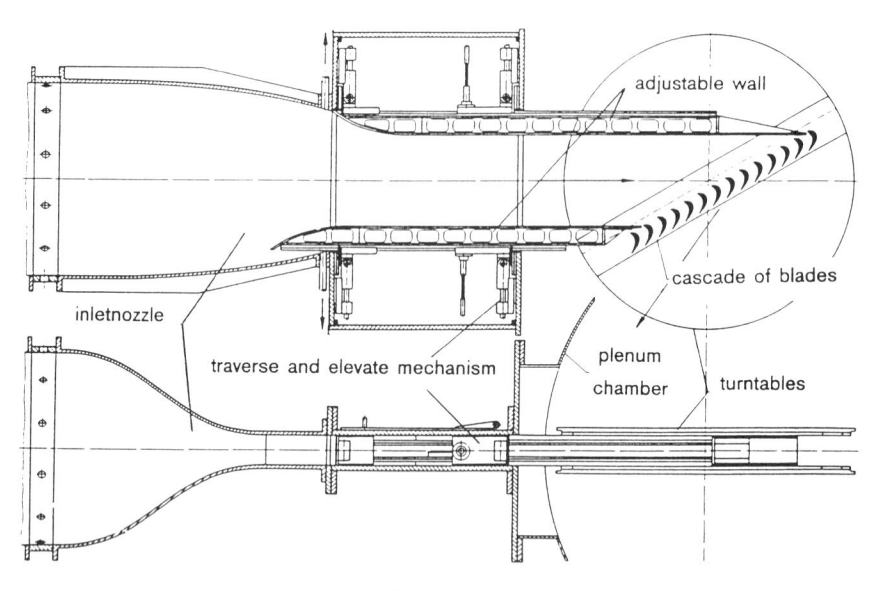

Figure 3.26 Test section.

(Figure 3.26) between two circular discs establishing the side walls of the flow channel. The inlet angle is adjusted by turning this assembly.

Of great importance is a homogeneous flow field at the inlet of the cascade. This is provided by a long upstream extension of the side walls up to the contraction and movable upper and lower walls. These can be adjusted in vertical and horizontal direction up to the desired location relative to the upper and lower blade. The optimum location must be found out empirically based on the static pressure distribution in a plane parallel to the cascade inlet.

Near the centre line of the flow channel the cascade support is equipped with exchangeable panes in which typically four blades (depending on the pitch–chord ratio) are fixed. For schlieren observations optical glass panes are installed in this location, but for special purposes (e.g. pressure distribution blades) steel panes are also in use.

Guided carriages driven by stepping motors via jacks are installed on both sides of the flow channel downstream of the cascade. They are utilized for the mounting of the probes and adjusting them parallel to the cascade outlet plane.

The flow downstream of the cascade is not guided. Flow angle and Mach number are only adjusted in mutual dependency by the plenum pressure (i.e. setting of the diffusor).

4.1.3 Coolant flow supply system

For the development of high-pressure turbines the investigation and optimization of cooled blades is of particular interest. Therefore the wind tunnel for straight cascades at DLR Göttingen is provided with all equipment needed for

Description of test facilities and measuring systems

the investigation of the aerodynamics of cooled turbine blades, using either air or carbon dioxide (CO_2) as coolant flow.

For tests with air ejection the fluid is provided by the central high-pressure system of the Göttingen research centre. From the minimum pressure level of the central supply system (70 bar) the pressure is reduced to 16 bar. Consecutively the air passes a standard measuring orifice (DIN 2071) and the remote-controlled valve for adjusting the desired mass flow rate. Finally the air is distributed by a manifold to both ends of the four blades equipped with ejection slots. The scheme is shown in Figure 3.27.

For the test with CO_2-ejection the fluid is provided by an isolated, mobile storage tank containing about 2000 kg CO_2 in liquid state. Normally enough of the CO_2 evaporates under ambient temperature conditions and for higher flow rates this process is supported by an electrical heater. But due to the heat required for the vaporation together with the Joule–Thomson effect, the CO_2 gas becomes very cold. Therefore a heat exchanger consisting of a pipe bundle (leading the CO_2 gas) in a water tank is installed downstream of the storage tank. The water is maintained at the desired gas temperature by an automatically controlled electrical heater. The size of the pipe bundle is designed

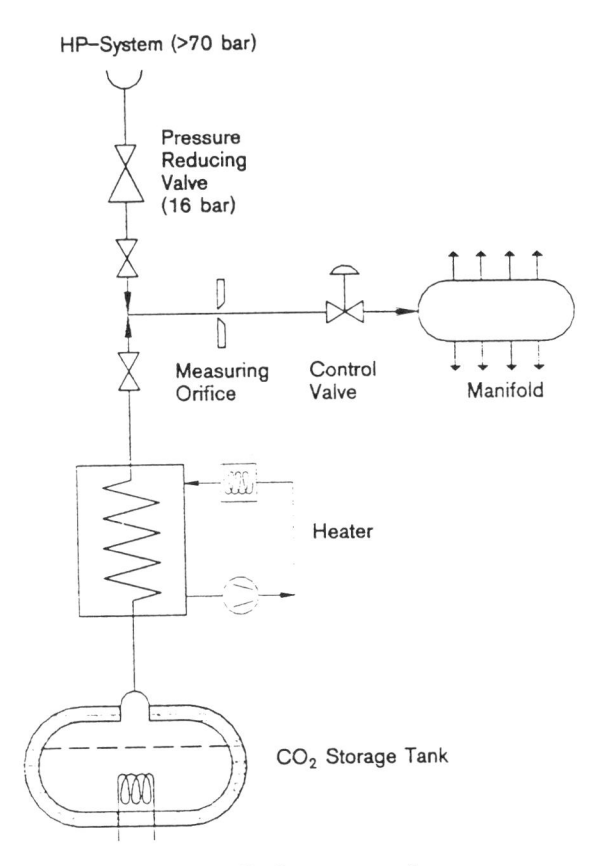

Figure 3.27 Coolant gas supply system.

such that the gas approaches the water temperature asymptotically until a negligible difference is achieved. This arrangement proved very reliable and avoids a complicated control system.

Further the gaseous CO_2 is guided to the test blades via the same installation as described above for the air ejection.

4.2 VKI compression tube annular cascade facility

4.2.1 Cascade tunnel

A detailed description of the facility, see Figure 3.28, is given in Reference [3.2].

The main elements of the facility are shown in Figure 3.29.

(1) a 1.6 m diameter and 8 m long compression tube containing a lightweight piston driven by air from the Institute's high-pressure air supply system;

Figure 3.28 Photograph of VKI Compression Tube Annular Cascade Facility CT-3.

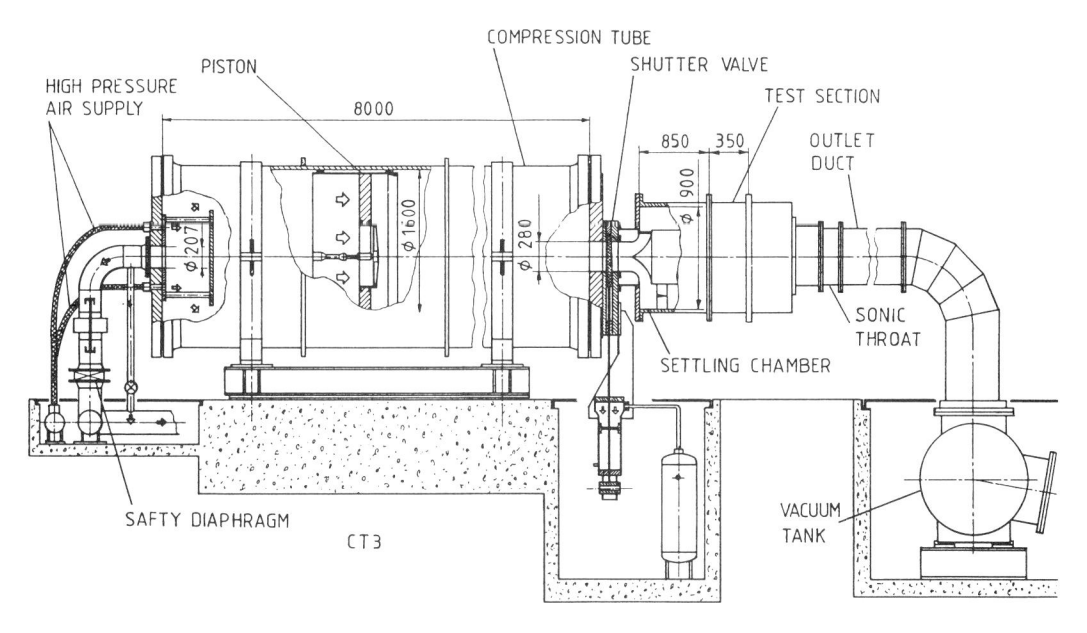

Figure 3.29 Main elements of facility.

(2) a vertically oriented fast-opening shutter valve closing the 280 mm diameter central vent hole in the end plate of the tube;

(3) a radial diffuser discharging the compressed air into an annular settling chamber of 500 mm inner diameter and 900 mm outer diameter;

(4) the test section, housing the nozzle guide vane and the probe traversing systems;

(5) a 15 m^3 vacuum tank separated from the test section through a variable sonic throat mounted in the test section outlet duct.

The operating sequence for one typical test is presented in Fig. 3.30. Before the test a vacuum pump decreases the pressure in the test section and the dump tank to a pressure level of about 0.1 bar. The initial pressure in the compression tube depends on the required test section total pressure during the test and the compression ratio required to increase the temperature in the tube from ambient conditions to the desired total temperature in the test section during the test. At t_1 cold high-pressure air is admitted into the tube behind the piston. The air in front of the piston is compressed and heated up until the tube pressure reaches the required pressure level. At that moment the shutter valve is opened, $t = t_2$, and air starts to flow through the test section. At t_3 the conditions in the test section have reached nominal conditions and remain constant as long as the downstream sonic throat is choked, t_4. At t_5 the piston has reached the end of the compression tube and, the mass flow drops

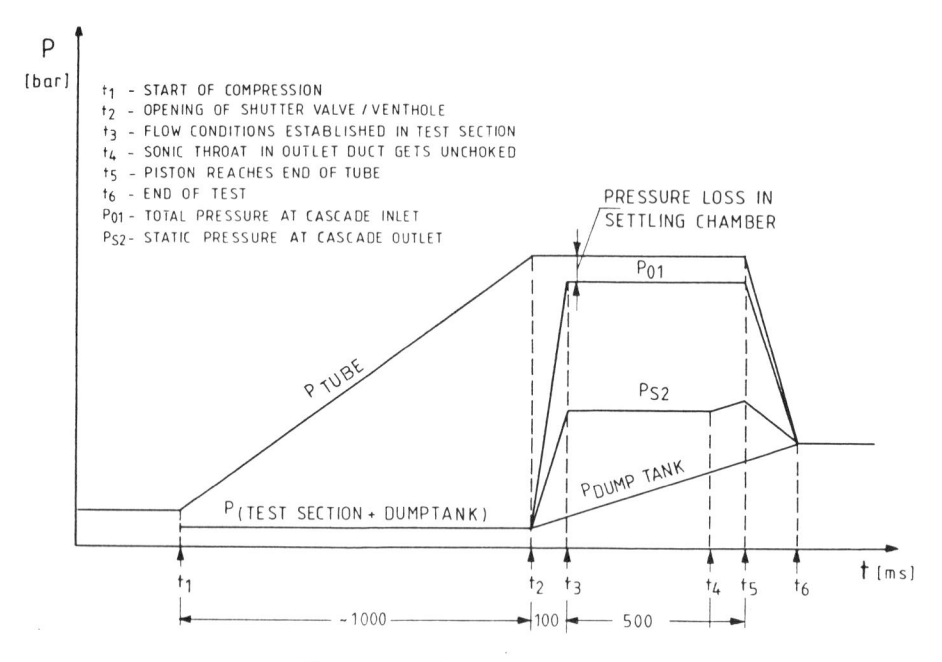

Figure 3.30 Operating sequence.

to zero. The running time is of the order of 1–0.5 s for mass flows of 15–30 kg/s.

The facility allows correct simulation of Mach number, Re, and temperature ratios T_{gas}/T_{blade} of advanced aircraft engines.

4.2.2 Test section

Figure 3.31 shows a view of the meridional plane of the test section. The contraction preceding the guide vane simulates the exit of a combustion chamber. A turbulence grid can be placed in the entrance of the contraction, but this was not the case for the present tests.

The guide-vane annulus is mounted in the guide-vane housing. The cooling air for the trailing edge cooling is injected into a large space between the guide-vane end wall and the guide-vane housing. From there the cooling air enters the hollow blades and is ejected through a pressure side trailing edge slot spanning the entire blade height. The hub and tip end walls are cylindrical with 689 and 790.4 mm diameter respectively. The cylindrical tip end wall continues downstream for about 450 mm, while the hub end wall extends only 100 mm downstream of the trailing edge. The downstream hub end wall is composed of a small fixed part, followed by a rotatable part carrying the probe. Spacer rings allow extension of the fixed part and shift the rotable part with the probe downstream, and thereby displace the measurement plane.

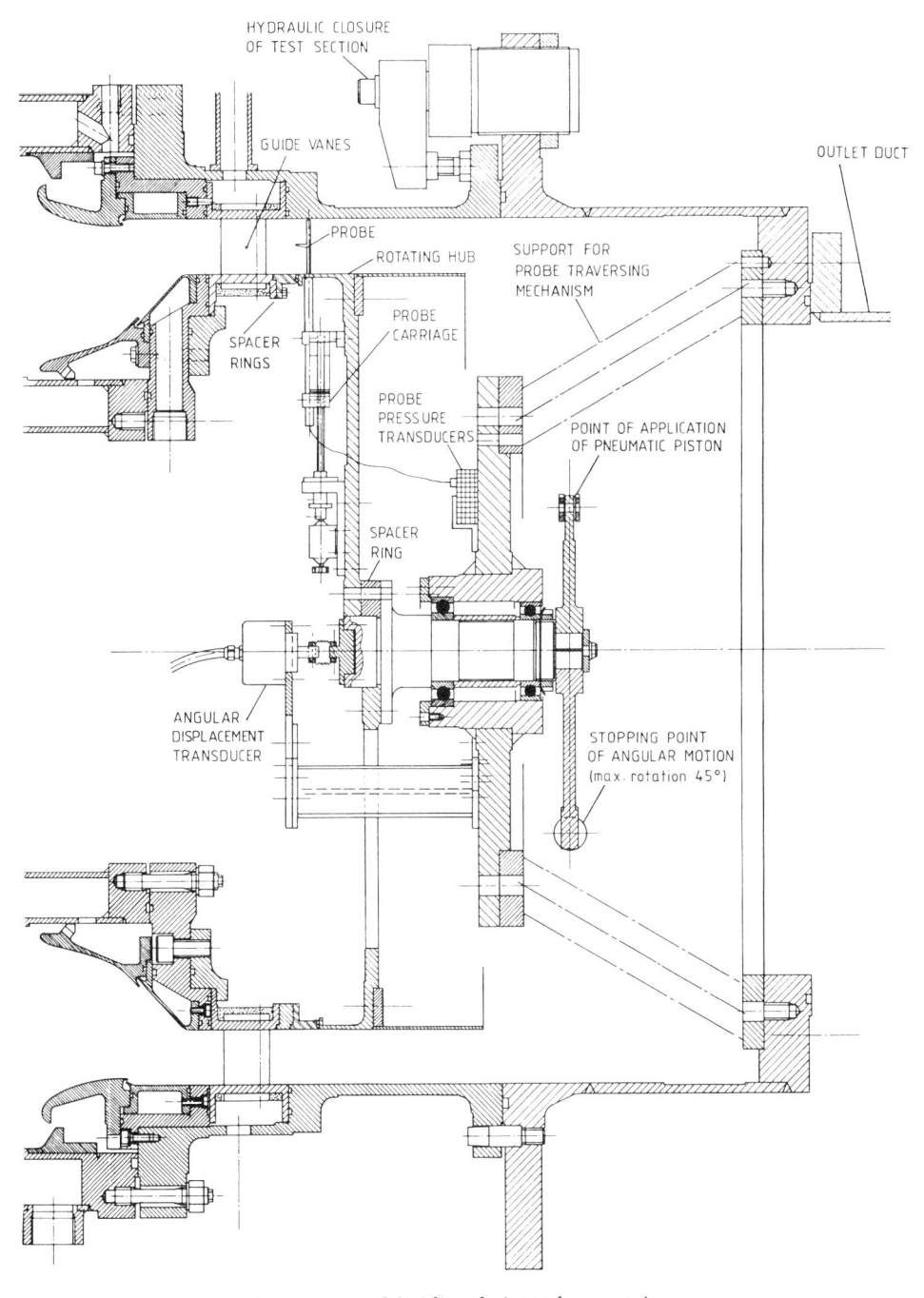

Description of test facilities and measuring systems

Figure 3.31 Meridional view of test section.

4.2.3 Measurement systems

The measurement system comprises:

- *upstream of the cascade*

  —a total pressure rake for the spanwise total pressure distribution;
  —a miniature thermocouple temperature probe;
  —a single hot-wire probe for turbulence measurements.

- *downstream of the cascade*

  —endwall static pressures tappings across three pitches positioned at a distance of 25%, 54% and 100% of the axial chord downstream of the trailing edge plane;
  —a slow-moving probe displacement mechanism allowing the traversing of probes over a limited annular sector of 35–40° covering 3 or 4 pitches, referred to hereafter as 'stationary probe measurements';
  —a high-speed rotating probe traversing mechanism with data transmission from the rotating to the stationary system, referred to hereafter as 'measurements in rotation'.

Stationary probe measurements The probe displacement speed is in most cascade tunnels of the order of a few millimeters per second. Due to the short testing time in compression-tube cascade facilities the probe displacement speed has to be much higher, usually of the order of 500 mm/s. Nevertheless the displacement speed remains small compared to the flow velocity and its effect on the measurements can be neglected. However, the probe–transducer system response time to a pressure step has to be of the order of $\Delta t \leq 10$ ms to guarantee correct wake traverse measurements. Extensive testing of various probe–pressure-line–transducer systems at VKI [3.3] have shown adequate response times for probes made of 1×0.8 mm diameter tubes connected via flexible pressure lines of the same inner diameter to Sensym differential pressure transducers located at 400 mm from the probe head. An opto-electronic encoder continuously monitors the circumferential position of the probe.

The immersion of a probe into a transonic or low supersonic flow inevitably alters the local flow conditions due to the blockage of the probe, in particular of the probe stem, if this stem is directed normal to the flow, which is typically the case when testing annular cascades. To prevent the blockage effect from varying with the probe immersion depth, the probe stem is backed up by a small cylinder which spans the whole passage height and has the same width as the probe stem.

Full details of the probe displacement system are shown in Figure 3.31. The probe moves with the rotatable part of the end wall, thus avoiding any probe traverse slot in the end wall. The probe nose always extends upstream of the rotatable end wall. The disc carrying the rotatable end wall is mounted on a

Description of test facilities and measuring systems

shaft set between two ball bearings. The shaft carries on its right side a lever which is actuated by a pneumatic piston.

Measurements in rotation The high-speed rotating probe system was designed to attenuate the probe blockage effects by spinning the probe at peripheral speeds sufficiently high to reduce the velocity relative to the probe to subsonic velocity. Besides the expected decrease of the blockage effect, the total pressure sensed by the probed does not need to be corrected for bow-wave shock losses as is the case for stationary measurements with the same flow conditions but supersonic flow to the probe. However, the calculation of the guide vane losses from the measurements in rotation require in addition to the measurement of the relative total pressure also the relative total temperature, see Section 6.3.2. Furthermore, the measured values have to be transmitted from the rotating to the stationary frame. To this end VKI developed a wide band width, high signal-to-noise ratio, IR (infra-red) optoelectronic data transmission system which is described in detail in Appendix 3. At present we limit ourselves to a short description.

A block diagram of the transmission system is presented in Figure 3.32. It consists of

- *on the rotating (emitter) side*:
  - —an amplifier;
  - —a voltage–frequency converter;
  - —a second amplifier driving the IR emitter diode.

- *on the stationary (receiver) side*:
  - —a receiver IR diode with pre-amplifier and operational amplifier;
  - —a frequency–voltage converter integrated in a PLL circuit;
  - —a low-pass filter used to filter the residual frequencies of the PLL.

The complete rotating probe traverse mechanism is depicted Figure 3.33. The probes are fixed on the disc carrying the rotatable hub end wall. The disc is

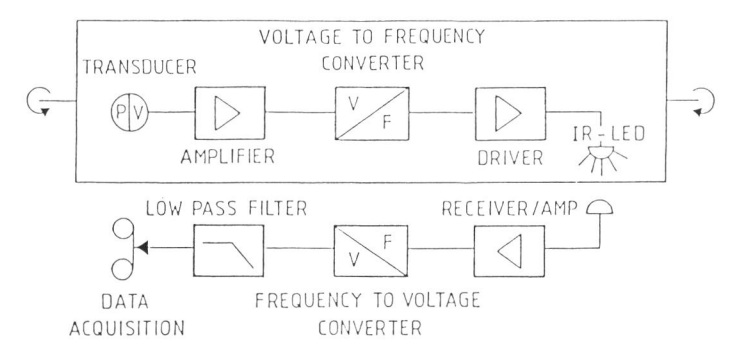

Figure 3.32 Block diagram of data transmission system.

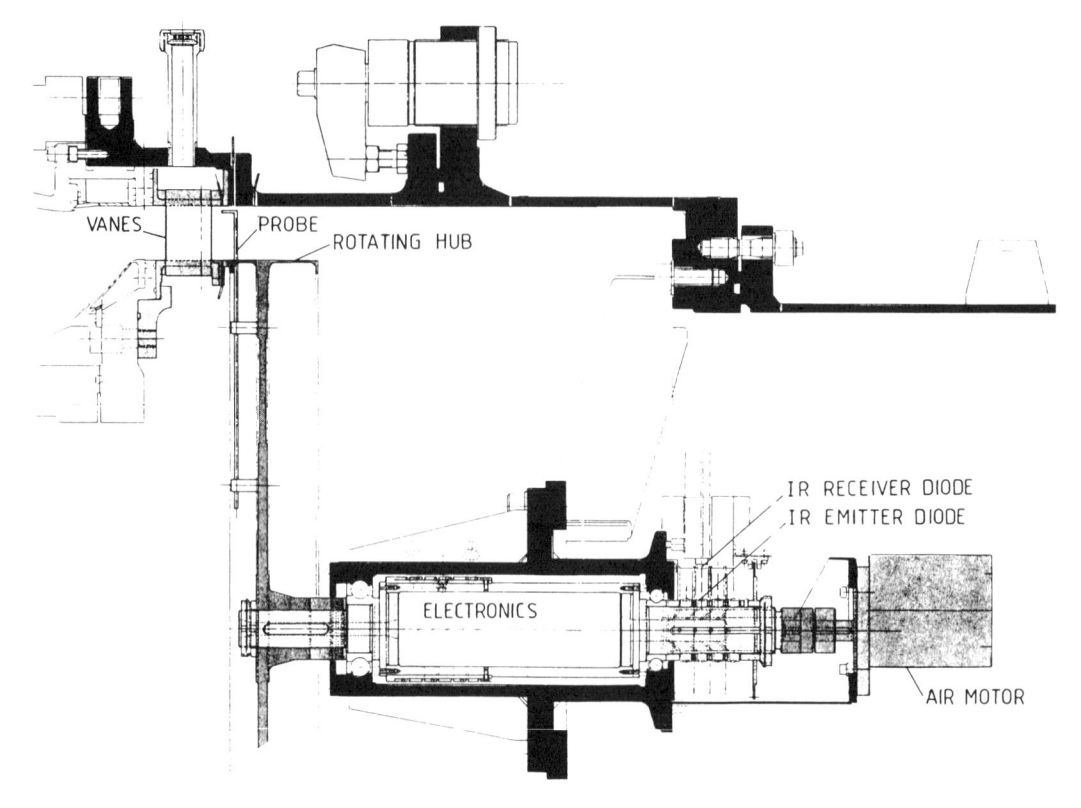

Figure 3.33 Rotating-probe traversing system.

fitted to a shaft driven by an air motor. The electronics for signal amplification and transmission are placed in the central part of the shaft in between the two bearings. The right end of the shaft carries the emitting diodes. For each data channel the signal is transmitted via four diodes positioned at $4 \times 90°$ around the shaft. The signal is captured by one fixed diode which is in permanent contact with at least one of the emitting diodes. The system in Figure 3.33 was designed for the simultaneous transmission of four data channels. Care has to be taken to avoid any interference between emitting and receiver diodes from different channels. The voltage supply for transducers, electronics and IR-diodes is transmitted via ordinary carbon sliprings (copper rings and carbon brushes).

It is worth while noting that this rotating-probe system is also an important step to guide vane performance measurements in a transonic turbine stage. The attachment of probes onto rotor blades appears in fact to be the only way to obtain information on the stator outlet flow field without causing unacceptable flow perturbations as caused by the immersion of probes into the small inter-blade row space.

A description of the probes used for the stationary and rotating measurements will be given in Section 6.

5 Straight cascade tests

5.1 Measurement programme

The test program was divided into two parts: a preliminary test program, in which air was used as coolant, and a main test program, in which carbon dioxide was the coolant. The downstream isentropic Mach number ranged from 0.7 to 1.2 with a Mach number increment of 0.1, whereas the design Mach number 1.05 was included as well. Coolant flow rates were adjusted to 0, 2, 3, i.e. the design coolant flow rate, and 4% for both programs. The cascade consisted of 8 inlet guide vanes, of which only the 4 centre blades were equipped with a coolant slot. Figure 3.34 shows the main characteristics of the cascade.

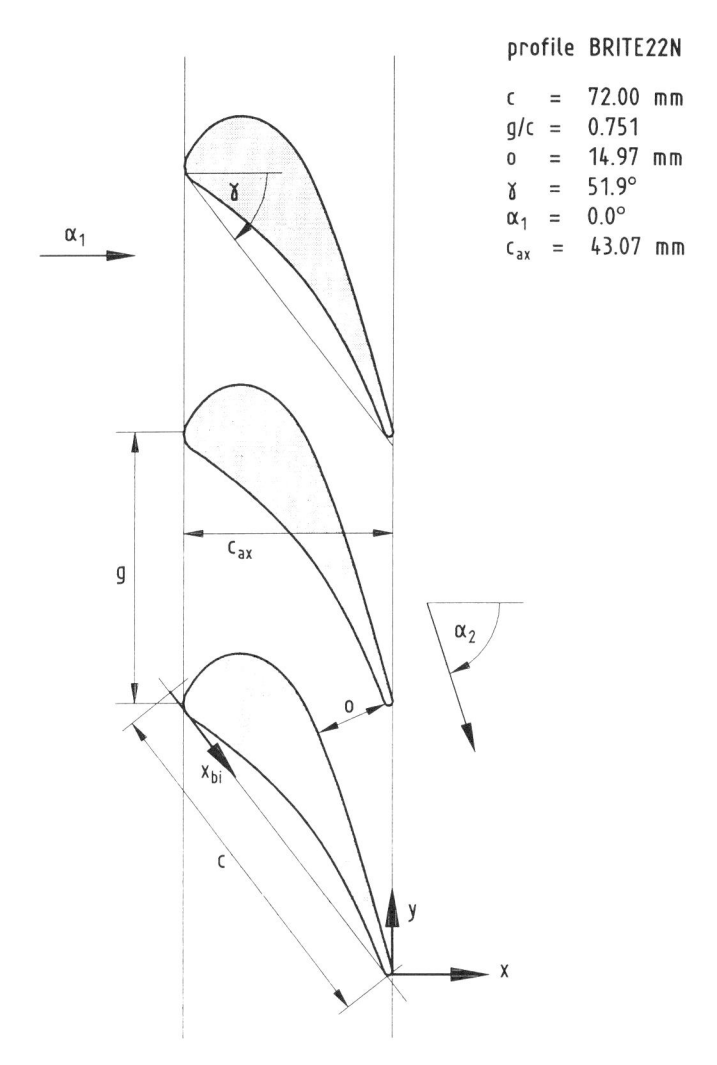

profile BRITE22N

c = 72.00 mm
g/c = 0.751
o = 14.97 mm
γ = 51.9°
α_1 = 0.0°
c_{ax} = 43.07 mm

Figure 3.34 Blade positions in plane cascade.

All measurements were carried out twice, that is, once for the blades with coolant ejection through the trailing edge slot and once for the blades with coolant ejection from the pressure side. The straight cascade program consisted of the following measurements:

- *Schlieren pictures*. Schlieren photographs were made with air as coolant, which was enabled by mounting the four centre blades in glass plates instead of the usually applied steel plates. The set-up of the optical instruments is shown in Figure 3.35.

- *Pressure distribution*. The pressure distribution over the blade surface was determined after inserting a specially prepared blade, which was equipped with 38 resp. 39 (depending on the blade type) pressure borings, into the cascade. These measurements were also made with air as well as with carbon dioxide as coolant.

- *Wake measurements*. Wake measurements of flow angle, total- and static pressures were carried out at 35 mm in axial direction downstream ($X/C_{ax} = 0.813$) of the trailing edges of the blades in the cascade with air as well as with carbon dioxide as coolant. The distance between adjacent measurement positions was set equal to 1.5 mm, which resulted in 36 data points per pitch.

- *Turbulence measurements*. The turbulence level in the flow inlet to the cascade was measured qualitatively using a hot wire probe and a wedge-type hot film probe. For stability reasons the turbulence level in the wake was measured qualitatively with the hot film probe only. As the constitution of the gas, in addition to the gas velocity, temperature and density, influences the readout of the hot film measurement device, the determination of the turbulence level in case of coolant ejection took place with ejection of air only.

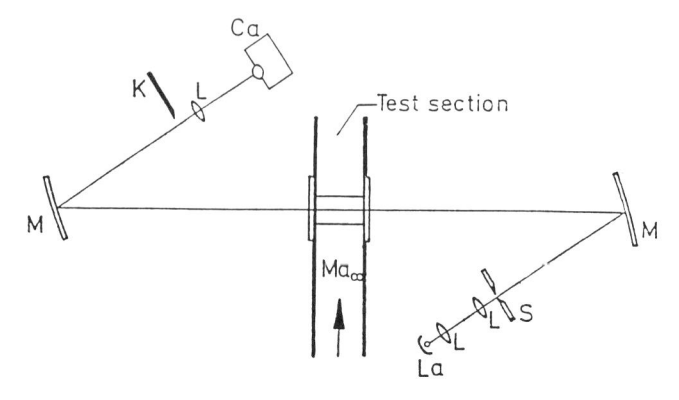

Figure 3.35 Set-up of the optical instruments used to take Schlieren pictures.

Straight cascade tests

- *Tests with carbon dioxide ejection*. The elucidation of the mixng process of the main flow and the coolant gas downstream of the cascade required wake traverses with the gas analyzer at more than one distance, as is usual for the standard wake measurements. Therefore the gas concentration of carbon dioxide was determined during 7 traverses at axial distances of 1, 2, 4, 7, 12, 20 and 35 mm from the plane containing the trailing edges of the blades. The distance between adjacent measurement positions ranged from 1.5 mm in the traverse at 35 mm distance to 0.1 mm in part of the traverse at 1 mm distance. The traverses at 7 mm and further downstream were attended with measurements of the flow angle and the total and static pressures by the wedge-type pressure probe as in the usual wake traverses. Closer to the trailing edges of the blades this probe could no longer be used, for dimensional reasons. Instead of this, the concentration measurements at 1, 2 and 4 mm were accompanied with total pressure measurements by a Pitot tube.

5.2 Overall flow conditions

In Figure 3.36 the inlet static pressure corresponding to $M_{2,is} = 1.05$ is drawn. The graph demonstrates that the inlet static pressure is uniform between upper and lower wall.

An example of the distribution of the total pressure in the flow entrance area is given in Figure 3.37. This picture is part of the work of Brück [3.4].

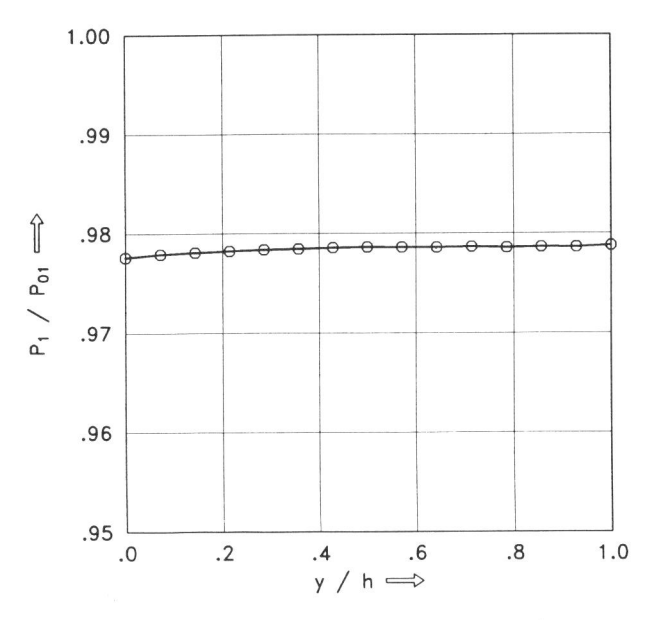

Figure 3.36. Inlet static pressure distribution between upper and lower test section walls at 85 mm before the cascade entry plane and under choking conditions in the cascade.

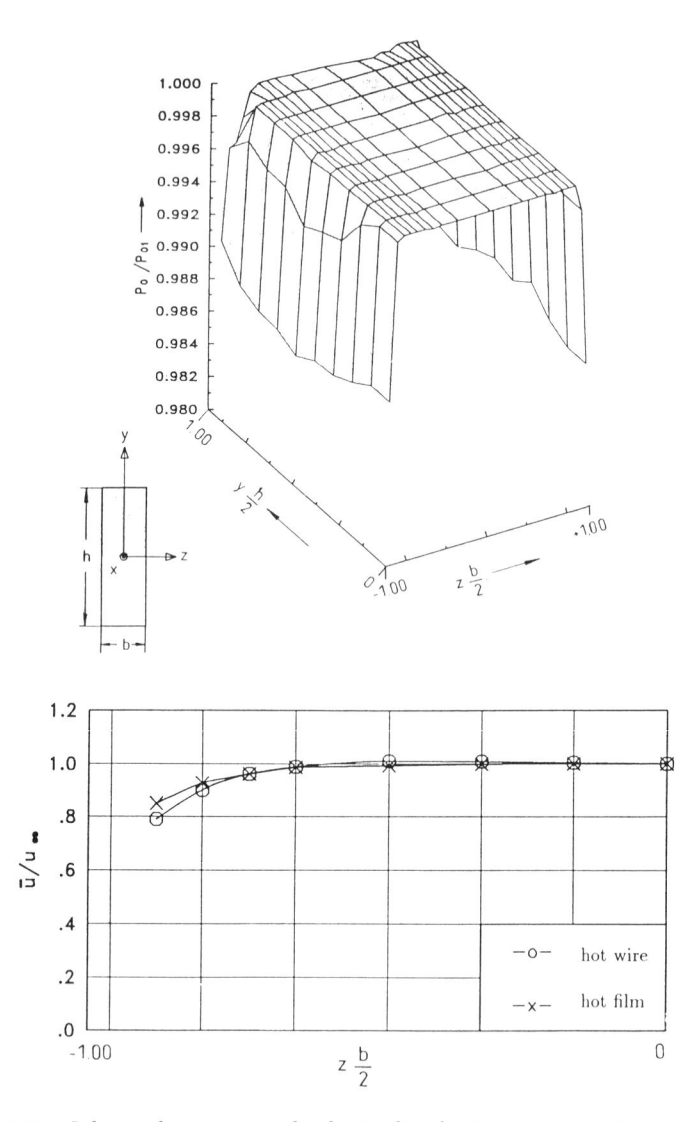

Figure 3.37 Inlet total pressure and velocity distribution, measured by S. Brück.

Typical wake traverses, carried out at $x/c_{ax} = 0.813$ were measured over 200 mm parallel to the cascade, covering almost 4 pitches ($g = 54.01$ mm), except for those measurements carried out during the repair time of the main compressor set, where the probe traverses were restricted to the two central pitches for reasons of limited available measurement time. The distance between succeeding probe positions was 1.5 mm. Figure 3.38 shows examples of the total pressure ratio, the exit flow angle and the static-to-total pressure ratio. These traverse data were recorded behind the cascade consisting of blades with coolant ejection slots in the trailing edge at $M_{2,is} = 1.05$, and $c_m = 0\%$. The four minima in the graphs denote the wakes of the four centre blades in the cascade, which are equipped with coolant slots.

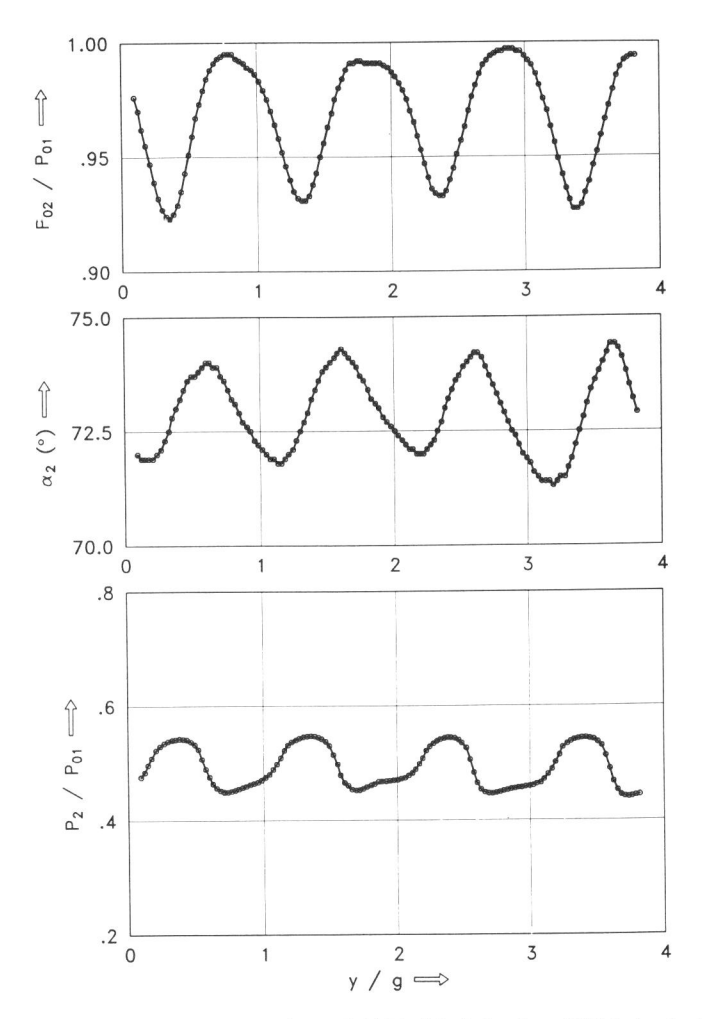

Figure 3.38 Local flow properties at $x/c_{ax} = 0.813$ behind the four TE-blades in the centre of the cascade at $M_{2,is} = 1.05$ and $c_m = 0\%$.

5.3 Tests with air ejection

5.3.1 Schlieren pictures

Figures 3.39–3.42 present schlieren pictures for both cascades at $M_{2,is} = 1.05$ and $c_m = 3\%$ as well as $c_m = 0\%$. The two completely visible blades in the photos are the two central blades of the four equipped with coolant slots.

The wake behind the blades is clearly distinguishable in all pictures, but the influence of the coolant ejection on the wake pattern is obscure. The effect of coolant ejection on the shock pattern is significant in case of the PS-ejection blades, see Figures 3.41 and 3.42. The presence of the step in the blade surface

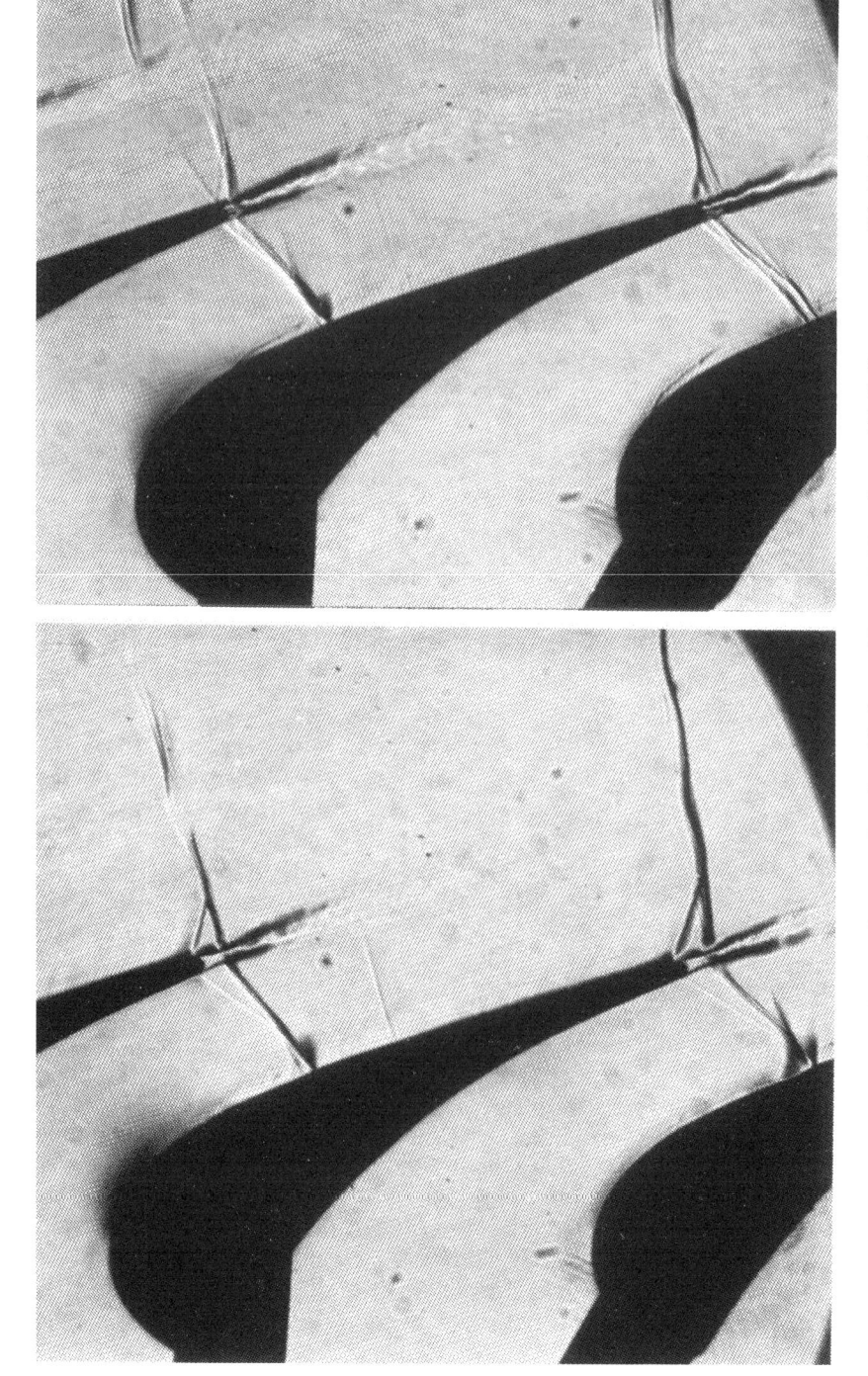

Figure 3.39 Schlieren photographs of TE-blades at $M_{2,is} = 1.05$ and $c_m = 0\%$ (left), resp. $c_m = 2\%$ (right).

Straight cascade tests 201

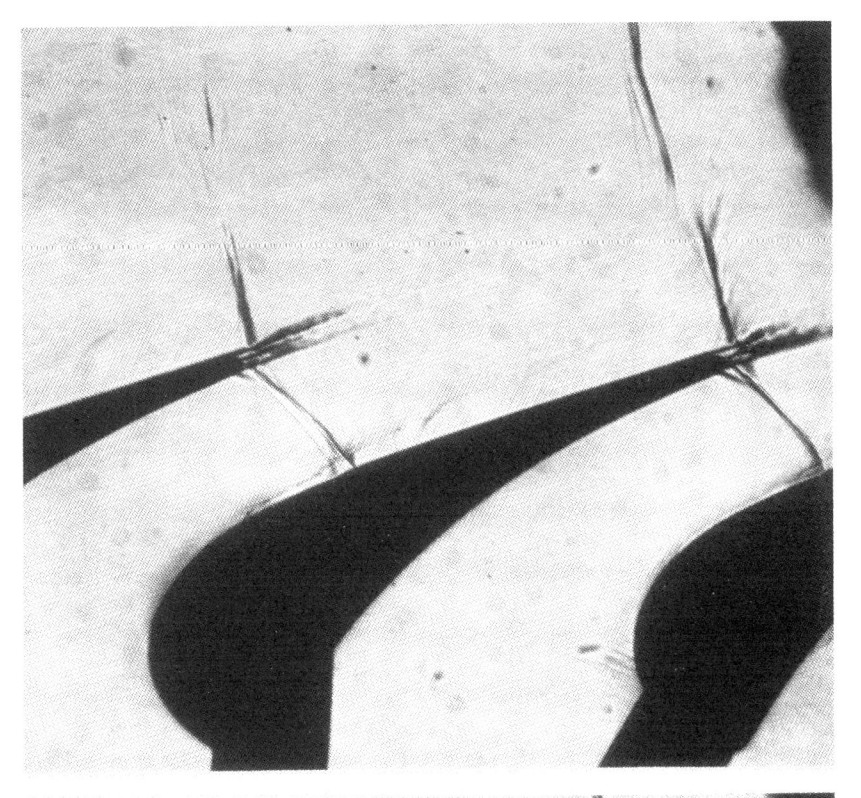

Figure 3.40 Schlieren photographs of TE-blades at $M_{2,is} = 1.05$ and $c_m = 3\%$ (left), resp. $c_m = 4\%$ (right).

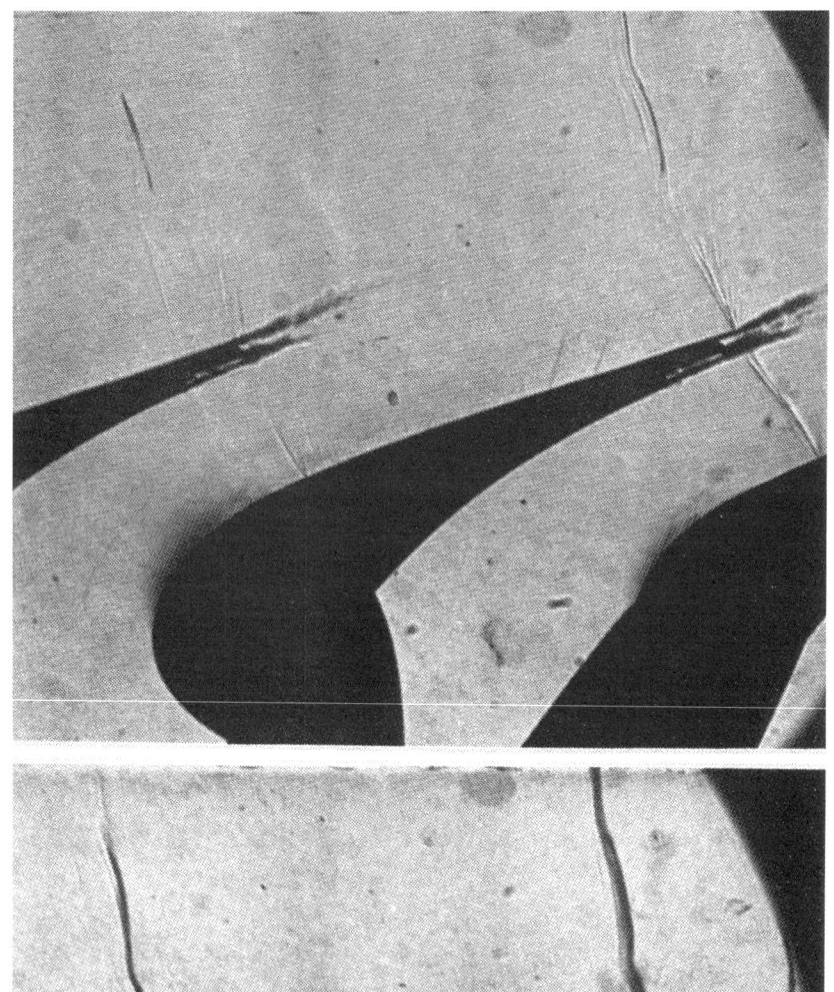

Figure 3.41 Schlieren photographs of PS-blades at $M_{2,is} = 1.05$ and $c_m = 0\%$ (left), resp. $c_m = 2\%$ (right).

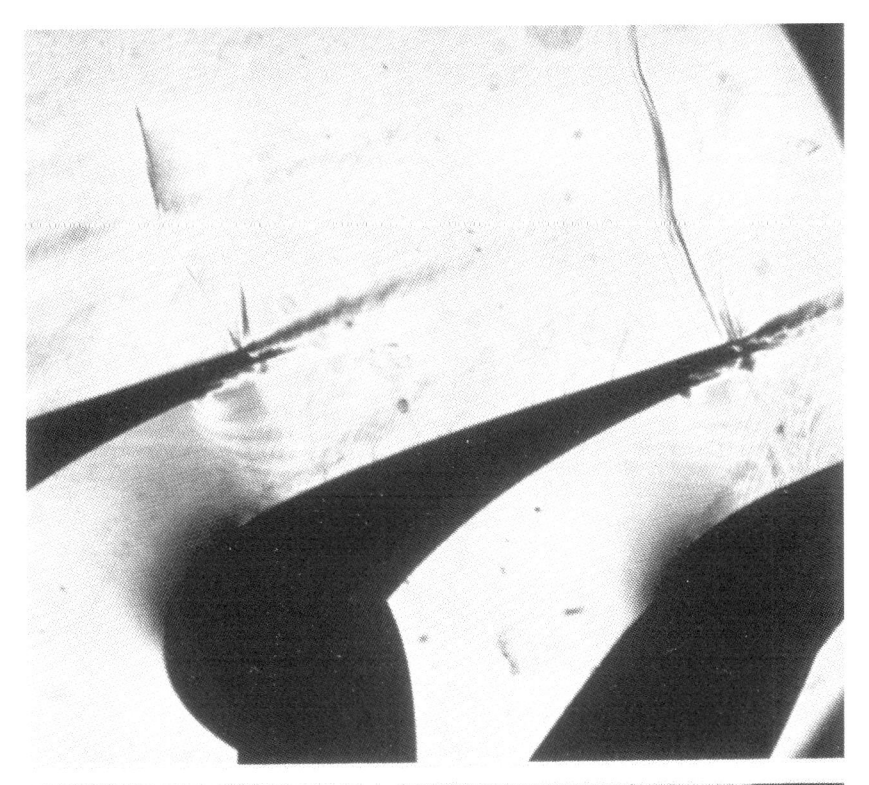

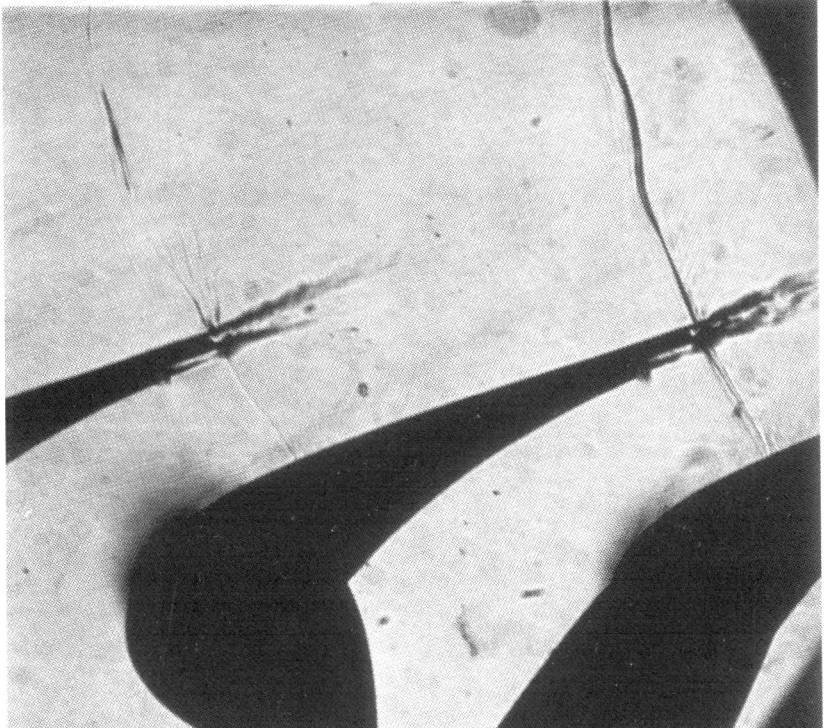

Figure 3.42 Schlieren photographs of PS-blades at $M_{2,is} = 1.05$ and $c_m = 3\%$ (left), resp. $c_m = 4\%$ (right).

at the pressure side generates an additional shock as long as coolant ejection is absent. This shock disappears at a coolant mass flow rate $c_m = 3\%$ due to displacement effects.

Based on the schlieren pictures it was decided to shift the blade equipped with pressure borings by one pitch in the downward direction, i.e. from the upper to the lower position in the photographs, to reduce interference effects from the free shear layer behind the top blade of the cascade. This shear layer can be seen in the upper right corner of the pictures.

5.3.2 Pressure distribution

The local Mach number along the blade profile is based on the ratio of the pressure measured at the blade surface through the pressure tap and the inlet total pressure in front of the cascade. The positions of the 39 pressure taps on the surface of the TE blade is shown in Figure 3.19, whereas Figure 3.20 shows the 38 taps of the PS blade. The number of pressure taps in the vicinity of the trailing edge could not be further increased for mechanical reasons. After

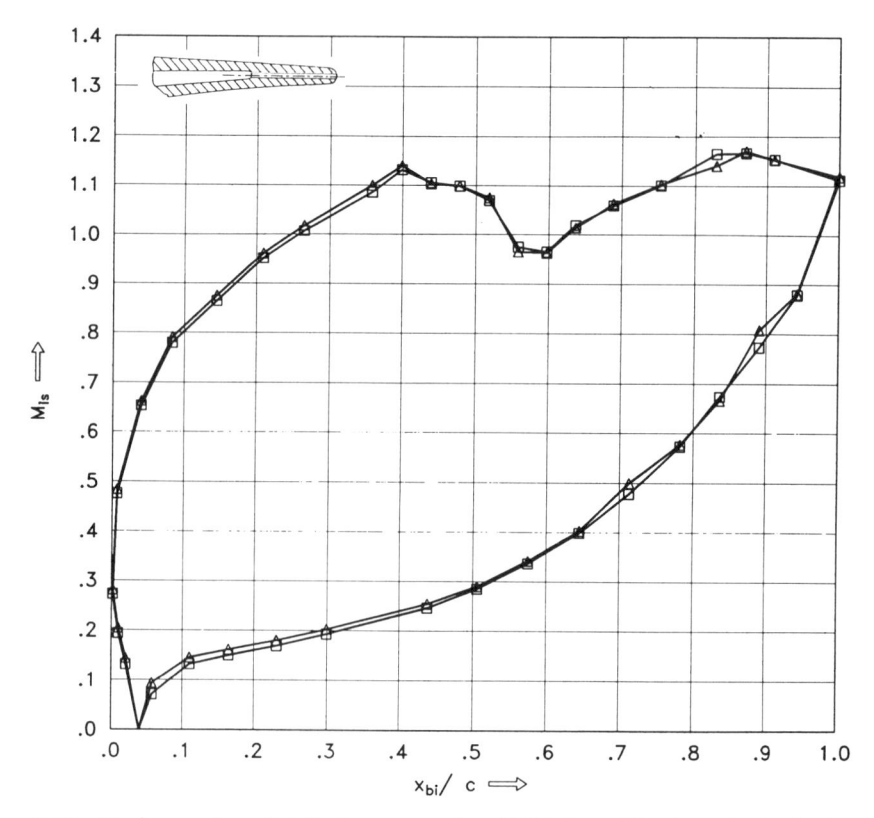

Figure 3.43 Mach number distribution over the TE-blade without coolant ejection at $M_{2,is} = 1.05$ with open ($\triangle$) as well as with closed ($\square$) slots.

instrumentation of the TE ejection blade it became clear that pressure tap no. 11 and 35 could not be used as the mounted tubes were squeezed flat. The position of tap no. 35 on the pressure side is not critical and, as the pressure rise on the suction side proved to be continuous till point 12, the measurements were accepted in this form.

For both blade types the velocity distribution in case of open coolant slots but without ejection has been compared with the distribution for a cascade with closed slots. The purpose of this comparison was the elucidation of three-dimensional effects in the flow through the cascade on the internal flow in the cavities of the blades.

The blade surface Mach number distribution at $M_{2,is} = 1.05$ for the blade with coolant slot in the trailing edge, see Figure 3.43, shows only very small differences between open and closed coolant slots. The differences are of the order of magnitude of the measurement errors. In Figure 3.44 the corresponding Mach number distribution for the PS ejection blades is drawn. Discrepancies are visible at the pressure side of the blade at the position of the coolant slot and at the trailing edge. At both points, especially at the coolant ejection slot, the measured pressure is smaller in case of a closed slot. At the suction

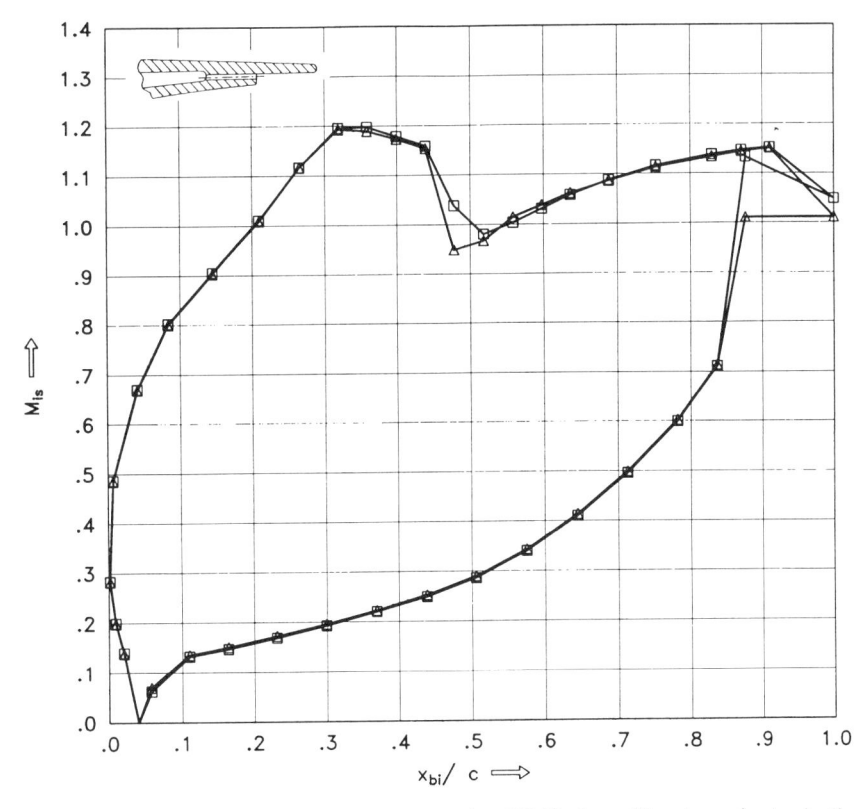

Figure 3.44 Mach number distribution over the PS-blade without coolant ejection at $M_{2,is} = 1.05$ with open ($\triangle$) as well as with closed ($\square$) slots.

side of the blade a deviation is measured at the position where the compression shock from the trailing edge of the above mounted blade reaches the boundary layer.

Figure 3.45 shows the Mach number distribution at $M_{2,is} = 1.05$ along the blade with coolant slot in the trailing edge at 4 coolant mass flow rates $c_m = 0\%$, 2%, 3% and 4%, with air as coolant. Near the position $x/c = 0.4$ on the suction side of the blade the influence of the incident compression wave, which propagates from the trailing edge of the neighbouring blade, gets noticeable, which results in the origin of a boundary layer separation bubble. At about $x/c = 0.55$ the boundary layer probably reattaches. At $x/c > 0.55$ the Mach number rises continuously farther downstream along the blade. The effect of coolant ejection on the Mach number distribution is rather small, reducing the local maxima and minima in the Mach number distribution on the suction side.

Figure 3.46 contains the corresponding data for the blade with pressure-side coolant ejection. In comparison with the already mentioned blade configuration the same phenomena are recognizable in the Mach number distribution in

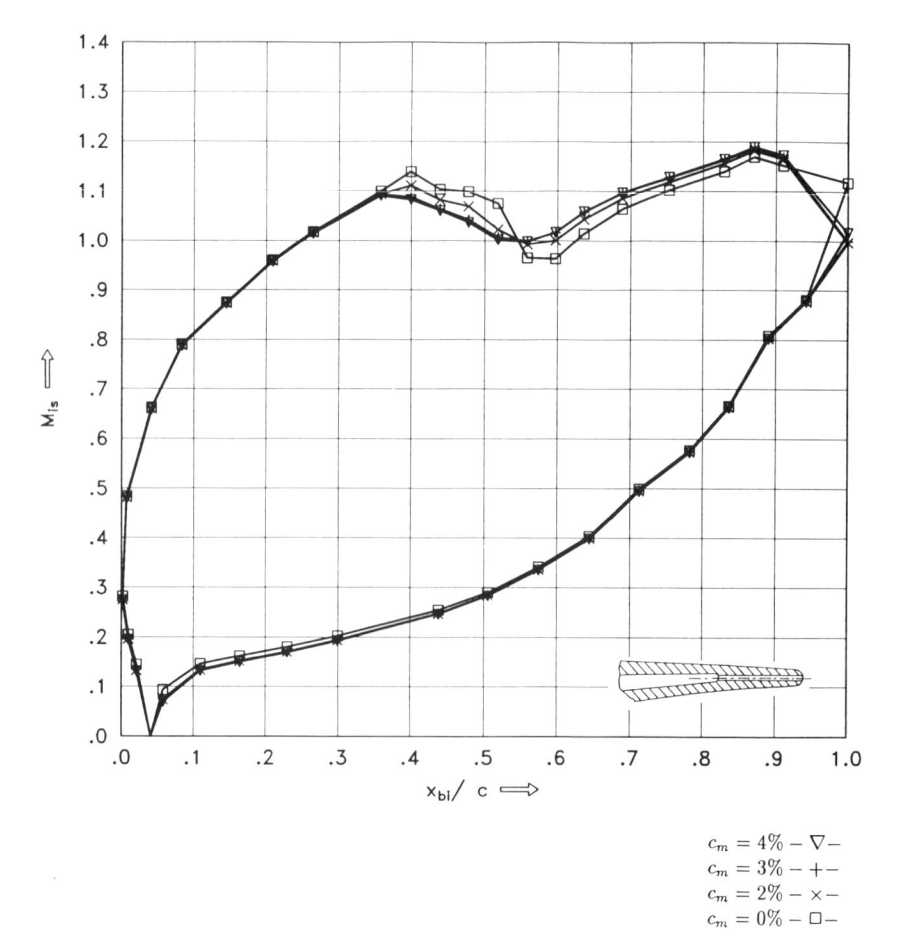

Figure 3.45 Mach number distribution over the TE-blade at $M_{2,is} = 1.05$ with air as coolant.

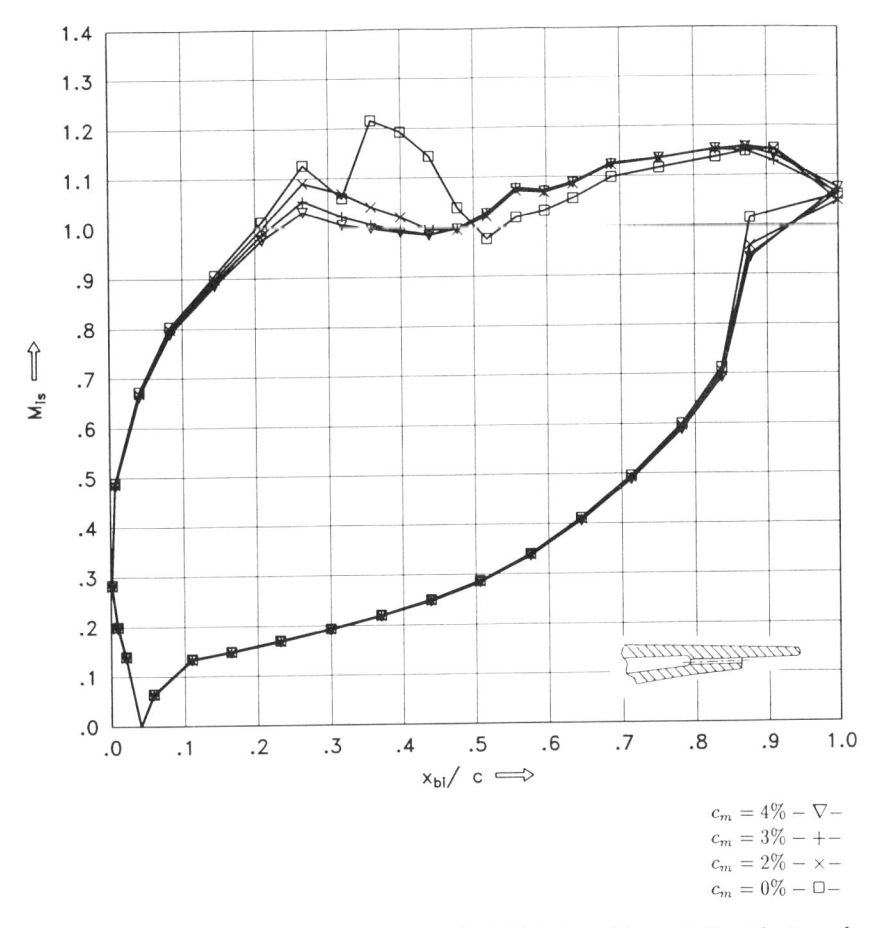

Figure 3.46 Mach number distribution over the PS-blade at $M_{2,is} = 1.05$ with air coolant.

$c_m = 4\% - \nabla -$
$c_m = 3\% - + -$
$c_m = 2\% - \times -$
$c_m = 0\% - \square -$

case of pressure-side ejection. However, the onset of the boundary-layer separation is at about $x/c = 0.3$ and reattachment takes place at about $x/c = 0.45$. The shift over 7 mm in upstream direction of the coolant ejection slot and, coupled to that, the throat between the blades, are probably responsible for this translation. The deviations in the area of boundary layer separation between Mach number distributions for flow with and without coolant ejection are remarkable. Especially the high Mach number values in the area of boundary-layer separation in case of zero coolant ejection are striking.

5.3.3 Base-pressure coefficients

The pressure tapping for measuring the base pressure P_b on the TE ejection blade is positioned in one of the struts in the coolant ejection slot (no. 25 in

Figure 3.19), which implies that the measurements point is not at mid-blade height. The blade with pressure-side ejection offers two possibilities to take P_b: (1) in the suction side trailing edge (no. 25 in Figure 3.20), (2) in the trailing edge of the pressure-side lip at the coolant ejection slot (no. 26 in Figure 3.20). In the latter case, the position is not at mid-span, similar to the base-pressure tapping for the TE ejection blade. Various base-pressure coefficients were deduced from the pressure measurements along the blade described in the previous section:

$$C_b = \frac{P_b - P_2}{P_{01}}. \tag{3.1}$$

Figure 3.47 contains C_b-data corresponding to (3.1) for the TE ejection blade as well as for the PS ejection blade:

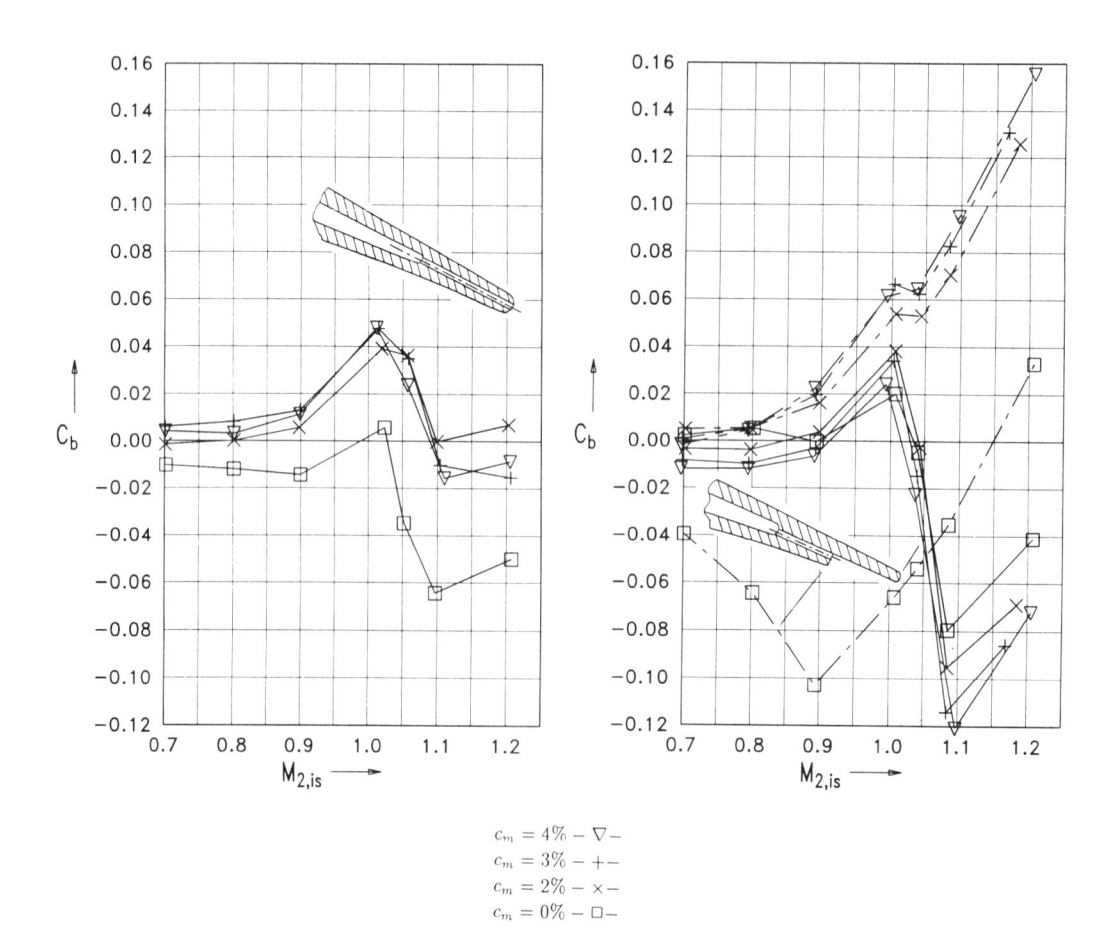

$c_m = 4\% - \triangledown -$
$c_m = 3\% - + -$
$c_m = 2\% - \times -$
$c_m = 0\% - \square -$

Figure 3.47 Base pressure coefficient on total pressure basis C_b for both blade types in case of air ejection.

Straight cascade tests

$$C_{p2} = \frac{P_b - P_2}{\frac{\rho_2}{2} \cdot v_2^2} \qquad (3.2)$$

C_{p2}-graphs according to (3.2) are presented in Figure 3.48 for both blade configurations:

$$C_{p^*} = \frac{P_b - P_b^*}{\frac{\rho_2}{2} \cdot v_2^2} \qquad (3.3)$$

P_b^* is the base pressure at the same downstream isentropic Mach number

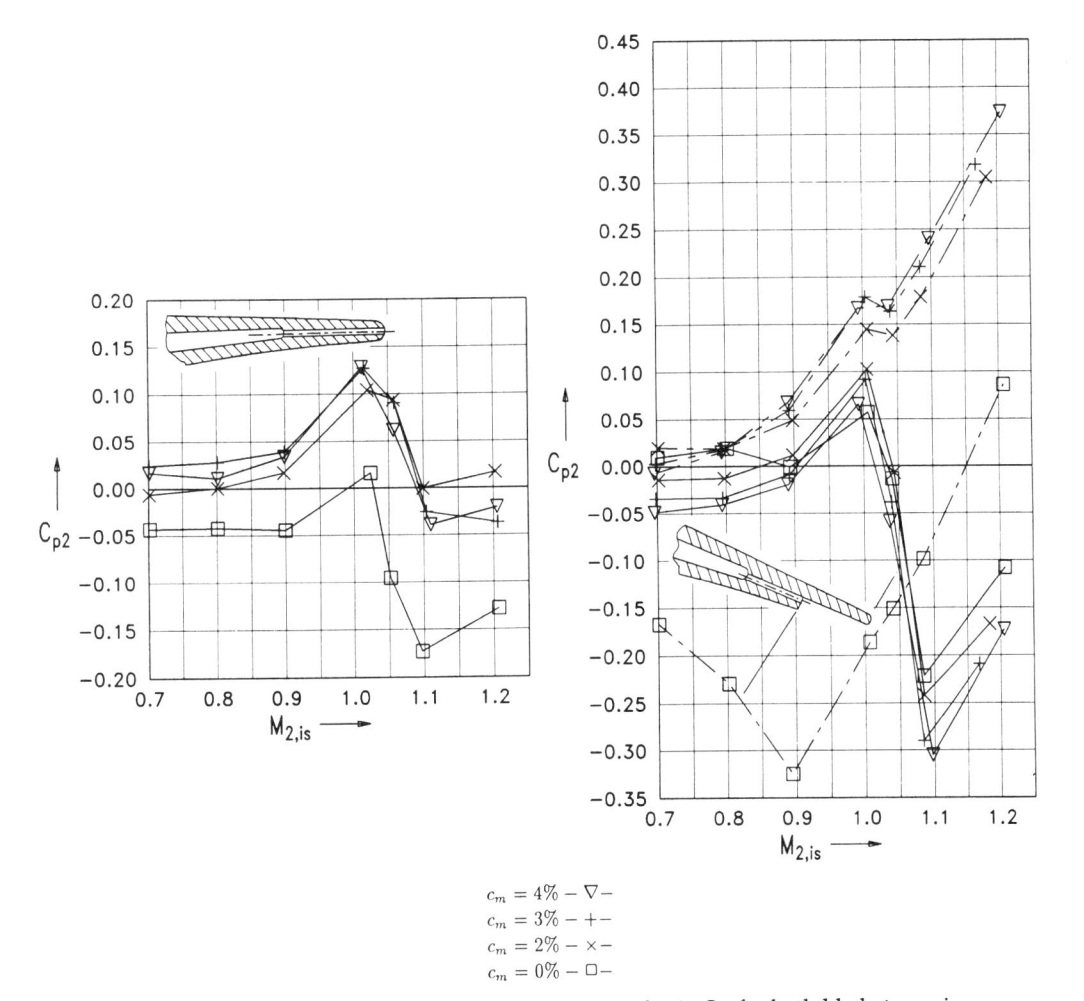

$c_m = 4\%$ – ▽ –
$c_m = 3\%$ – + –
$c_m = 2\%$ – × –
$c_m = 0\%$ – □ –

Figure 3.48 Base pressure coefficient on kinetic energy basis C_{p2} for both blade types in case of air ejection.

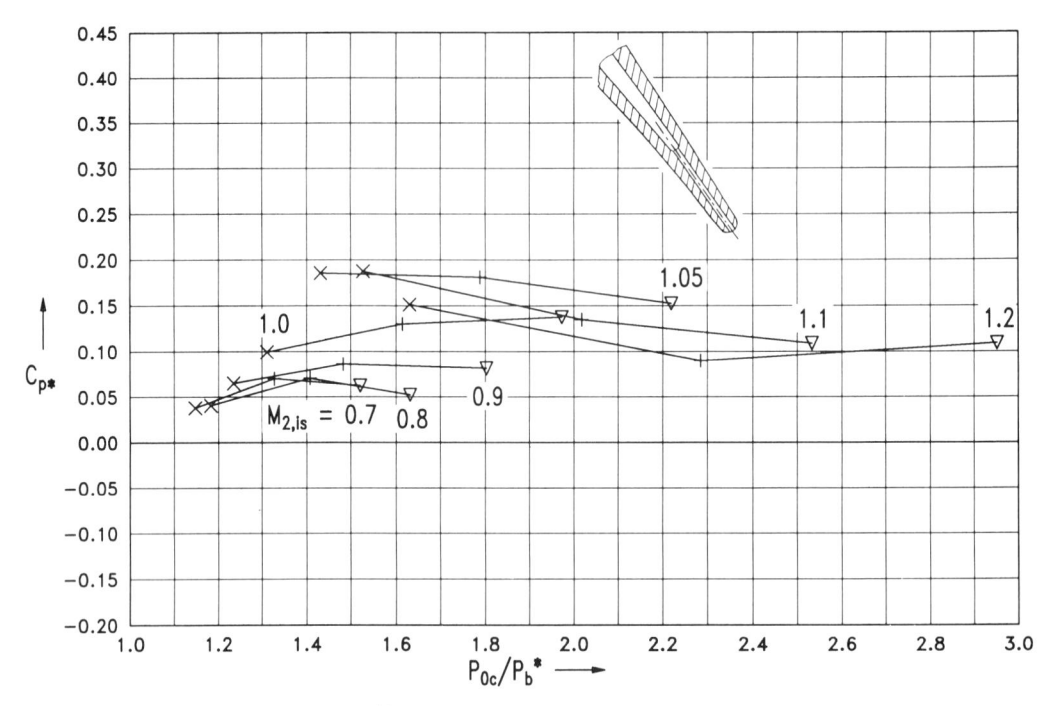

Figure 3.49 Base pressure coefficient on kinetic energy basis C_{p^*} for the TE-blade in case of air ejection. The base pressure for flow without ejection is denoted P_b^*.

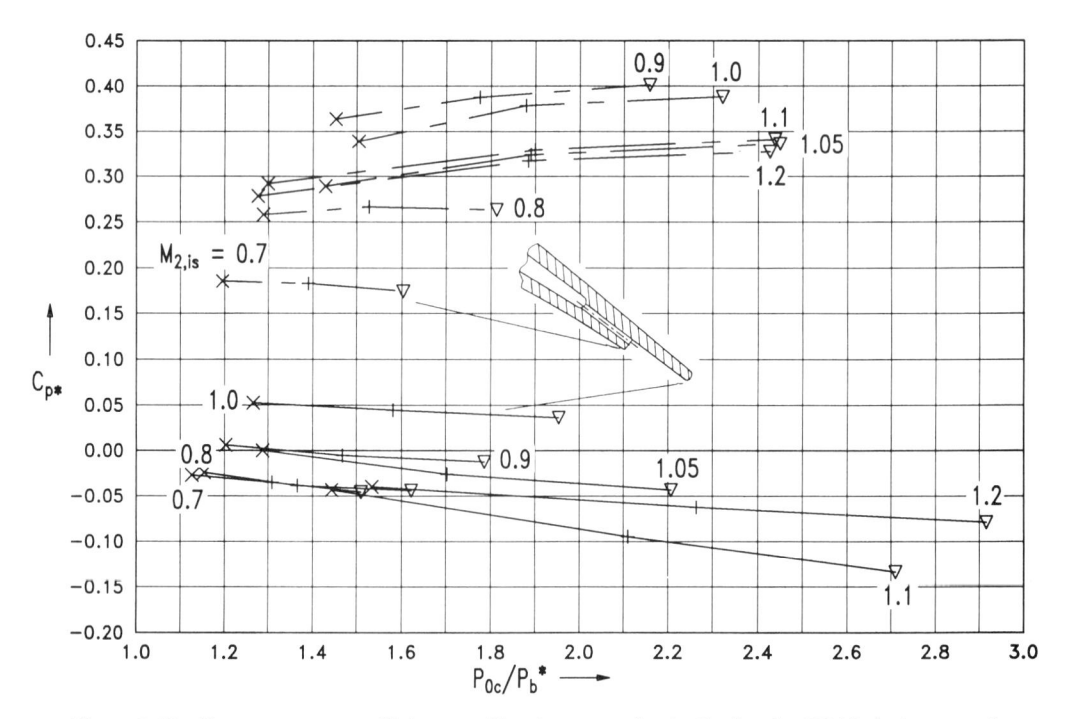

Figure 3.50 Base pressure coefficient on kinetic energy basis C_{p^*} for the PS-blade in case of air ejection. The base pressure for flow without ejection is denoted P_b^*.

without coolant ejection. Figures 3.49 and 3.50 show C_{p^*}-values according to (3.3) for the blade with TE and PS ejection respectively.

At the trailing edge itself both types of blade show similar tendencies for the variation of C_b/C_{p2} with Mach number; i.e. a maximum near $M_{2,is} = 1.0$ and a minimum at $M_{2,is} = 1.1$. The strong decrease in C_b from $M_{2,is} = 1.0$ to $M_{2,is} = 1.1$ can be explained by the shock movement from a position in front of the trailing edge to a position behind the trailing edge.

In contrast to this the base pressure behind the step of the PS-ejection blade exhibits a steady increase with Mach number.

As expected coolant ejection increases the base pressure, because coolant is diluted to the wake.

The coefficient C_{p^*} gives the difference between the base pressure with coolant to the datum base pressure without coolant. The abscissa P_{0c} is the total pressure of the coolant in the ejection slot. This total pressure increases monotonically with coolant flow rate. Accordingly the curves in Figures 3.49 and 3.50 show the influence of cooling flow rate on the base pressure. These curves show no strong dependency upon the total pressure of the coolant. When interpreting these graphs one should bear in mind that these C_{p^*}-values should all go to zero at P_{0c}/P_b^* equal to 1.0. The difference with respect to the point without coolant flow demonstrates that there is in fact an effect of coolant flow ejection, which is more or less confined to small coolant flow rates.

5.3.4 Wake measurements

The measurements of total pressure, static pressure and flow angle in the wake of the cascade were carried out at an axial distance of 35 mm to the cascade exit plane, i.e. at $x/c_{ax} = 0.813$. The length of the traverse parallel to the cascade exit plane was about 200 mm, which corresponds to nearly 4 pitches. Data were recorded every 1.5 mm. The graphs in this section present the flow data belonging to the pitch in which the wake of the second cooled blade, counted in the cascade from bottom to top, is contained. The Mach number in the rectangular inlet section (M_1), the axial velocity density ratio (AVDR), the homogeneous downstream flow angle (α_2) and the energy loss coefficient (ζ) are plotted as a function of the isentropic downstream Mach number ($M_{2,is}$) with the coolant flow ratio as parameter for both blade types in Figures 3.51–3.54.

The measured axial velocity density ratio in case of choking conditions in the cascade exceeds the theoretical maximum, marked by the horizontal line in Figure 3.51. This phenomenon is caused by side wall boundary layers, which lead to three-dimensional effects in the wind tunnel.

According to Figure 3.53 the deflection of the flow reaches its maximum near $M_{2,is} = 1.0$ for both blade configurations. The influence of coolant ejection on the flow deflection α_2 differs for both blade types: for the TE-blade α_2 decreases with increasing coolant flow rate c_m. Ejection of coolant from the PS-blade on the other hand enhances the flow deflection except for the maximum Mach number, i.e. $M_{2,is} = 1.2$.

The energy loss coefficient ζ in Figure 3.54 is almost constant in the Mach

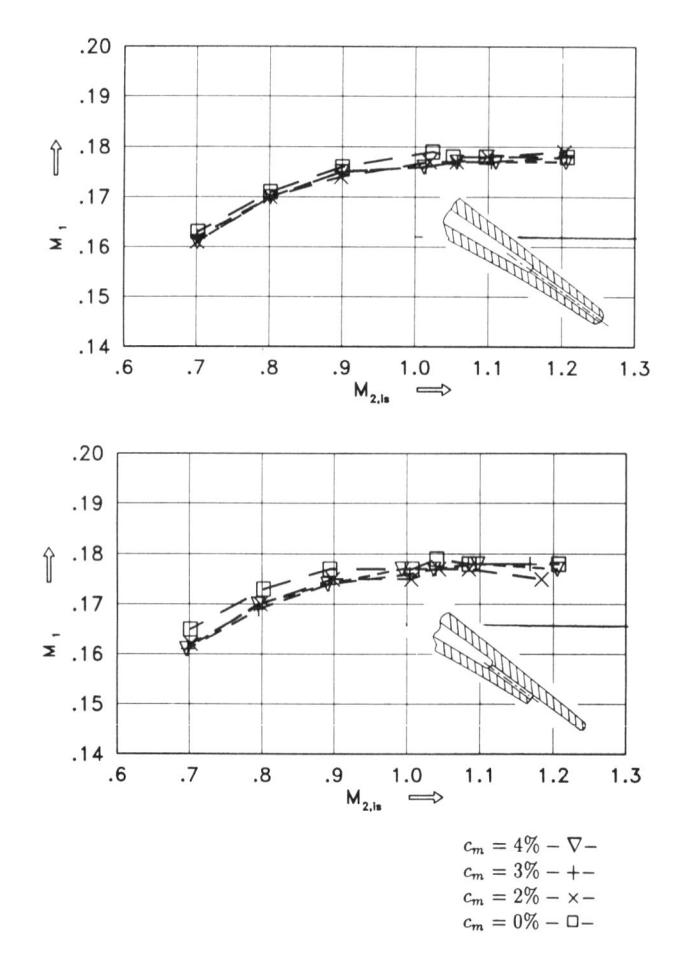

Figure 3.51 Mach number in the rectangular inlet section at choking conditions in the cascade for both blade types.

number region up to $M_{2,is} = 1.0$. At higher Mach numbers, however, the losses rapidly rise. At design conditions, i.e. $M_{2,is} = 1.05$, ζ has already clearly risen. Generally the losses of the PS-blade are higher than for the TE-blade, with the exception of the design point, where they happen to be equal. Coolant ejection for the lower Mach numbers reduces the losses. This tendency is reversed for the highest Mach number.

Tables A.1 and A.2 in Appendix 2 present an overview of the flow data deduced from the wake measurements with air ejection from the TE blade and the PS blade respectively.

Wake measurement data, i.e. total pressure ratio, flow angle and static- to total-pressure ratio, are drawn in Figure 3.55 for the TE as well as for the PS configuration at $M_{2,is} = 1.05$, with air ejection at a rate $c_m = 3\%$. In Figure 3.56 the same quantities are presented at an outlet Mach number $M_{2,is} = 0.80$.

Straight cascade tests

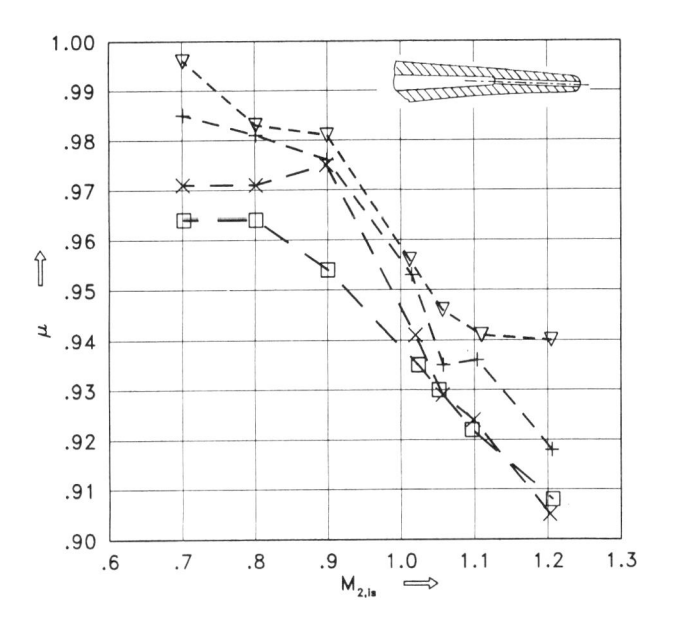

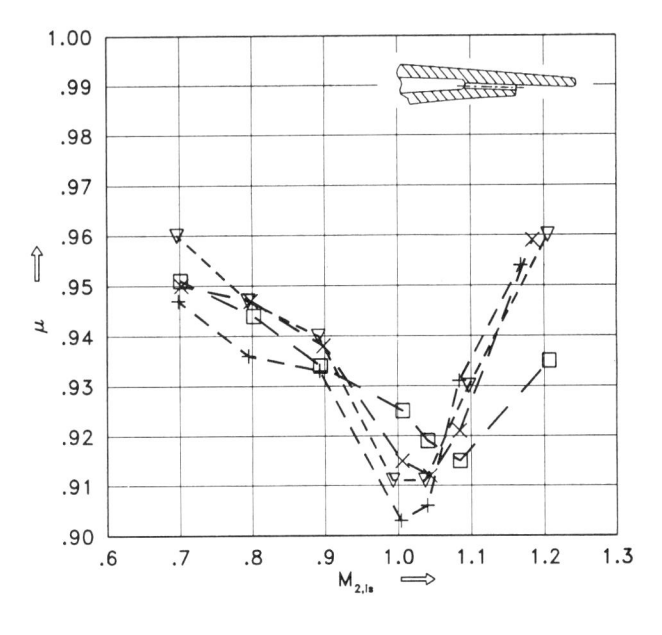

$c_m = 4\% - \triangledown -$
$c_m = 3\% - + -$
$c_m = 2\% - \times -$
$c_m = 0\% - \square -$

Figure 3.52 Axial velocity density ratio for both blade types.

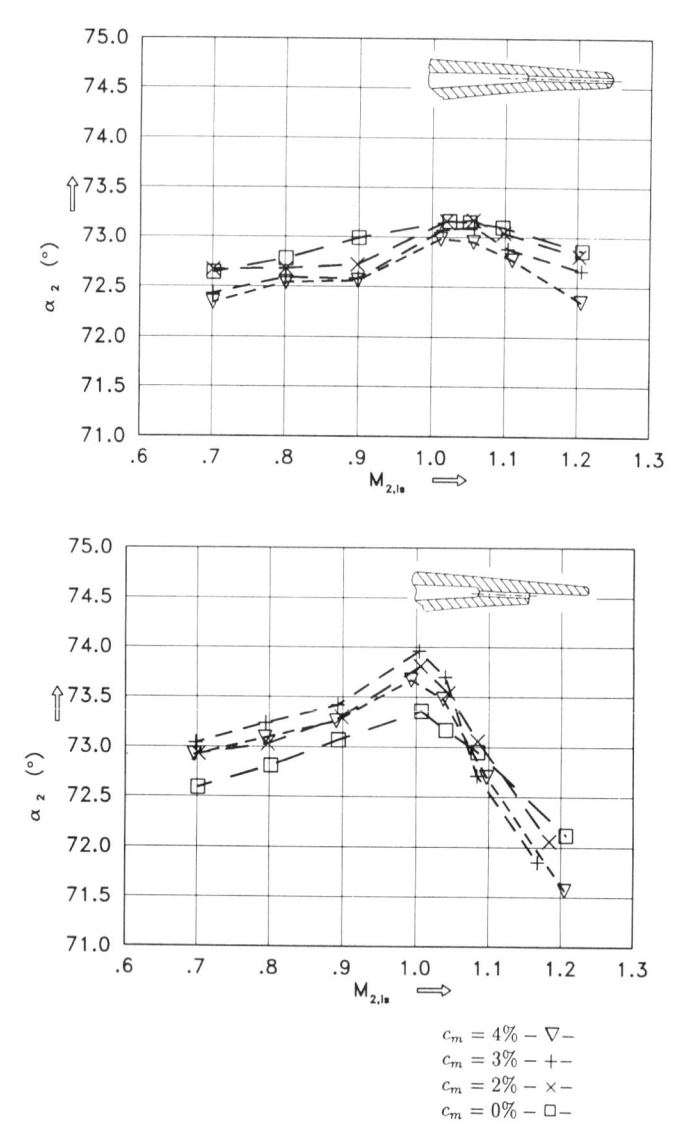

Figure 3.53 Homogeneous downstream flow angle for both blade configurations with air as coolant.

5.3.5 Turbulence measurements

Turbulence measurements were carried out in the settling chamber and in the rectangular inlet section upstream of the cascade both with a hot wire probe and a hot film probe. For reasons of mechanical probe stability downstream of the cascade only the hot film probe was used. During the measurements in the inlet section, $M_{2,is}$ was held at 0.8. Typical corresponding flow velocities in the rectangular inlet section are 60 m/s.

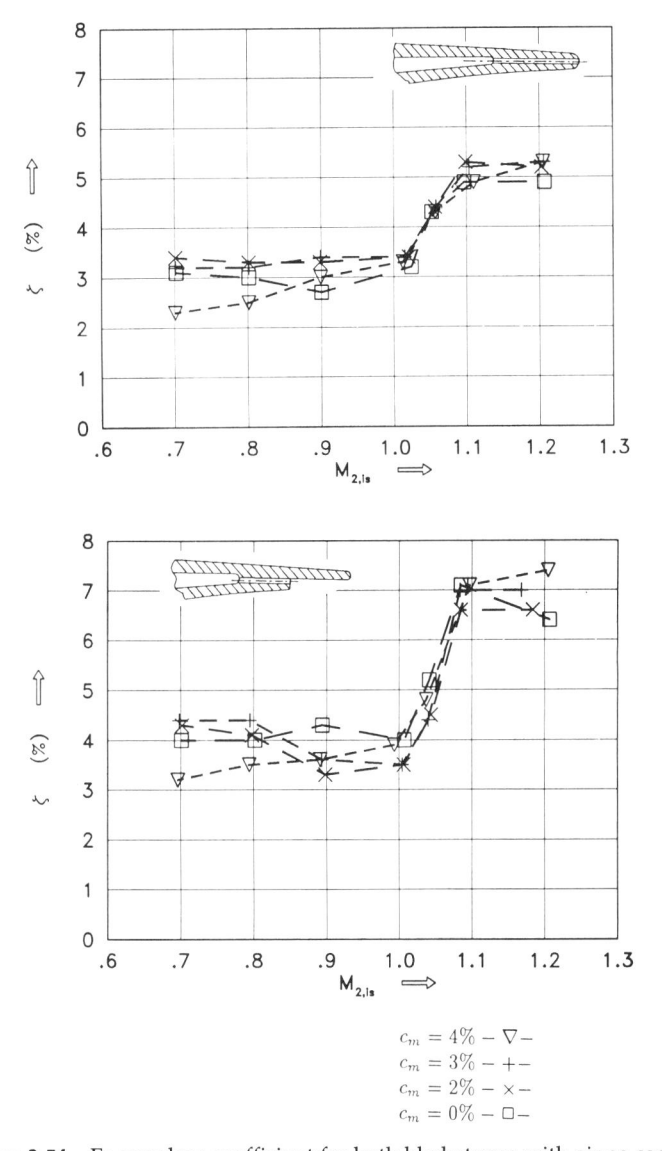

Figure 3.54 Energy-loss coefficient for both blade types with air as coolant.

$c_m = 4\% - \nabla -$
$c_m = 3\% - + -$
$c_m = 2\% - \times -$
$c_m = 0\% - \square -$

The velocity component u is assumed to be in the direction of the time mean velocity. Upstream of the cascade u is therefore usually in the axial direction; however, downstream of the cascade this no longer holds. The three velocity components are split into their time mean values and fluctuations in time:

$$u = \bar{u} + u' \tag{3.4}$$

$$v = \bar{v} + v' \tag{3.5}$$

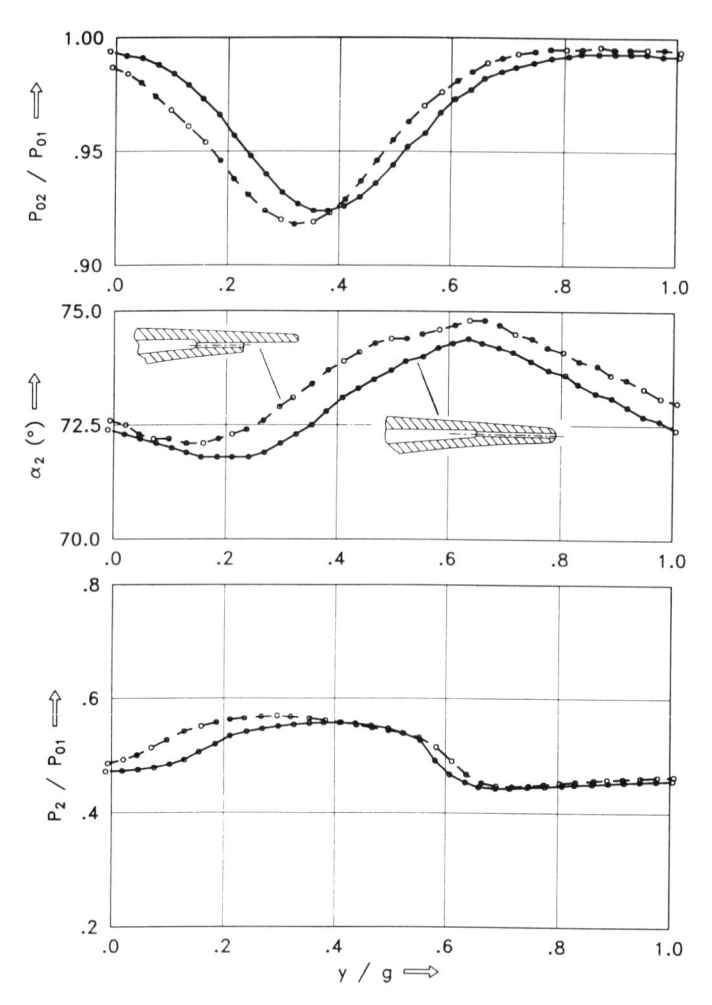

Figure 3.55 Local flow properties for both blade configurations at $x/c_{ax} = 0.813$, $M_{2,is} = 1.05$ and $c_m = 3\%$ (air).

$$w = \bar{w} + w' \tag{3.6}$$

with $\bar{v} = \bar{w} = 0$. Further it is assumed that the RMS-values of the velocity fluctuations in all three directions are identical:

$$\sqrt{u'^2} = \sqrt{v'^2} = \sqrt{w'^2} \tag{3.7}$$

Then the turbulence degree Tu may be defined as

$$Tu = \frac{\sqrt{u'^2}}{u_\infty} \tag{3.8}$$

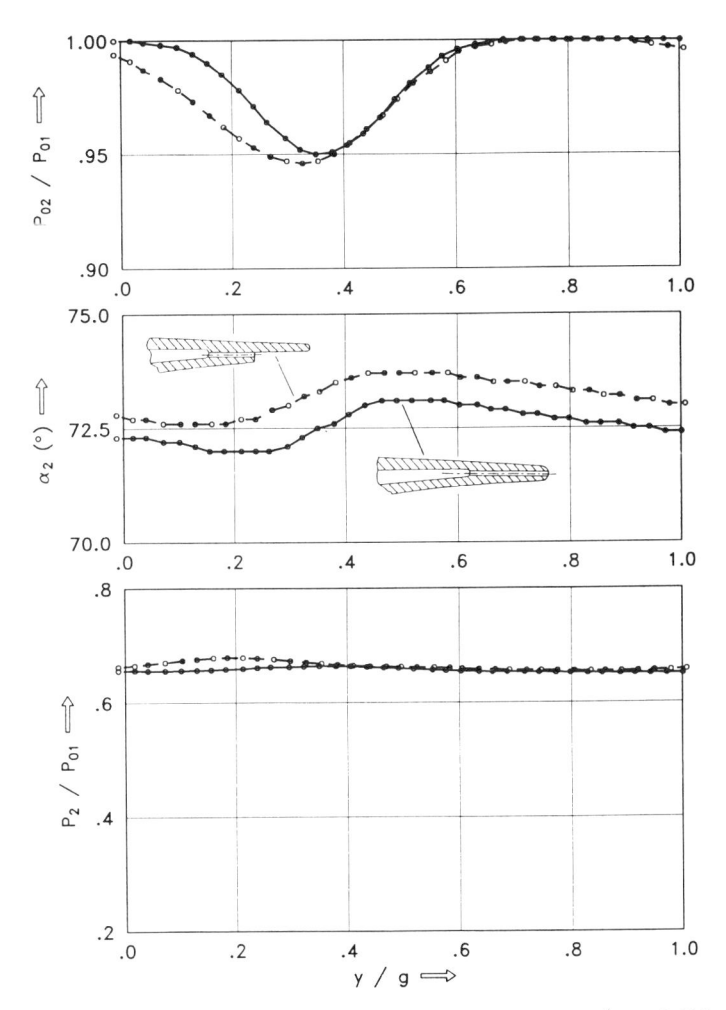

Figure 3.56 Local flow properties for both blade configurations at $x/c_{ax} = 0.813$, $M_{2,is} = 0.80$ and $c_m = 3\%$ (air).

where u_∞ presents the mean velocity undisturbed by a boundary or a wake. The evaluation of the data, sampled during the turbulence measurements, was carried out in accordance with the procedure described by Höhler and Baumgarten [3.5]. In case of isothermal, incompressible flow King's law is reduced to the following equation:

$$\frac{\bar{e}^2 - \bar{e}_0^2}{\Delta T} = B\bar{u}^{1/n} \tag{3.9}$$

where ΔT is the overheat temperature of the wire or film with respect to the flow, $\bar{u}$ is the time mean velocity, $\bar{e}$ denotes the anemometer output voltage, $\bar{e}_0$

is the output voltage when the flow velocity is zero. B and n are calibration constants, $n \simeq 2$. The velocity fluctuations are assumed to be small, leading to the approximations:

$$\sqrt{u'^2}\,\frac{d\bar{e}}{d\bar{u}} = \sqrt{e'^2} \tag{3.10}$$

$$\frac{\bar{u}}{u_\infty} = \left(\frac{\bar{e}^2 - \bar{e}_0^2}{\bar{e}_\infty^2 - \bar{e}_0^2}\right)^{1/n} \tag{3.11}$$

After application of these approximations the following equation for the turbulence degree is deduced:

$$Tu = \frac{\sqrt{e'^2}\,2n\bar{e}}{\bar{e}^2 - \bar{e}_0^2}\left(\frac{\bar{e}^2 - \bar{e}_0^2}{\bar{e}_\infty^2 - \bar{e}_0^2}\right)^n \tag{3.12}$$

Figure 3.57 presents turbulence values in the rectangular inlet section. the hot wire probe indicates a higher turbulence level than that of the hot film probe. In the centre of the inlet section the turbulence degree amounts to 0.5% for the hot wire and 0.4% for the hot film. A possible reason for the discrepancy between the results of the two measurement systems might be the smaller frequency range of the hot film probes.

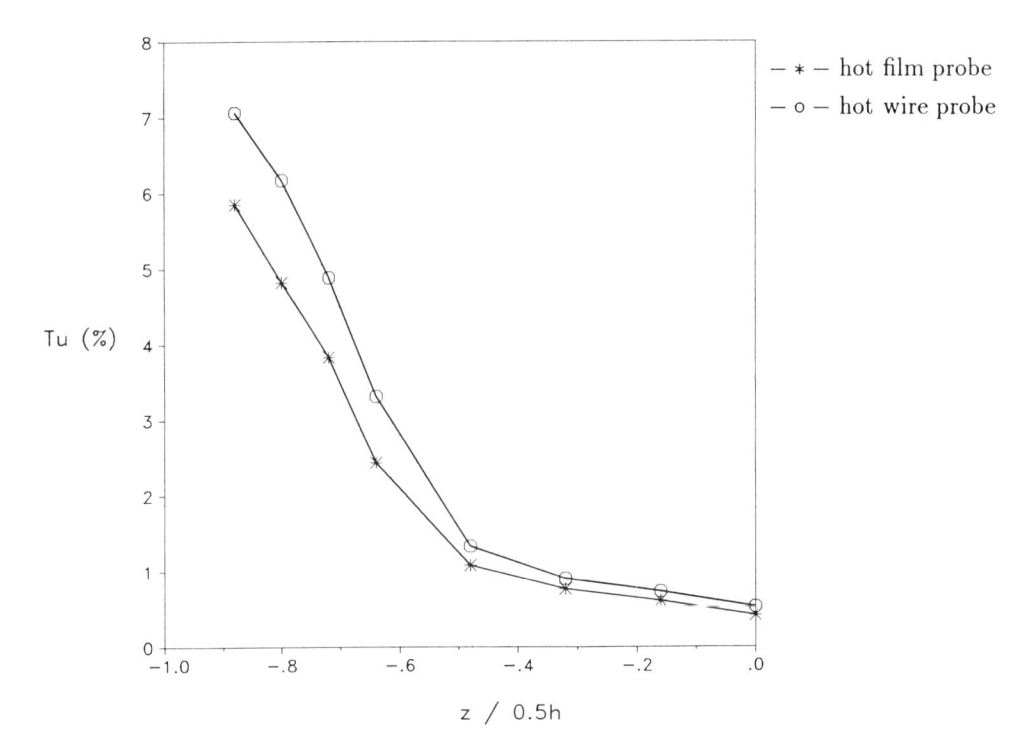

Figure 3.57 Turbulence degree in the rectangular inlet section.

Downstream of the cascade the interpretation of the measurement data is at the moment restricted to a qualitative interpretation. Figure 3.58 presents the ratio of the turbulence levels according to (3.13) between the data belonging to the measurement position and those at the position with the maximum mean voltage $\bar{e}$. These measurement data were recorded behind the trailing edge ejection blade at $X/C_{ax} = 0.813$:

$$\frac{Tu}{Tu_{ref}} = \frac{\bar{e}}{\bar{e}_{ref}} \frac{\sqrt{e'^2}}{\sqrt{e'^2_{ref}}} \left(\frac{\bar{e}^2 - \bar{e}_0^2}{\bar{e}^2_{ref} - \bar{e}_0^2}\right)^{n-1} \qquad (3.13)$$

At $M_{2,is} = 0.80$ the influence of coolant ejection is clearly distinguishable in the wake. The actual position of the maximum in this graph is not directly comparable with the maxima of the wake losses measured with the wedge-type probe, as the probe mounting devices are not identical. In the case of choked flow conditions a distinct effect of the coolant flow ejection on the turbulence distribution is scarcely visible as demonstrated by the turbulence data recorded at $M_{2,is} = 1.05$.

5.4 Tests with carbon dioxide ejection

5.4.1 Data reduction of wake flow measurements

The standard data reduction system for the evaluation of cascade wake measurements takes into account the compressible flow field of a homogeneous perfect gas with constant specific heat. It is based on the conservation laws for mass, momentum and energy and is described in detail in [3.6]–[3.8].

Usually the total temperature in the entire flow field is assumed to be constant. To comply with this assumption also for the present test series the total temperature of the coolant flow is adjusted exactly to the total temperature of the main flow. This resolution avoids measuring the temperature in the wake flow field, which is generally burdened with strong errors. Nevertheless the evaluation of tests with a variable (measured) total temperature is possible by application of the extended data reduction method by Oldfield and Schultz [3.9].

In the actual research programme additional tests were performed with the ejection of carbon dioxide (CO_2) in order to study the wake-mixing process. Therefore the gas concentration was measured in the wake in addition to the standard aerodynamic values (P_0, P, α) and consequently the data reduction system had to be extended to gas mixtures.

The applied analyser yields the concentration of CO_2 as relation of volumes. This value can be transformed into the mass share of the gas concentration with help of the molecular mass of components:

$$\vartheta_m = \frac{\vartheta_v M_{CO_2}}{(1 - \vartheta_v) M_{AIR} + \vartheta_v M_{CO_2}} \qquad (3.14)$$

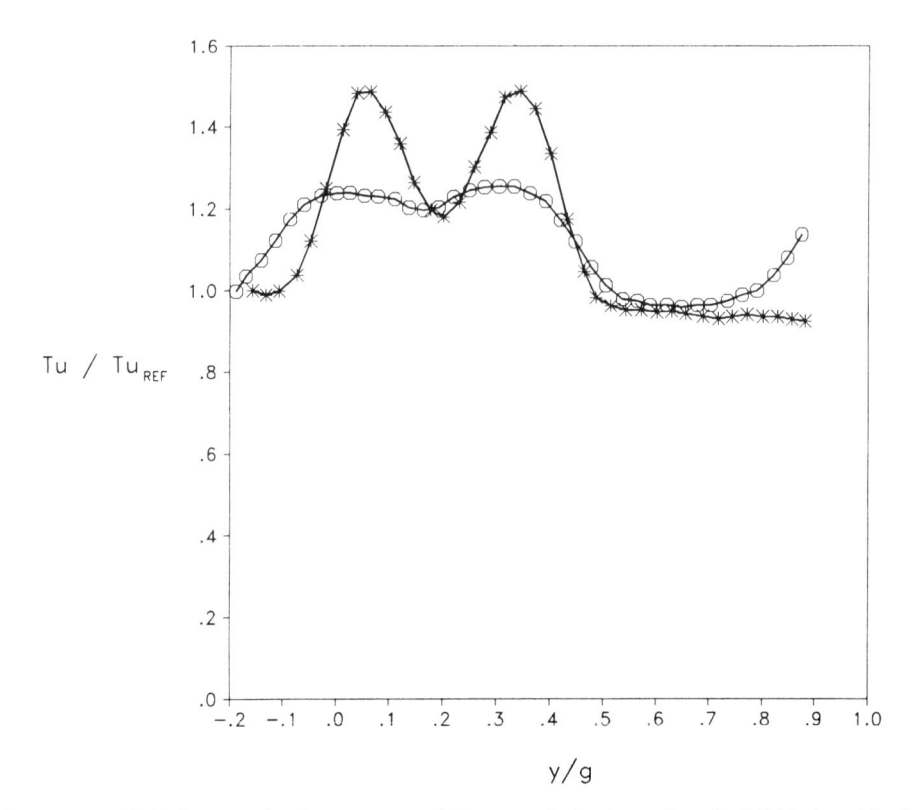

Figure 3.58 Turbulence ratio downstream of the cascade in the wake of a TE-blade with air ejection at $M_{2,is} = 0.80$ (left) and at $M_{2,is} = 1.05$ (right).

The gas constant of a mixture is evaluated from the gas constants of the components according to their mass shares:

$$R = (1 - \vartheta_m)R_{\text{AIR}} + \vartheta_m R_{\text{CO}_2} \tag{3.15}$$

The isentropic exponent—also required for the transformation of the local values in the measuring plane—is defined as ratio of the specific heats at constant pressure and constant volume:

$$\kappa = \frac{c_p}{c_v} \tag{3.16}$$

For the specific heats of mixtures the same law as for the gas constant:

$$\kappa = \frac{(1 - \vartheta_m)c_{p,\text{AIR}} + \vartheta_m c_{p,\text{CO}_2}}{(1 - \vartheta_m)c_{v,\text{AIR}} + \vartheta_m c_{v,\text{CO}_2}} \tag{3.17}$$

Hence it follows that the isentropic exponent itself is not composed according to the mass shares. But after some treatment we can transform the above

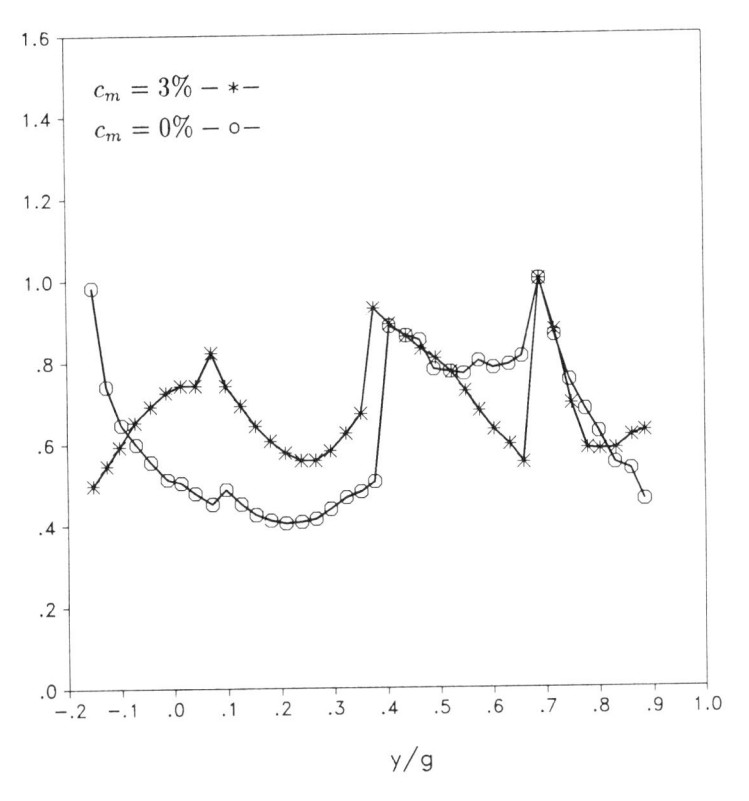

Figure 3.58 (continued).

formula into

$$K_m = \frac{(1-\vartheta_m)\kappa_{AIR}}{(1-\vartheta_m)+\vartheta_m\dfrac{c_{v,CO_2}}{c_{v,AIR}}} + \frac{\vartheta_m \kappa_{CO_2}}{(1-\vartheta_m)\dfrac{c_{v,AIR}}{c_{v,CO_2}}+\vartheta_m} \quad (3.18)$$

This equation demonstrates that the mixing law applied above for the gas constant and the specific heat holds also approximately for the isentropic exponent, because

$$c_{v,CO_2} \approx c_{v,AIR} \quad (3.19)$$

The thermodynamic properties of air and CO_2 are listed in Table A.3 in Appendix 2. The calculation of the homogeneous flow field from the local values will be outlined below as supplement to the original derivation of the wake reduction method in [3.8]. The starting point is the continuity equation transformed into a dimensionless form by dividing it by the values of the critical state of the homogeneous downstream flow (equation (16) in [3.8]):

$$\frac{\rho_2 v_2}{\rho_2^* a_2^*}\cos\alpha_2 = \int_{(y/g)}^{(y/g)+1} \frac{\rho_{2y} v_{2y}}{\rho_2^* a_2^*}\cos\alpha_{2y}\,d\left(\frac{y}{g}\right) \qquad (3.20)$$

For perfect gases with constant specific heat the following holds:

$$\rho^* a^* = \frac{p^*}{RT^*}\sqrt{(\kappa RT^*)} \qquad (3.21)$$

$$\rho^* a^* = \sqrt{\left[\kappa\left(\frac{2}{\kappa+1}\right)^{\frac{\kappa+1}{\kappa-1}}\right]}\frac{p_0}{\sqrt{(RT_0)}} \qquad (3.22)$$

For simplification we can introduce

$$\mathcal{K} = \frac{1}{\kappa}\left(\frac{\kappa+1}{2}\right)^{\frac{\kappa+1}{\kappa-1}} \qquad (3.23)$$

and obtain

$$\rho^* a^* = \frac{p_0}{\sqrt{(\mathcal{K} RT_0)}} \qquad (3.24)$$

Finally the continuity equation becomes, with this simplification for constant temperature,

$$\frac{p_{02}/p_{01}\Theta_2\cos\alpha_2}{\sqrt{\left(\frac{\mathcal{K}_2}{\mathcal{K}_\infty}\frac{R_2}{R_1}\right)}} = \int_{(y/g)}^{(y/g)+1} \frac{(p_{02y}/p_{01})\Theta_{2y}\cos\alpha_{2y}}{\sqrt{\left(\frac{\mathcal{K}_{2y}}{\mathcal{K}_1}\frac{R_{2y}}{R_1}\right)}}\,d\left(\frac{y}{g}\right) \qquad (3.25)$$

The local gas constant and the isentropic exponent can be calculated from the measured concentration. The corresponding data of the homogeneous flow must be calculated from the measured main flow rate and the ejection flow rate.

The treatment of the additional term in the derivation as presented in [3.8] leads finally to the following system of equations as a solution of the extended conservation equations:

$$I_1 = \int_{(y/g)}^{(y/g)+1} \frac{\dfrac{p_{02y}}{p_{01}}\Theta_{2y}\cos\alpha_{2y}}{\sqrt{\dfrac{\mathcal{K}_{2y}}{\mathcal{K}_1}\dfrac{R_{2y}}{R_1}}}\,d\left(\frac{y}{g}\right) \qquad (3.26)$$

The other integrals remain unchanged.
Finally the following system of equations is obtained:

$$M_2^{*2} = \left(\frac{\kappa+1}{2}\right)^{\frac{2}{\kappa-1}} \frac{I_2^2}{I_1^2} \frac{\mathcal{K}_1}{\mathcal{K}_2} \frac{R_1}{R_2} \left\{ \frac{1}{2} - \left(\frac{2}{\kappa+1}\right)^{\frac{2}{\kappa-1}} \frac{I_1^2}{I_2^2} \frac{\mathcal{K}_2}{\mathcal{K}_1} \frac{R_2}{R_1} \right.$$

$$\left. + \frac{\kappa+1}{2\kappa} \frac{I_3^2}{I_2^2} \pm \sqrt{\left[\frac{1}{4} - \left(\frac{2}{\kappa+1}\right)^{\frac{2}{\kappa-1}} \frac{I_1^2}{I_2^2} \frac{\mathcal{K}_2}{\mathcal{K}_1} \frac{R_2}{R_1} + \frac{\kappa^2-1}{4\kappa^2} \frac{I_3^2}{I_2^2}\right]} \right\} \quad (3.27)$$

$$\cos\alpha_2 = \frac{I_3}{I_1} \frac{\Theta_2}{2\frac{q_2}{p_{0_2}} \sqrt{\left(\frac{\mathcal{K}_2}{\mathcal{K}_1} \frac{R_2}{R_1}\right)}} \quad (3.28)$$

$$\frac{p_{0_2}}{p_{0_1}} = \sqrt{\left(\frac{\mathcal{K}_2}{\mathcal{K}_1} \frac{R_2}{R_1}\right)} \frac{I_1}{\Theta_2 \cos\alpha_2} \quad (3.29)$$

All other equations remain unchanged.

The comparison of the integrated CO_2 mass flow rate in the wake with the CO_2 mass flow rate measured in the supply line allows to control the accuracy of the wake flow measurement. Therefore the integral in (3.25) must be rewritten as follows:

$$\dot{m}_{CO_2} = \frac{p_{0_1}}{\sqrt{(\mathcal{K}_1 R_1 T_0)}} \int_{(y/g)}^{(y/g)+1} \vartheta_{m_{2y}} \frac{\frac{p_{0_{2y}}}{p_{0_1}} \Theta_{2y} \cos\alpha 2y}{\sqrt{\left(\frac{\mathcal{K}_{2y}}{\mathcal{K}_1} \frac{R_{2y}}{R_1}\right)}} d\left(\frac{y}{g}\right) \quad (3.30)$$

5.4.2 Wake-mixing profiles

Wake measurements of the carbon dioxide volume concentration were carried out using the probe traversing mechanism with a pitot tube instead of the wedge type probe normally employed for wake pressure measurements. Figure 3.59 illustrates the positions of the measurement planes in the wake where concentration measurements as well as pressure measurements were carried out. A vacuum pump took gas samples through this pitot tube, the carbon dioxide volume concentration of which was determined by a CO_2-analyser. The layout of this analysing equipment is shown in Figure 3.60.

Because of the dimensions of the wedge-type probe, the comparison of the incoming mass flow (through the ejection slot) and the mass flow in the measurement plane can only be carried out at axial distances ≥ 7 mm, $(x/c_{ax} \geq 0.163)$, to the trailing edge of the blade. The results for the blade with coolant ejection through the trailing edge are listed in Table 3.2 for $M_{is} = 1.05$ and in Table 3.3 for $M_{is} = 0.80$.

The wake-mixing profiles measured at the design coolant flow rate $c_m = 3\%$ and at Mach numbers $M_{2,is} = 0.8$ and $M_{2,is} = 1.05$ are shown in Figures 3.61–3.64. These measured concentration curves are approximated by Gaussian curves.

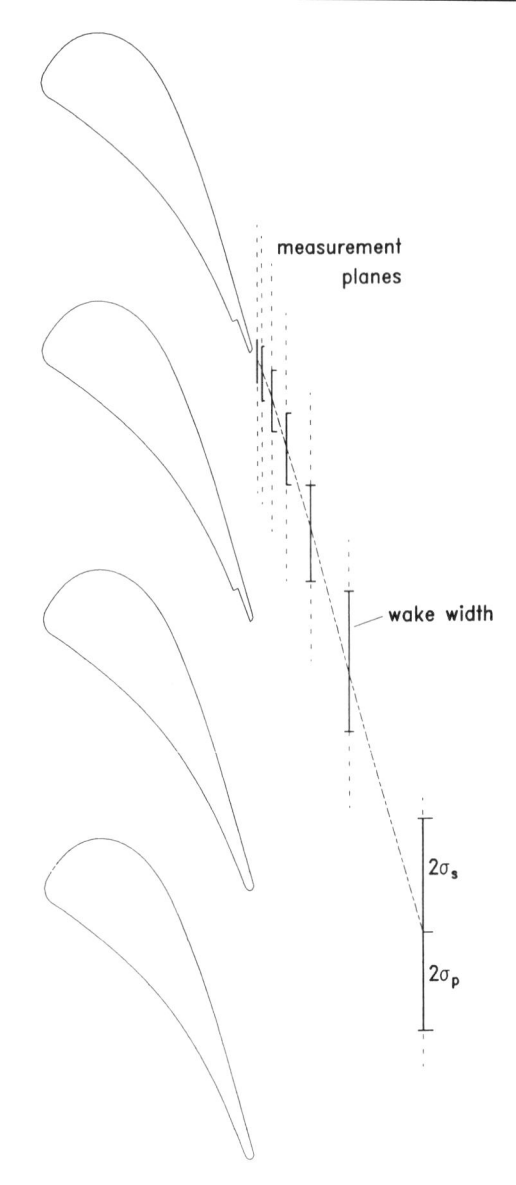

Figure 3.59 Total pressure wake minima and wake width behind the PS-blade with CO_2-ejection.

Table 3.2 Comparison of ejected (ej.) and integrated (int.) mass flow of CO_2 in case of TE-ejection at $M_{is} = 1.05$ and $c_m = 3\%$

x (mm)	c_m (%)	Ej.mass (g/s)	Int.mass (ch.1, g/s)	Ratio (%)	Int.mass (ch.2, g/s)	Ratio (%)
35	3.08	11.45	11.38	99.4	11.17	97.5
20	3.08	11.45	11.30	98.7	11.13	97.2
12	3.11	11.57	10.97	94.8	10.78	93.1
7	3.13	11.42	9.94	87.1	9.81	85.9

Straight cascade tests

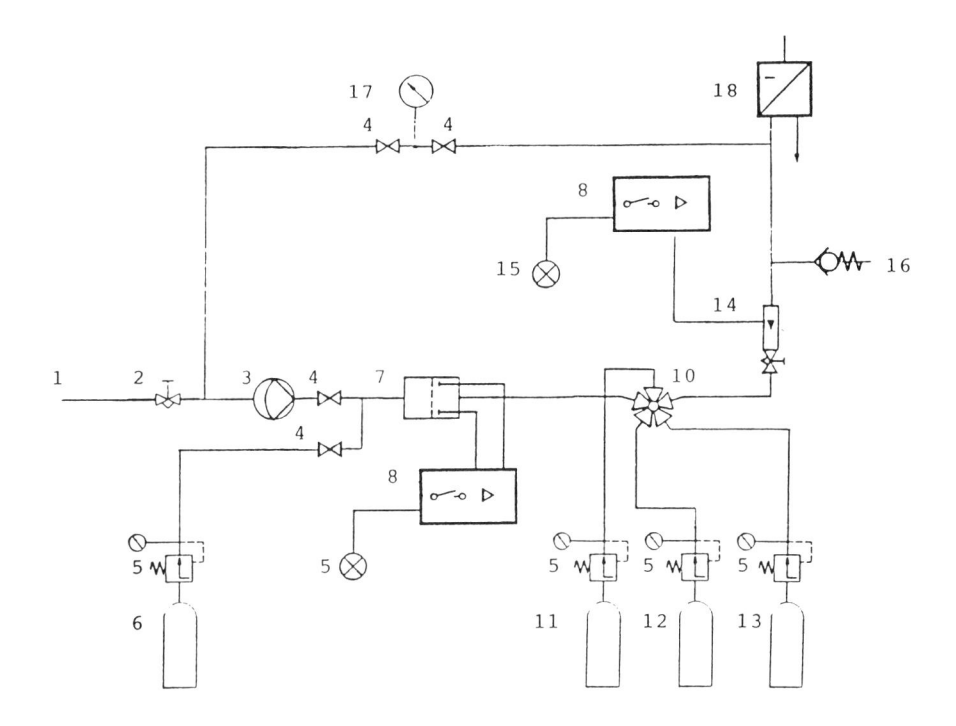

Figure 3.60 Layout of carbon dioxide measurement system.

1. gas sample tube
2. needle valve
3. vacuum pump
4. valve
5. pressure reducer
6. test gas bottle containing 4.75 vol. % CO_2
7. condensation sensor/filter
8. amplifer/relay
9. warning lamp, indicating condensation
10. multiway valve
11. calibration gas bottle containing N_2
12. calibration gas bottle containing 97.5 vol. % CO_2
13. calibration gas bottle containing 9.00 vol. % CO_2
14. flowmeter
15. warning lamp, indicating low gas flux
16. safety valve, saving the analyzer from overpressure
17. pressure gauge
18. infrared analyzer, measuring CO_2 volume concentration

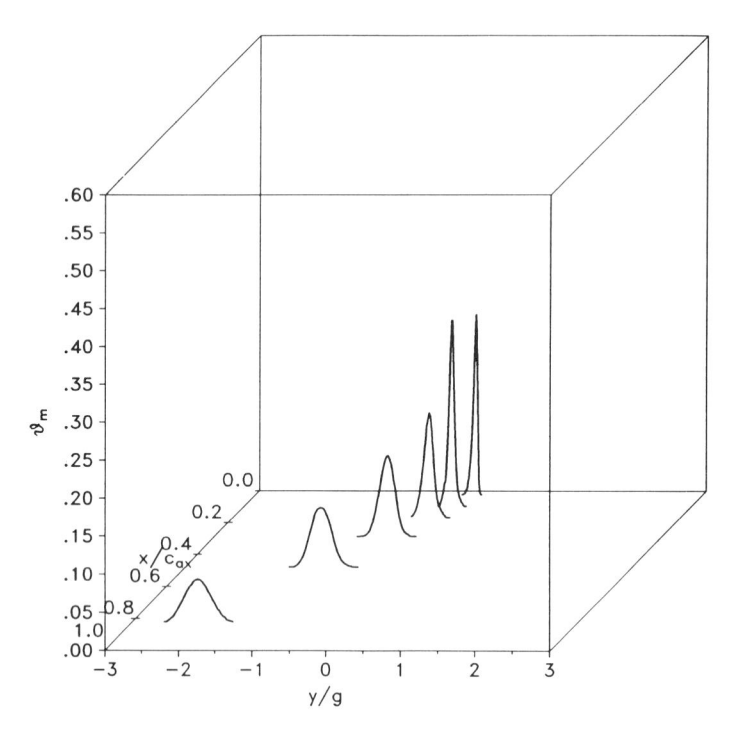

Figure 3.61 Measured CO_2-concentration profiles in the wake of the blade with TE-ejection at $M_{2,is} = 1.05$ and $c_m = 3\%$.

In statistics, the Gaussian curve gives a theoretical probability density distribution. It is given by the normal distribution function defined by

$$\vartheta(y) = \frac{1}{\sqrt{(2\pi\sigma)}} \exp\left(-\frac{(y - \bar{y})^2}{2\sigma^2}\right) \qquad (3.31)$$

In terms of statistics, $\vartheta(y)$ is the probability for the variable to take the value y, $\bar{y}$ being the average value of y and σ the standard deviation. The lesser the standard deviation, the more the data is concentrated around the mean value which results in a narrow distribution with a big amplitude. In the present case

Table 3.3 Comparison of ejected (ej.) and integrated (int.) mass flow of CO_2 in case of TE-ejection at $M_{is} = 0.80$ and $c_m = 3\%$

x (mm)	c_m (%)	Ej.mass (g/s)	Int.mass (ch.1, g/s)	Ratio (%)	Int.mass (ch.2, g/s)	Ratio (%)
35	3.04	11.52	11.08	96.2	10.57	91.8
20	3.04	11.52	12.44	108.0	11.90	103.3
12	3.04	11.52	12.76	110.7	12.36	107.3
7	3.04	11.52	11.73	101.8	11.40	99.0

Straight cascade tests 227

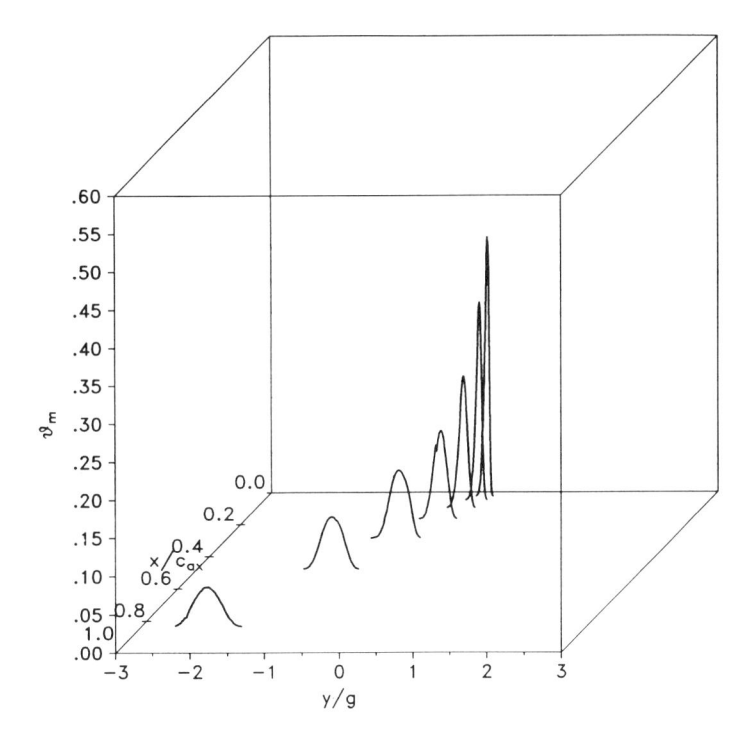

Figure 3.62 Measured CO_2-concentration profiles in the wake of the blade with PS-ejection at $M_{2,is} = 1.05$ and $c_m = 3\%$.

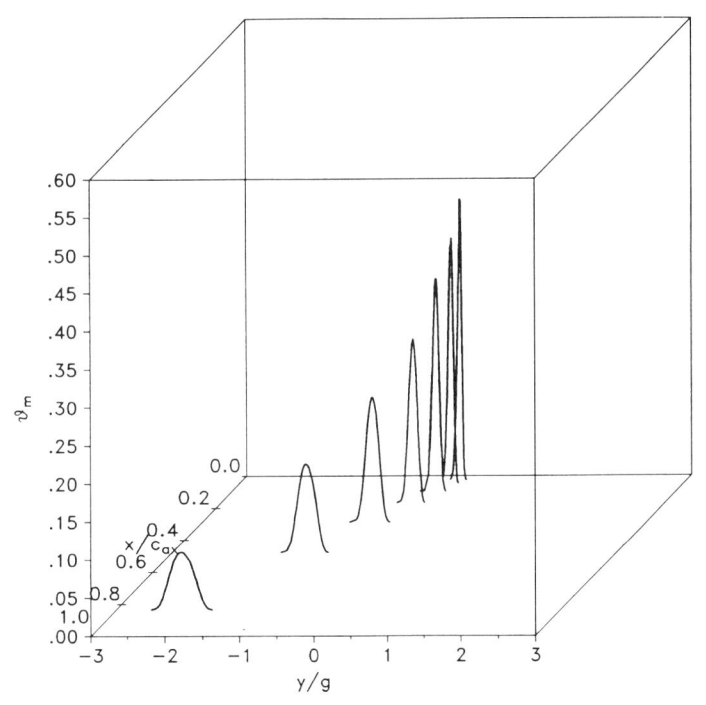

Figure 3.63 Measured CO_2-concentration profiles in the wake of the blade with TE-ejection at $M_{2,is} = 0.80$ and $c_m = 3\%$.

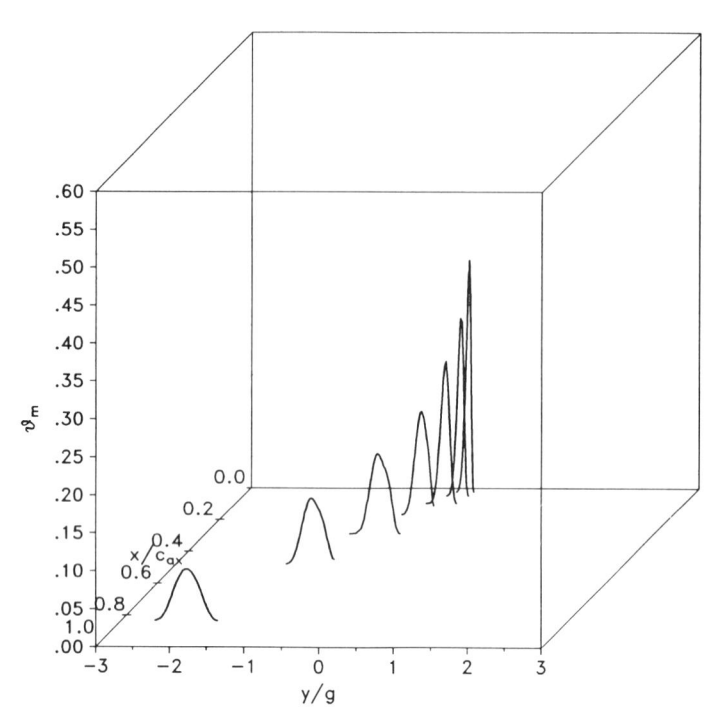

Figure 3.64 Measured CO_2-concentration profiles in the wake of the blade with PS-ejection at $M_{2,is} = 0.80$ and $c_m = 3\%$.

y is the pitchwise direction, $\vartheta(y)$ is the CO_2 concentration and σ characterizes the shape of the distribution. A big σ indicates a large and flat distribution, a low σ value and a narrow and deep distribution. Plotting σ as a function of the axial distance quantifies the axial evolution of the concentration.

A better fit can be obtained taking two Gaussians: one for the pressure side and one for the suction side of the wake.

The Gaussian distribution has several interesting properties. Integrating the normal function between $-\infty$ and $+\infty$ will result in 1 whatever σ and $\bar{y}$ are. To fit a wake profile with a Gaussian, $\vartheta(y)$ has to be multiplied by A, the area under the curve $f(y) = \vartheta$. The distribution functions are then

$$\vartheta_{ps} = \frac{A_{ps}}{\sqrt{(2\pi\sigma_{ps})}} \exp\left(-\frac{(y-\bar{y})^2}{2\sigma_{ps}^2}\right) \quad (3.32)$$

and

$$\vartheta_{ss} = \frac{A_{ss}}{\sqrt{(2\pi\sigma_{ss})}} \exp\left(-\frac{(y-\bar{y})^2}{2\sigma_{ss}^2}\right) \quad (3.33)$$

for the pressure and suction sides of the wake, respectively.

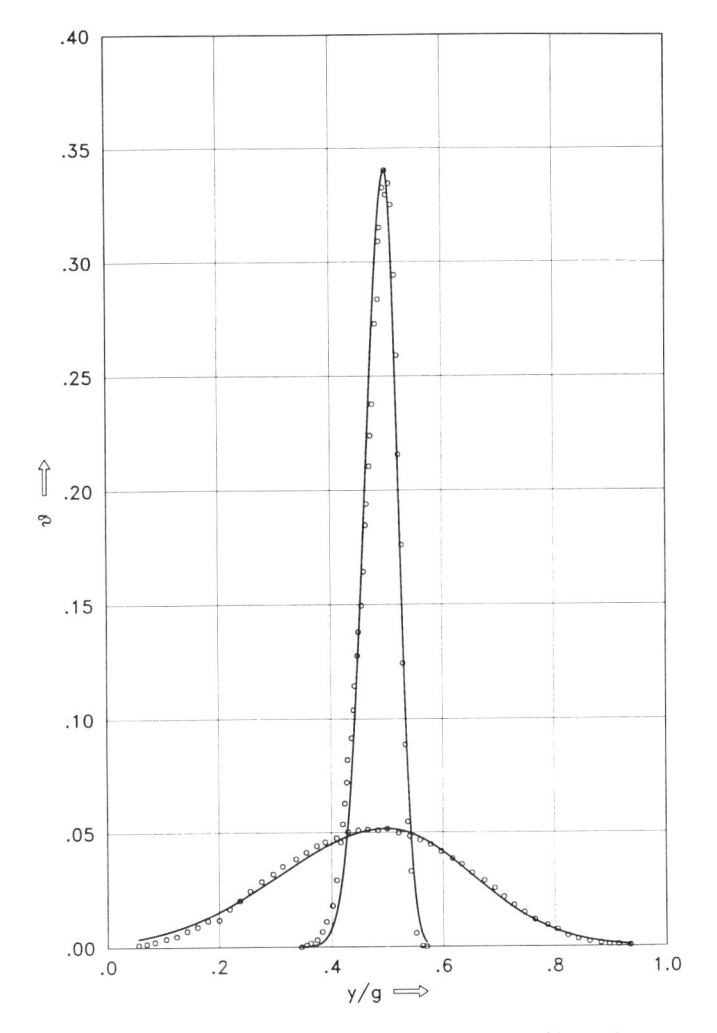

Figure 3.65 Comparison of measured CO_2-concentration profile with asymmetrical Gaussian curve adapted to it for the wake of a PS-blade at $M_{2,is} = 1.05$, $c_m = 3\%$ at measurement positions $x/c_{ax} = 0.023$ and $x/c_{ax} = 0.812$.

The abscissa of the curves is parallel to the cascade and not, as would seem obvious, normal to the flow downstream of the blade. The transformation required to obtain the preferable orientation would have resulted in loss of information due to the large angle of rotation and the restricted number of measurement curves. The Gaussian curves fit reasonably well at axial distances from the cascade exit plane equal to or larger than 12 mm, i.e. $x/c_{ax} \geq 0.279$; examples are presented in Figure 3.65.

Figure 3.66 shows the maximum carbon dioxide concentration $\vartheta_{m,max}$ for $M_{2,is} = 0.80$, resp. $M_{2,is} = 1.05$. The odd value of $\vartheta_{m,max}$ at $x/c_{ax} = 0.023$ in

Figure 3.66 at $M_{2,is} = 1.05$ is caused by the analysing system, which was operated at the limit of its momentary possibilities, especially due to a comparatively large pressure difference between the wake area just behind the blade at $M_{2,is} = 1.05$ and the atmosphere. The limitations are due to the capacity of the vacuum pump and the compromise between the dimensions of the gas sample probe, which has to be small to not disturb the flow, and the required gas sample time to get reliable measurement data.

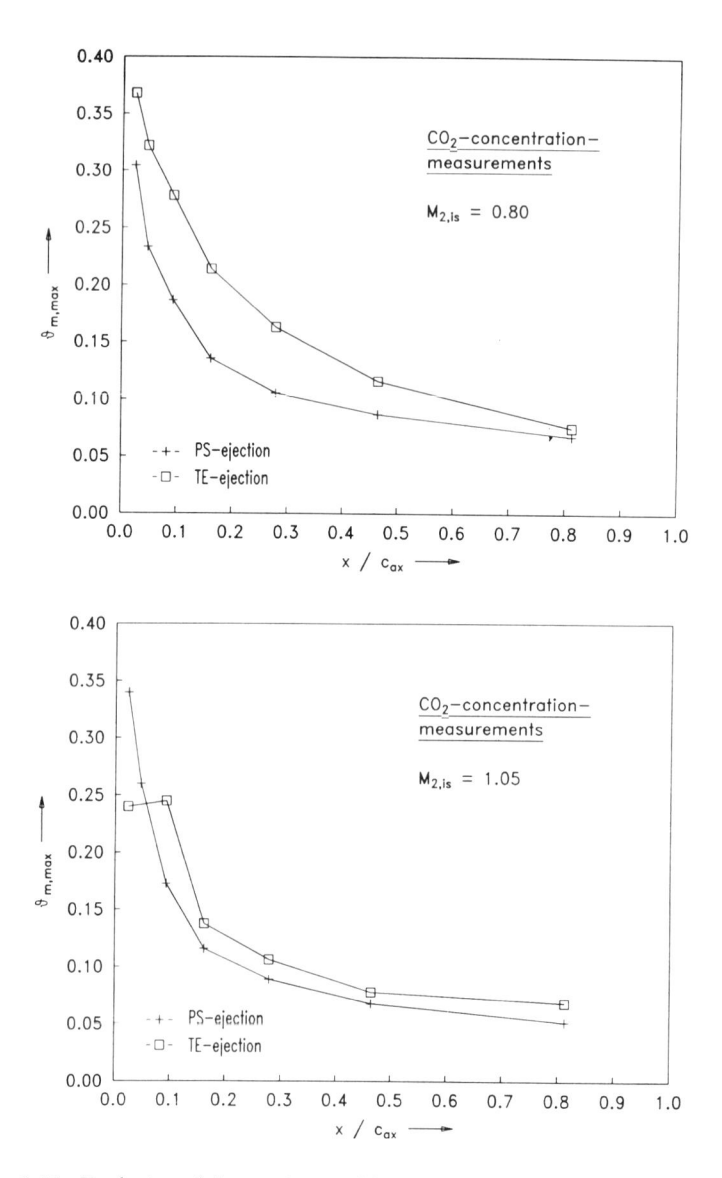

Figure 3.66 Evolution of the maximum CO_2-concentration in the wake at $c_m = 3\%$.

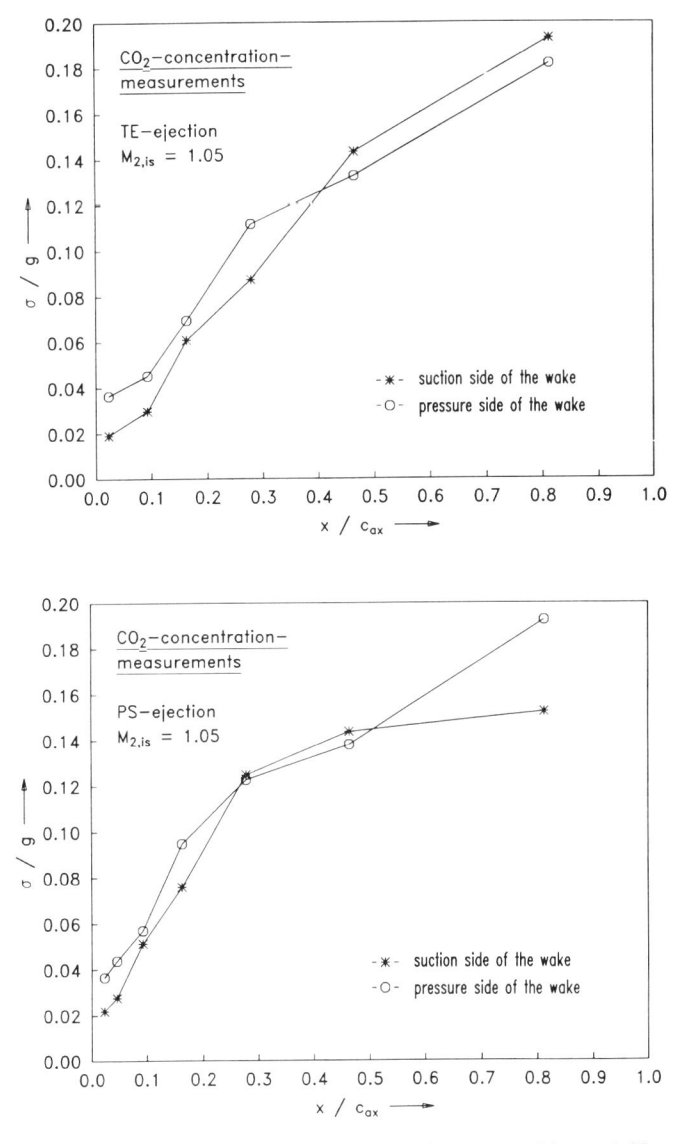

Figure 3.67 Width of the CO_2-concentration distributions at $M_{2,is} = 1.05$ and $c_m = 3\%$.

The width of the wake, indicated by σ_{ps}/c and σ_{ss}/c, is displayed in Figures 3.67 and 3.68.

Figures 3.67 and 3.68 demonstrate the continuous spread of carbon dioxide in the lateral direction with increasing distance to the trailing edge. This process is accompanied by a reduction of the maximum concentration, shown in Figure 3.66.

The fact that the ejection slot of the blade with PS-ejection is shifted in the

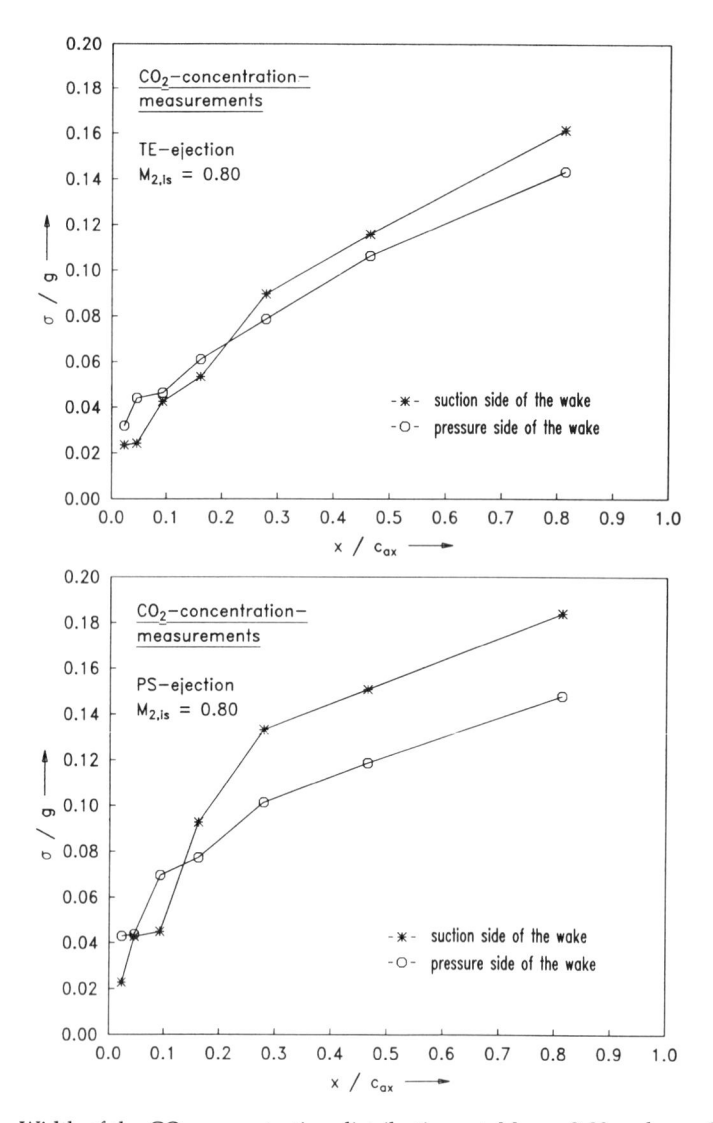

Figure 3.68 Width of the CO_2-concentration distributions at $M_{2,is} = 0.80$ and $c_m = 3\%$.

upstream direction with respect to the slot position of the TE-blade configuration (the difference amounts to $x/c_{ax} = 0.049$) leads to the expectation that at identical probe positions the mixing process might have further progressed in the case of pressure-side ejection. The maximum concentration curves in Figure 3.66 confirm this assumption, although the support from the side of the σ-curves is questionable.

A procedure similar to that for the concentration measurements was carried out for the measurements of the total pressure in the wake.

The total pressure ratio at $M_{2,is} = 1.05$ as measured with the wedge-type probe at distances $x/c_{ax} \geqslant 0.049$ and as measured with a pitot-tube at smaller

Straight cascade tests

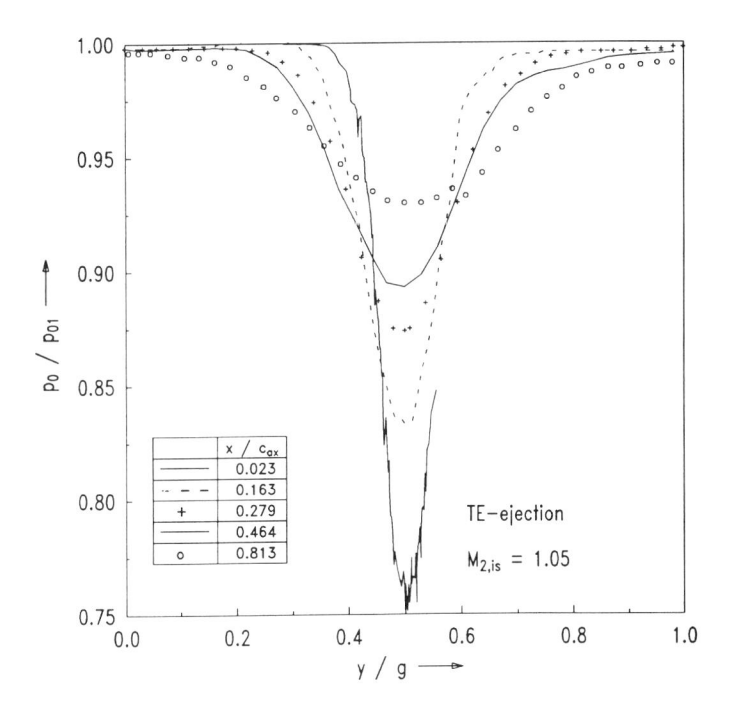

Figure 3.69 Total pressure measurements in the wake of the TE-blade with CO_2-ejection.

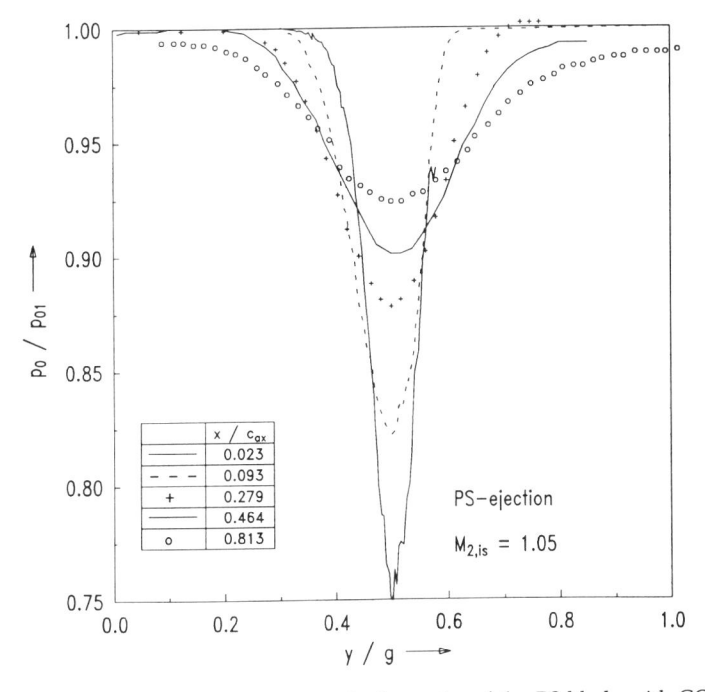

Figure 3.70 Total pressure measurements in the wake of the PS-blade with CO_2-ejection.

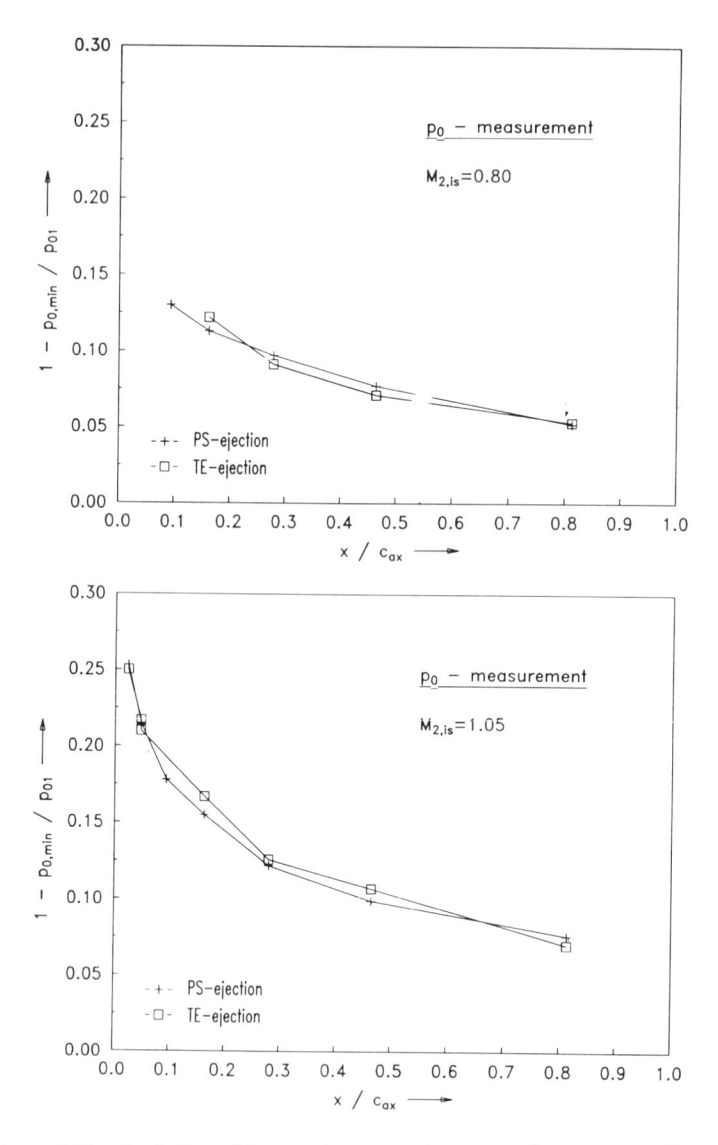

Figure 3.71 Evolution of the maximum total pressure loss at $c_m = 3\%$ (CO_2).

distances is presented in Figures 3.69 and 3.70 for the TE-blade, resp. the PS-blade.

The best-fitting asymmetrical Gaussian curves through these measured data resulted in a maximum of the total pressure loss and wake width on the suction and pressure sides for every measurement plane. Figure 3.71 shows the maximum total pressure loss for $M_{2,is} = 0.80$ and for $M_{2,is} = 1.05$. The wake widths at $M_{2,is} = 0.80$ and at $M_{2,is} = 1.05$ are drawn in Figures 3.72 and 3.73 respectively.

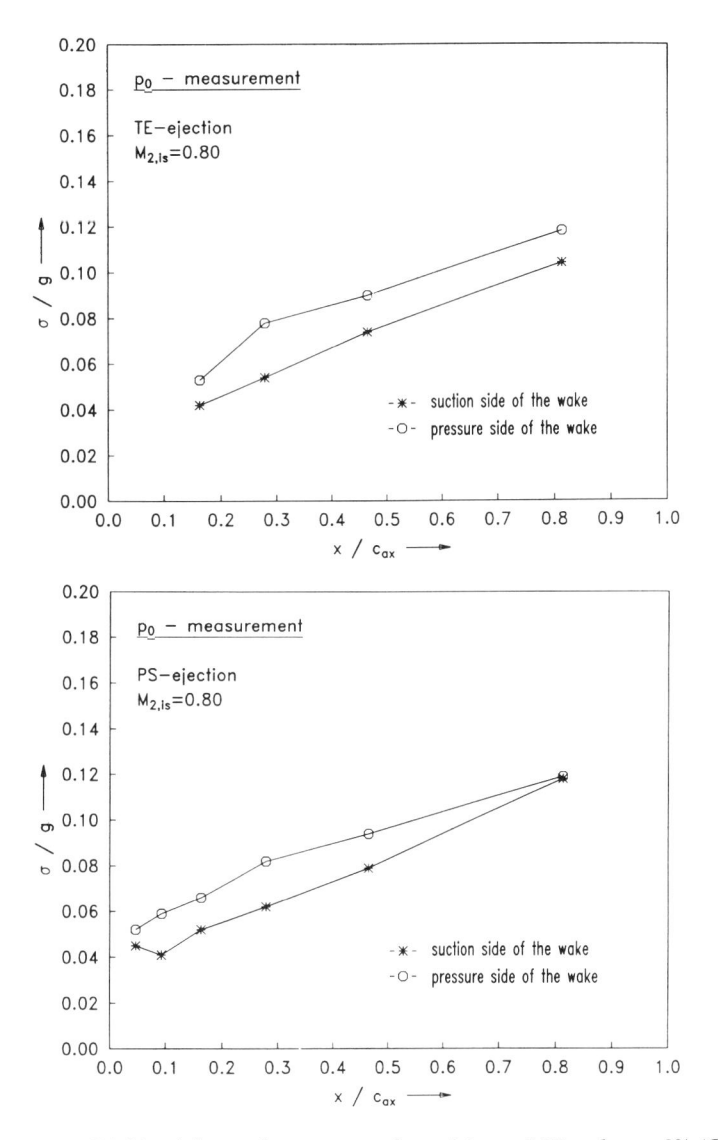

Figure 3.72 Width of the total pressure wake at $M_{2,is} = 0.80$ and $c_m = 3\%$ (CO_2).

6 Annular cascade tests

6.1 Testing program

The cascade configuration selected for the annular cascade tests is the nozzle guide vane with coolant flow ejection through a slot on the rear pressure side (PS-ejection blade). Tests are run at design Mach number and Reynolds

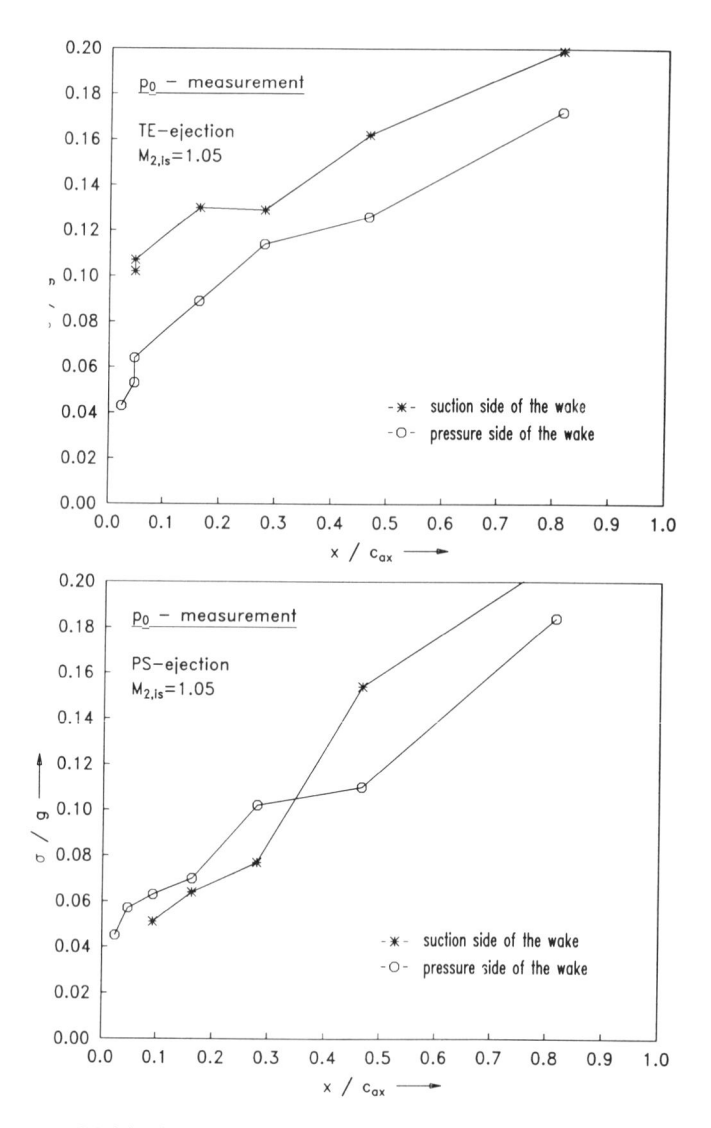

Figure 3.73 Width of the total pressure wake at $M_{2,is} = 1.05$ and $c_m = 3\%$ (CO_2).

number and nominal coolant flow rate. The test conditions are summarized below:

- isentropic Mach numer at mid-span $\quad M_{2,is} = 1.05$
- Reynolds number based on chord
  length and outlet velocity $\quad RE = 10^6$
- degree of turbulence $\quad Tu = 1\%$
- coolant flow rate $\quad \dot{m}c/\dot{m}g = 3\%$
- inlet total pressure $\quad P_{01} = 1.62$ bar
- inlet total temperature $\quad T_{01} = 440$ K

- coolant flow temperature $\qquad T_{oc} = 287$ K
- blade temperature $\qquad T_{bl} = 287$ K

The coolant flow to main flow temperature ratio is $T_{oc}/T_{og} = 0.65$. Ideally this ratio should have been close to 0.5 to correctly simulate the conditions in advanced turbo engines. However, this would have required to cool the coolant air down to ≈ 220 K which would have caused considerably more complications in the experimental setup and longer testing times. Due to the fundamental character of this study, this was not deemed to be necessary.

The bulk of the experimental data consists of loss and angle measurements at three axial planes $X/C_{ax} = 0.25$, 0.54 and 1.0 over the entire blade height. The measurements are obtained with combined total-directional probes using the stationary measurement system (see Section 4.2.3 and Figure 3.31). These aerodynamic data are completed by the measurement of the thermal wake profile at selected radial and axial positions. Turbulence data are measured at mid-span at $X/C_{ax} = 0.25$.

The measurements in rotation are limited to fast response total pressure measurements at mid-span at $X/C_{ax} = 0.25$. A long testing program was carried out for measuring also the time-varying relative total temperature with the high-speed rotating probe system using the double hot-wire aspirating probe by Ng [3.10]. The research effort concentrated on the minimization of the probe dimensions and the development of appropriate calibration techniques. The aerothermal characteristics of the probe proved to be much more complex than expected and brought this part of the project to a temporary deadlock. Near the end of the project period it was found that the cause was most probably due to unexpected heat conduction effects between hot wires and probe head. This finding came too late to be exploited for this project but it shows the direction to be taken for future developments of appropriate techniques.

6.2 General flow conditions

6.2.1 Upstream flow conditions

Due to their particular operating mode, compression tube cascade facilities are prone to small total pressure fluctuations, Figure 3.74. They are caused by slight oscillations of the piston and the compressed mass of air upon the sudden release of the air through the vent hole in the compression tube. The amplitude of these quasi-sinusoidal variations decreases rapidly with time. The mean pressure variation over the time period in which the wake traverses are made, typically between $t = 300$ and $t = 500$ ms, is of the order of $\pm 0.7\%$, which is remarkably small for this type of facility.

As far as the total temperature evolution is concerned, it remains nearly constant (within 3°) during the useful running time, Figure 3.74. The temperature is measured with a K-type bare thermocouple probe with 0.2 mm wire diameter.

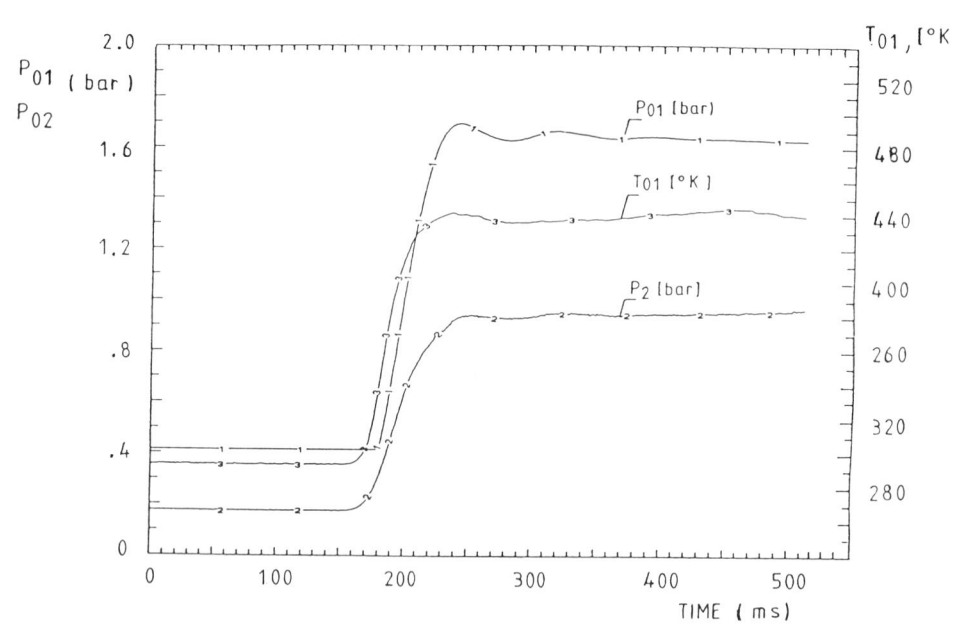

Figure 3.74 Variation of overall flow conditions for typical test run in the VKI Compression Tube Annular Cascade Facility.

Figure 3.75 shows the radial variation of the inlet total pressure as measured with a total pressure rake, each pressure tube being connected to a separate Sensym absolute pressure transducer. The boundary-layer thickness is very small on both end walls—around 5% of the blade height. At the hub the first measurement point is at 0.5 mm wall distance. The ratio of the total pressure at this point to the free-stream total pressure is $P_{ol}/P_{01} = 0.993$.

The inlet flow angle is not measured, but considering the construction of the tunnel with its axisymmetric settling chamber preceding the test section, it can reasonably be assumed that the inlet flow direction is axial.

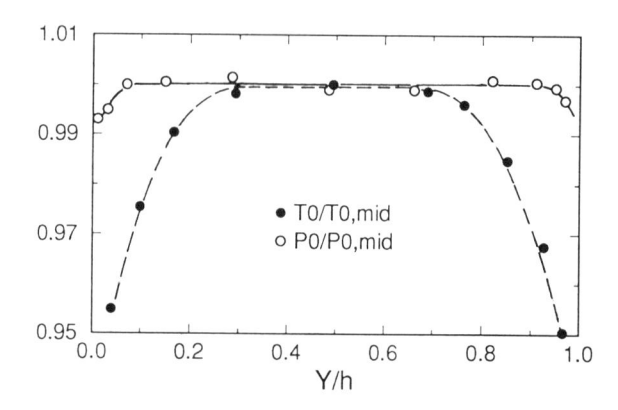

Figure 3.75 Spanwise variation of upstream total pressure and temperature.

The turbulence at mid-span, measured with a hot-wire probe, is $Tu \approx$ 1.6–2%.

6.2.2 Outlet flow conditions

The opening of the downstream throttling valve is regulated for setting the mid-span isentropic Mach number at the axial plane $X/C_{ax} = 0.54$ at the design Mach number $M_{2,is} = 1.05$. However, the static pressure at mid-span is not measured directly. It is derived by linear interpolation between the pitch-averaged Mach numbers at hub and tip end wall. Table 3.4 compares the experimental conditions with those predicted with a three-dimensional Euler code (see Section 2.5.).

The Mach number differences between hub and tip agree closely between predictions and experiments.

The good periodicity of the circumferential isentropic Mach number distribution is demonstrated in Figure 3.76 for the hub flow conditions in the measurement plane $X/C_{ax} = 0.54$. These data were taken in an early stage of the test program at an outlet Mach number slightly lower than the nominal conditions, i.e. $M_{2,is,Hub} = 1.10$ instead of 1.12. Figure 3.76 is representative for the flow conditions in all axial planes at both hub and tip end walls.

Although only qualitative in nature, the oil flow visualization of the blade suction side flow in Figure 3.77, taken at design Mach number and blowing

Table 3.4 Comparison between three-dimensional Euler code and experiment

	$M_{2,is} = f(P_2/P_{01})$ 3D Euler code	Experiments
hub	1.13	1.12
mean	1.05	1.05
tip	0.987	0.98

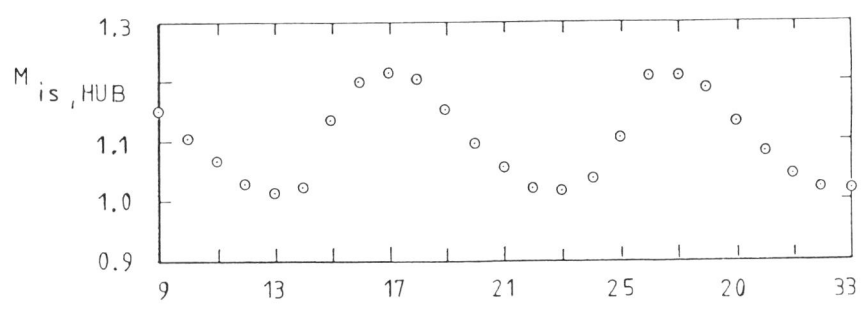

Figure 3.76 Downstream hub end-wall isentropic Mach number distribution.

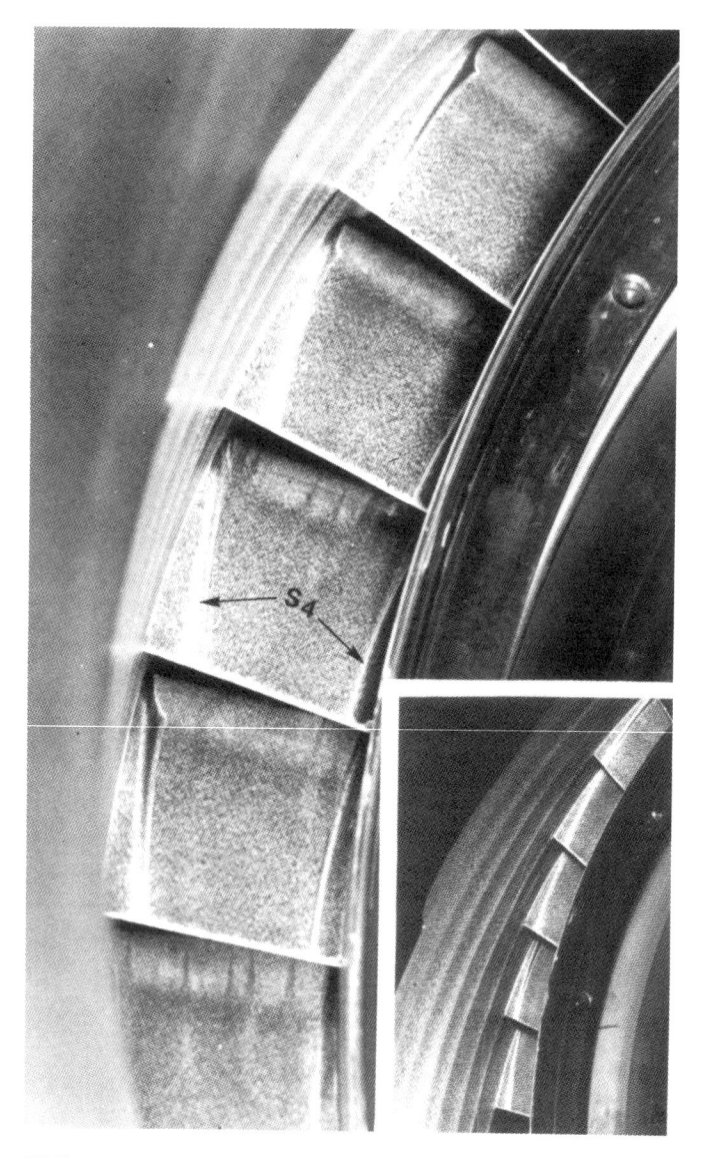

Figure 3.77 Oil-flow visualization of nominal outlet Mach number and coolant flow ejection.

rate, allows some conclusions to be drawn as regards the outlet flow conditions:

(a) the main flow remains attached up to the trailing edge, indicating low blade losses;
(b) the radial pressure gradient significantly affects the spanwise extension of the secondary flow regions at hub and tip as witnessed by the corresponding passage vortex lift off lines S4; the accumulation of white paint along

the line S4 results from the interaction of the end wall boundary-layer cross-flow, driven by the passage vortex, and the blade suction side flow*;

(c) there appear to exist some slight differences in the spanwise position of the tip S4 line at the trailing edge, over several blades.

6.2.3 Coolant flow

The spanwise distribution of the coolant flow total pressure P_{oc} at the trailing edge coolant slot exit was measured with five flattened total pressure tubes immersed 2–3 mm into the slot. The measurements were taken at nominal coolant mass flow conditions but in absence of the main flow. Figure 3.78 shows the pressure distribution for two blades which differ by their position with respect to the coolant air admission tubes feeding the coolant air into the annular chamber on top of the blades. The maximum pressure difference between the two measurement series is $\Delta P/P_{01} = 0.02$. One observes also a slight spanwise variation of the pressure, with slightly higher pressures near the hub region ($\approx 3\%$). In presence of the main flow, however, the coolant flow total pressure will change somewhat, because of the different static pressure conditions at the trailing edge. The trailing-edge static pressure is in fact lower than the atmospheric pressure conditions for the measurements in Figure 3.78. In addition the coolant mass flow will increase from tip to hub due to the important decrease of the static pressure from tip to hub resulting in much higher coolant jet velocities at the hub.

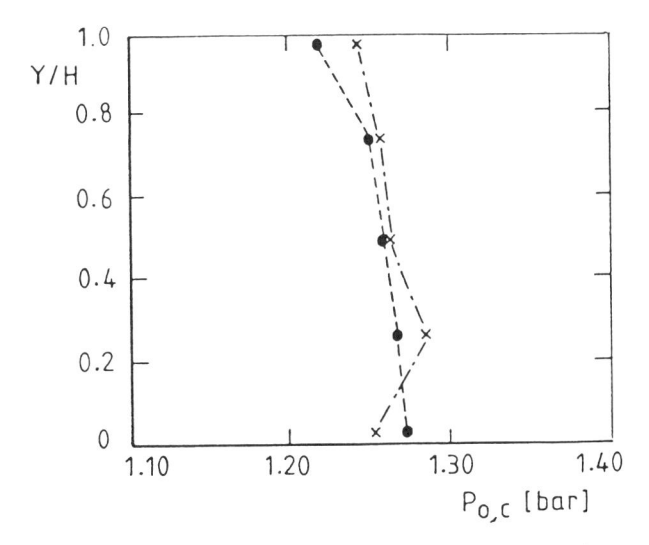

Figure 3.78 Total pressure distribution along coolant slot.

*Terminology adapted from Sieverding [3.11].

6.3 Stationary measurements

6.3.1 Pressure probes

Types of probes The stationary measurement system is described in detail in Section 4.2.3. At present we shall discuss the design of the pressure probes. The main constraints for the probe are

(a) a sufficient short response time (see Section 4.2.3);

(b) the use of the probes in strong pressure-gradient fields;

(c) the minimization of the probe-blockage effect,

Point (a) determines the minimum dimensions of the probe head, point (b) determines to a large extent the arrangement of the differential tubes for the angle measurements, point (c) affects in particular the probe-stem size and the distance between probe stem and probe nose.

In Section 4.2.3 it was concluded that the inner diameter of the metal tubes for the pressure probes should be at least 0.8 mm. Hence, the diameter of the probe head of a five-hole probe would be typically 3 mm.

Figure 3.79 shows various three-hole probe configurations for the simultaneous measurement of the total pressure and the yaw angle (flow angle in blade-to-blade plane). The alignment of the pressure tubes in a plane parallel to the wake profile for probe type A leads inevitably to large local angle errors during the wake traverse, if the distance between the two differential pressure tubes is significant compared to the wake width. Calculations showed that traversing the mid-span wake in the axial plane $X/C_{ax} = 0.25$ with probe A, made of three tubes each of 1 mm diameter, would have resulted in local errors of up to $\pm 1.8°$. The traversing of a shock wave would of course entail even

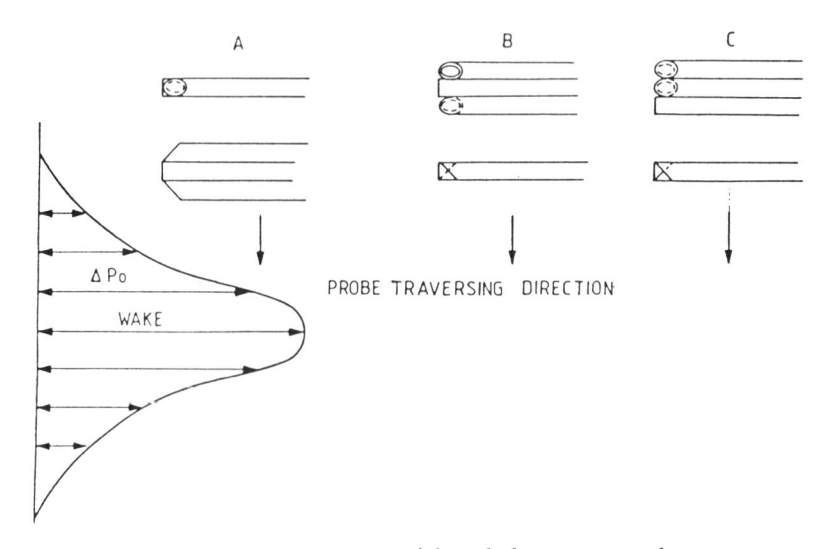

Figure 3.79 Various types of three-hole pressure probes.

bigger errors. On the contrary, probe type B with the tubes arranged in a plane normal to the wake profile avoids these errors to a large extend.

In five-hole probes, the arrangement of the differential pressure tubes for the yaw angle measurement is equivalent to probe type A and hence the same comments apply. Therefore, the simultaneous measurement of total pressure, static pressure and flow angles is excluded. Rather than using a separate probe for the static pressure, it was preferred to derive it by linear interpolation between hub and tip end-wall pressure measurements. The linearity of the spanwise pressure distribution was checked by three-dimensional Euler calculations.

The use of type B probe heads for the yaw angle measurements makes it impossible to also measure with the same probe the radial flow angle which is needed to correct the yaw angle measurements for any radial flow angle effects. Three-dimensional NS calculations showed, however, that the radial flow angle varies at most by 3 to 4 degrees. Calibration tests showed that the yaw angle probe was insensitive to such angle variations.

The use of type B three-hole probes, however, is also problematic near the end walls, where the spanwise total pressure gradients associated with end-wall boundary layers and secondary flows are even bigger than the pitchwise gradients of the wake profiles. This is evident from the carpet plots for the blade losses in Figure 3.87. Calculations showed that angle measurements with probe B in the region of the secondary loss peak at the hub would be in error by as much as $\pm 2°$ and angle measurements within the end-wall boundary layer should be disregarded altogether.

Type C probe heads are better suited in the end-wall regions because of the reduced distance between the differential pressure tubes. With this probe the maximum local error in the region of the secondary loss peak drops to $\pm 1°$; however, within the end-wall boundary layers the errors are still high and angle measurements for wall distances $y < 2.5$ mm (5% of blade height) are not to be trusted.

The wake traverses between 15% and 85% blade height were carried out with probes of type A; for traverses closer to the walls use was made of probes of type C.

Probe blockage effect The best way of evaluating the importance of the probe blockage effect on the downstream measurements is to compare the end-wall static pressure distributions in the measurement plane in the absence and in the presence of the probe. Ideally the probe stem should be so small and far removed that the flow in the measurement plane is not at all effected by its presence. This appears to be unrealistic in transonic annular cascades because it implies either unreasonably long probe heads or too small probe stem dimensions, either of both leading to undesirable probe vibrations. A less stringent, but sufficient requirement is to ensure that the end-wall static pressure at the point of measurement and upstream of it is not affected, i.e., the wall pressure tappings to the right of the probe nose in Figure 3.80, while the flow downstream of the probe nose, i.e., to the left of it, may by disturbed.

The standard VKI wake traverse probes are of type B in Figure 3.79, with a

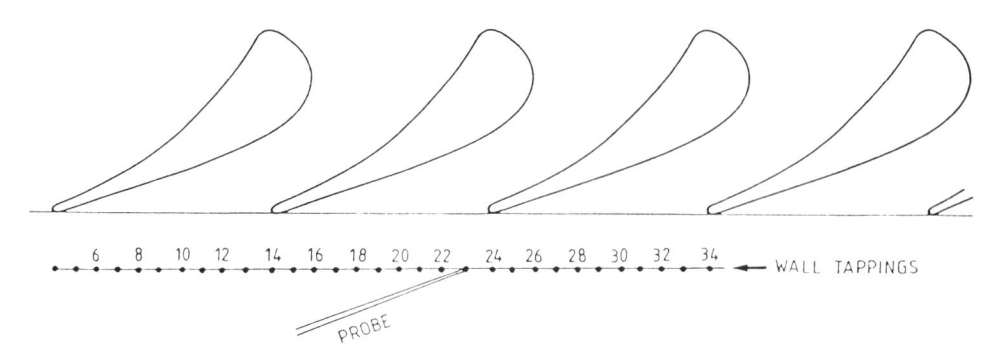

Figure 3.80 Probe in downstream measurement plane.

probe head length of 35 mm and a probe stem width of 2.2 mm. In the case of the BRITE cascade the ratio of probe stem width to cascade throat width amounts to 15%. It is reminded that the blockage effect is made independent from the probe immersion depth by putting behind the movable probe stem a fixed cylinder which spans the entire flow channel height (see Section 4.2.3). Measurements with the standard VKI probe at $X/C_{ax} = 0.54$ distance from the trailing edge demonstrated an unacceptably high blockage effect. The isentropic Mach numbers, based on the hub end-wall static pressures at the position of the probe nose (probe head at mid-span height), were low by 5–10% compared to measurements in absence of the probe. Although the effect on the blade flow is probably much smaller, one cannot expect reliable blade loss measurements under such conditions.

After several vain attempts to modify the standard probe for reduced blockage effect, an entirely new probe was designed, Figure 3.81. The end of the rectangular-shaped probe stem (7 × 3.3 mm) is bent laterally at 90°. A 110 mm probe head is attached to this transverse piece. A cross-arm provides the required stiffness for the probe head. The probe head is curved to align it with the mid-span cylindrical stream surface. Aligning the probe head with the mean flow direction puts the probe stem at more than two pitches away from the probe nose. Figure 3.82 compares the hub end-wall isentropic Mach number distributions at $X/C_{ax} = 0.54$, recorded at two different circumferential positions of the probe, with the Mach number distributions in absence of the probe. The mean Mach number in these tests was slightly higher than for nominal conditions, i.e. 1.175 instead of 1.12. The numbers on the horizontal axis in Figure 3.82 indicate the position of the wall pressure tappings. The spacing of the tappings equals 10% of the pitch. In position A the probe nose is aligned with wall tapping 23. Tappings with a number smaller than 23 indicate a downstream position with respect to the probe nose, while higher numbers indicate an upstream position. For probe position A the first significant influence of the probe stem is felt five tappings downstream of the probe nose, i.e. at 50% of the pitch; for probe position B the influence is felt at four tappings or 40% of the pitch downsteram of the nose. These results demonstrate clearly that the new probe design has overcome the important probe-blockage problem for the performance measurements of the BRITE cascade.

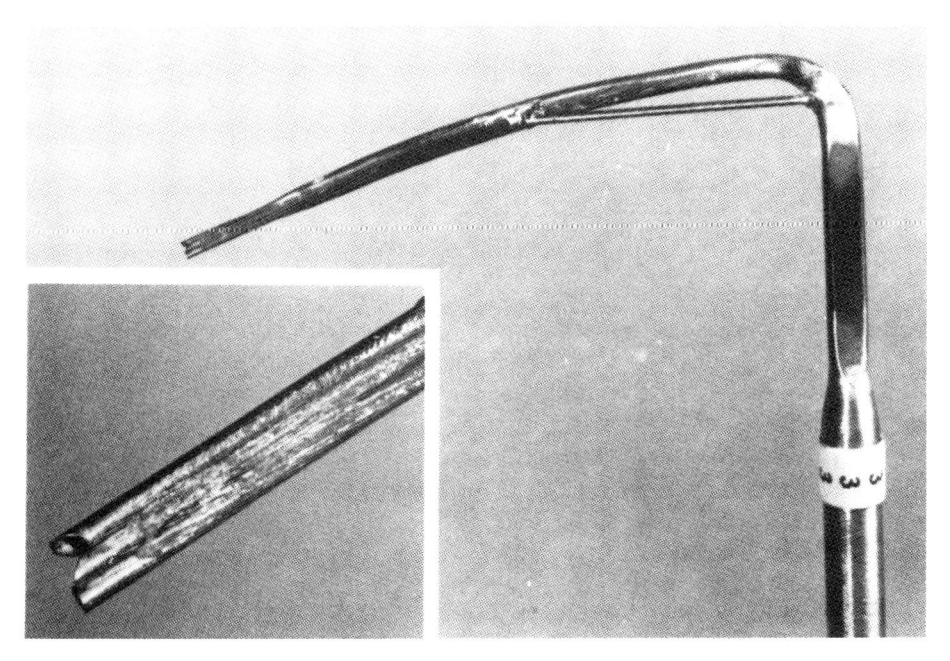

Figure 3.81 New low-blockage pressure probe.

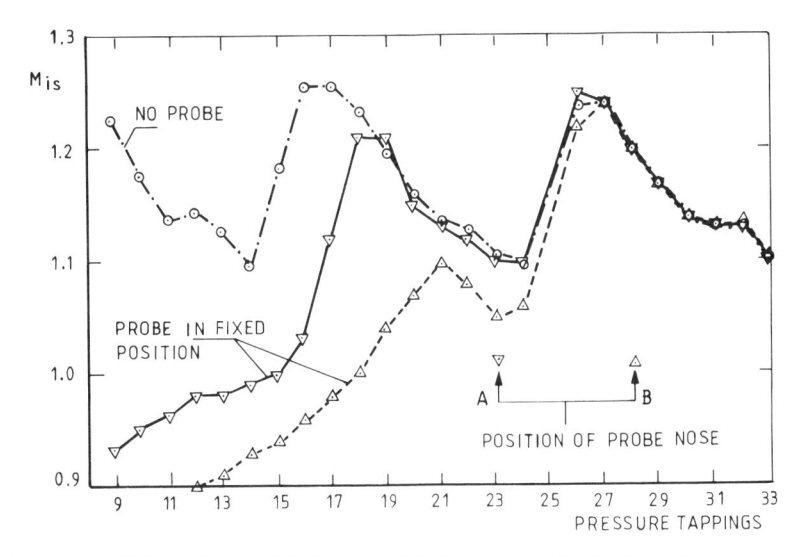

Figure 3.82 Effect of probe blockage on Mach-number distribution at hub end wall.

6.3.2 Performance measurements

Measurement errors, repeatability, non-periodicity The measurement errors are dependent on the linearity and stability of pressure transducers and the accuracy of the calibration procedures. The uncertainty on the blade losses is affected by the individual errors in the measurement of the upstream and downstream total pressures P_{01} and P_{02} and the downstream static pressure P_2. (The influence of temperature gradients in the outlet flow field is not taken in consideration.) The measurement errors for these quantities is estimated to be of the order of $\Delta P/P = 0.2$–0.3%. The uncertainty on the local static pressure may actually be higher because it is derived from a linear interpolation between the static pressures at hub and tip. The uncertainty of the blade kinetic energy loss is evaluated as $\delta\zeta/\zeta = 10\%$.

The uncertainty on the angle measurements is much more difficult to evaluate, because it relies additionally on the calibration of the directional probe over a wide range of Mach numbers and flow angles and on its behaviour in strong gradient fields. In weak pressure gradient fields the flow angle uncertainty is estimated to $\pm 0.8°$, but in high spanwise gradient fields the local errors may be considerably higher, see p. 243.

Repeat tests are also an excellent mean for the assessment of measurement errors. Figure 3.83 compares the blade losses of two test series in plane $X/C_{ax} = 0.25$, separated by several months, with complete dismantling of the probe-traversing mechanism in between. For both test series the losses were averaged over three wakes. Out of the 8 repeat tests, the difference in blade losses at five spanwise positions is less than 5% between the two test series. For the other three positions the difference amounts to ≈ 10–12%. These data corroborate the values of the uncertainty on the losses of $\delta\zeta/\zeta = 10\%$ quoted above.

Non-periodic downstream conditions arise from non-uniform inlet conditions and blade-to-blade variations in pitch, throat, stagger angle and blade surface roughness. The inlet uniformity is disturbed by three struts in the settling chamber placed at 0°, 120° and 240° with respect to the vertical direction in the upper half of the test section. The downstream traverses are made across

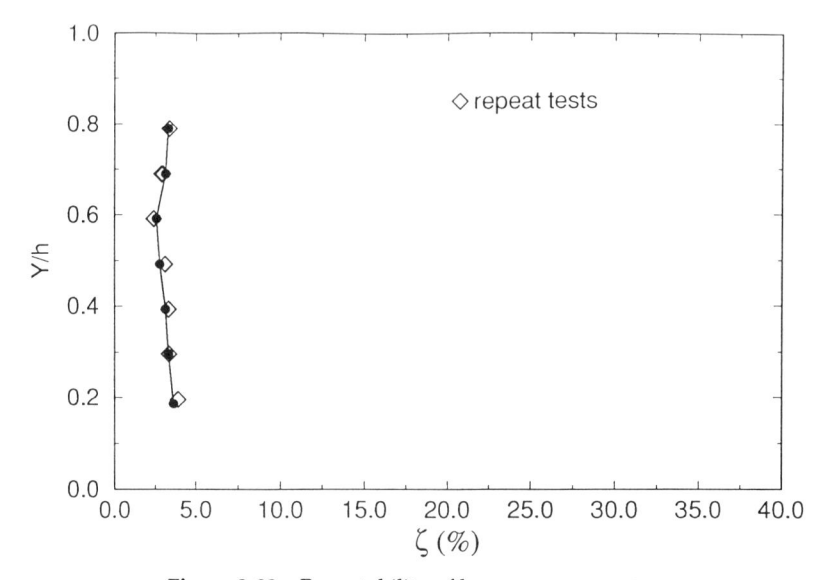

Figure 3.83 Repeatability of loss measurements.

Annular cascade tests

the wakes of the blades in an annular sector extending from 35° to 70° and should not be affected by the wakes of the struts. In any event the total pressure losses induced by the struts are very low since the velocity in the settling chamber is only of the order of 10 m/s. As regards the blade-to-blade variations, the manufacturing accuracies for pitch and stagger angle are respectively 0.1 mm and 0.1°. Nothing can be said about the variation of the blade surface roughness. The blades are manufactured by electro-erosion.

An example of the loss variations over several blades is presented in Figure 3.84. The blade-to-blade variations change, however, along the span and from one axial plane to the other. To evaluate globally the blade-to-blade loss variations, the results of some 60 traverses from all three measurement planes, covering each three wakes, have been examined. For each traverse the loss of each individual blade was compared with the loss value obtained by averaging the losses over three wakes. The differences are summarized in Table 3.5.

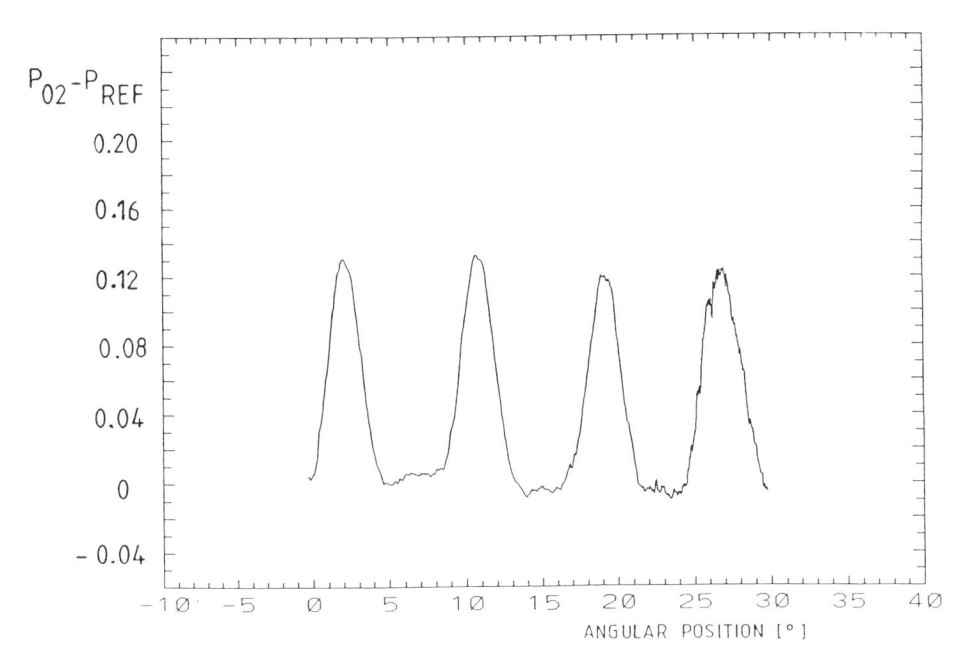

Figure 3.84 Blade-to-blade loss variations.

Table 3.5 Individual losses compared with losses over three wakes

$\dfrac{\zeta_{blade} - \zeta_{3\,blades}}{\zeta_{3\,blades}}$ (%)	Percentage of traverses
0–10	60
10–20	28
>20	12

Note that the highest differences are recorded in the secondary and end-wall boundary-layer regions, where span-wise gradients are predominant.

The differences are surprisingly high, considering the expected 'inherent periodicity' of blade flows through annular cascades. Obviously, minor differences in blade and cascade geometry can considerably affect the boundary-layer development on the blades and end walls.

Presentation of results Blade losses and outlet flow angles are presented in form of carpet plots and spanwise distributions of pitch-averaged values. Except a few, all data are averaged data over three pitches. In view of the comparison with numerical predictions, losses and angles are presented under the form of area-averaged data. However, the blade losses were also calculated under the form of mixed-out values using the conservation equations for mass, momentum and energy between the measurement plane and a plane infinitely far downstream. For all measurement points the difference between area-averaged and mixed-out losses was $\Delta\zeta < 0.2\%$.

The blade losses are expressed by the kinetic loss coefficient

$$\zeta = 1 - \frac{1 - (P_2/P_{02})^{(k-1)/k}}{1 - (P_2/P_{01})^{(k-1)/k}}$$

The differences between upstream and downstream mass flows and the differences between the potential energies of mean and coolant flows have not been taken into account.

Losses The spanwise evolution of the losses is the result of the combined effect of secondary flows, of Mach number variation along the blade height and radial pressure gradient. The variation of the pitch-to-chord ratios g/c along the blade height may also play an effect, but the optimum g/c is Mach number dependent as shown by Dejc and Trojanovskij [3.12]. It turns out that the variation of g/c along the span of ~15% corresponds approximately to the change in the optimum g/c suggested by Dejc for the Mach number variation of $M_{2,is} = 0.98 \rightarrow 1.12$ from tip to hub. Therefore, it appears reasonable to neglect this parameter in the following discussion.

Near the end walls, Mach-number effects are masked by secondary flow effects. As indicated by the oil flow visualizations in Figure 3.77, the extension of the secondary flow zone is larger at the tip than at the hub. At the trailing edge this zone covers about 10% of the span at the hub and 20% of the span at the tip. With increasing distance the secondary flows will continue to expand radially thanks to the continuously growing end-wall boundary layer, the radial pressure gradient and the combined action of passage and trailing edge vortex (sum of trailing filament vortices and trailing edge shed vorticity). The overall aspect of the loss distribution in the planes $X/C_{ax} = 0.25$ and 0.54, Figure 3.85(a,b), follows the typical pattern: at the end wall high losses due to the end-wall boundary layer; at some distance from the end wall a relative loss maximum (secondary loss core); beyond that, a slow decrease down to the level of profile losses. The secondary loss core at the tip is not very prominent and the position of its maximum is not well defined. The secondary loss maximum exceeds the profile losses by hardly 1 point. The maximum moves from about $Y/h = 0.85$ in the axial plane $X/C_{ax} = 0.25$ to $y/h = 0.8$ and 0.75 at the downstream planes $X/C_{ax} = 0.54$ and 1.0, Figure 3.85(d). Contrary to the tip, at the hub one observes a very prominent loss core. Its extension grows from $y/h = 0.15$ at the plane $X/C_{ax} = 0.25$ to $y/h = 0.25$ at $X/C_{ax} = 0.54$. The big difference between hub and tip is explained at least partially through the action of the radial pressure gradient, which causes a radial migration of low momentum boundary layer from tip to hub through the trailing-edge base-flow region. Such a radial migration was already observed by Sieverding et al. [3.13] in a low-speed annular cascade. In the plane $X/C_{ax} = 1.0$, Figure 3.85(c), the loss distribution at the hub takes a parabolic shape, extending from the hub end wall to midspan. Obviously the end-wall boundary growth has become the dominating factor. One of the reasons for the strong change in the hub loss profile may be looked for in the evolution of the static pressure field, see Table 3.6.

Between $X/C_{ax} = 0.25$ and 0.54, the pitch-averaged pressure at the hub remains nearly identical, but between $X/C_{ax} = 0.54$ and 1.0 there is a significant pressure rise. At the tip the pressure change is much smaller over the same axial distance. The reason for the pressure rise at the hub is possibly the upstream influence of the constant pressure field at the exit of the cylindrical flow duct, see Figure 3.31.

The effect of Mach number variation on the spanwise loss distribution is best observed in the measurement planes $X/C_{ax} = 0.25$ and 0.54. Minimum profile losses occur at the spanwise position $y/h = 0.6$. This position of minimum losses is confirmed also by other tests e.g. measurements in plane $X/C_{ax} = 1.0$ and measurements without coolant flow ejection at $X/C_{ax} = 0.54$. At this spanwise position, the mean isentropic Mach number is $\overline{M_{2,is}} = 1.03$, the losses amount to $\zeta = 2.8\%$, Figure 3.85(a). From $y/h = 0.6$ downwards, there is a slow linear increase in losses up to $\zeta = 4\%$ at $y/h = 0.15$. From $y/h = 0.6$ upwards, the losses rise also slowly, but when accounting for some secondary losses, the profile losses will never exceed $\zeta = 3.2\%$. At no place does one observe a sudden strong increase of the losses with Mach number as in the straight cascade tests.

Figure 3.86 presents the blade losses in plane $X/C_{ax} = 0.54$ between 20 and 80% blade height with and without coolant flow ejection. Except at $Y/h = 0.7$ and 0.8, the ejection of coolant flow results in a reduction of the profile losses,

Table 3.6 Mean pressure ratios and isentropic Mach number

Plane	X/C_{ax}	$\overline{P_2/P_{01}}$ Tip	Hub	$\overline{M_{2,is}}$ Tip	Hub
1	0.25	0.550	0.4545	0.967	1.124
2	0.54	0.5426	0.4562	0.977	1.121
3	1.0	0.5345	0.4824	0.987	1.076

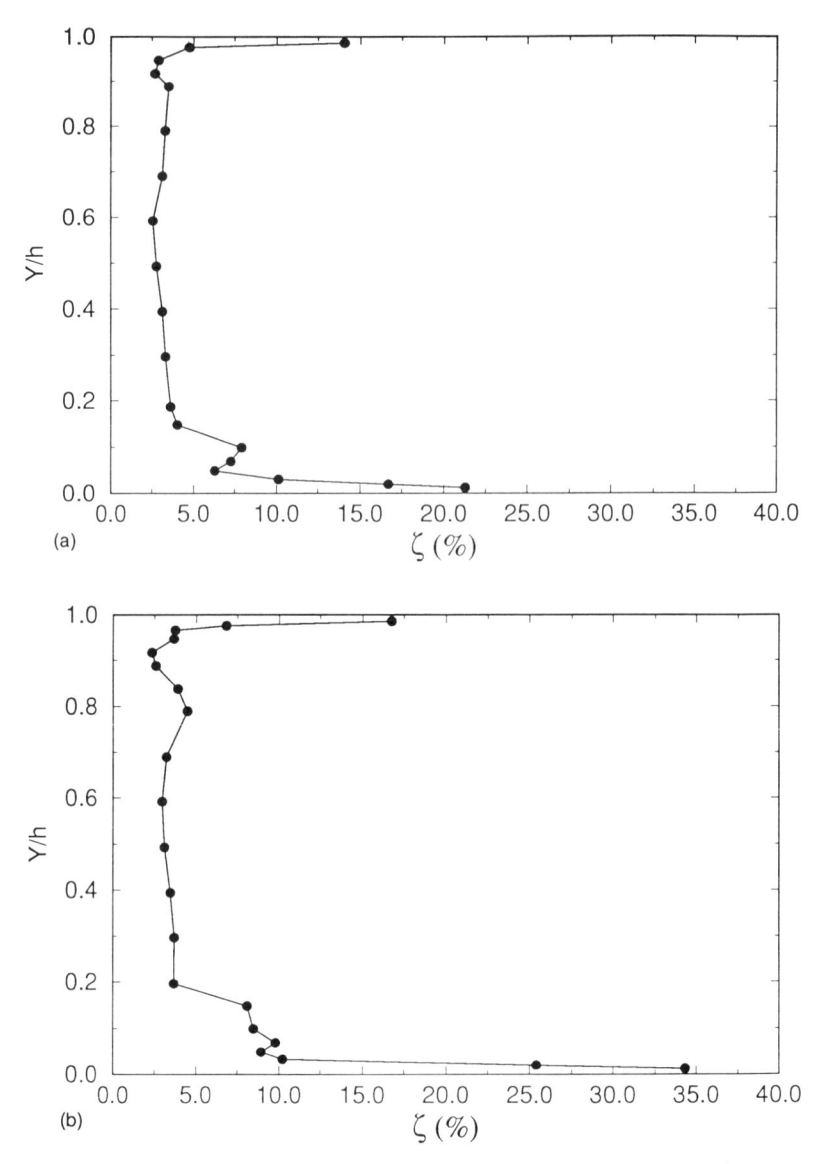

Figure 3.85 Span-wise loss distribution at downstream planes (a) $X/C_{ax} = 0.25$, (b) $X/C_{ax} = 0.5$, (c) $X/C_{ax} = 1.0$; (d) comparison of spanwise loss distribution in three downstream planes.

and this reduction increases towards the hub, i.e. to the region of higher Mach numbers. The higher losses at zero coolant flow rate are to be attributed to the backwards facing step on the rear pressure side. At choked conditions, the corner will produce a strong Prandtyl–Meyer expansion, followed by a reattachment shock at a short distance from the corner. On the suction side the flow will experience first a strong acceleration due to the Prandtl–Meyer expansion from the pressure side corner and then undergo, one after the other,

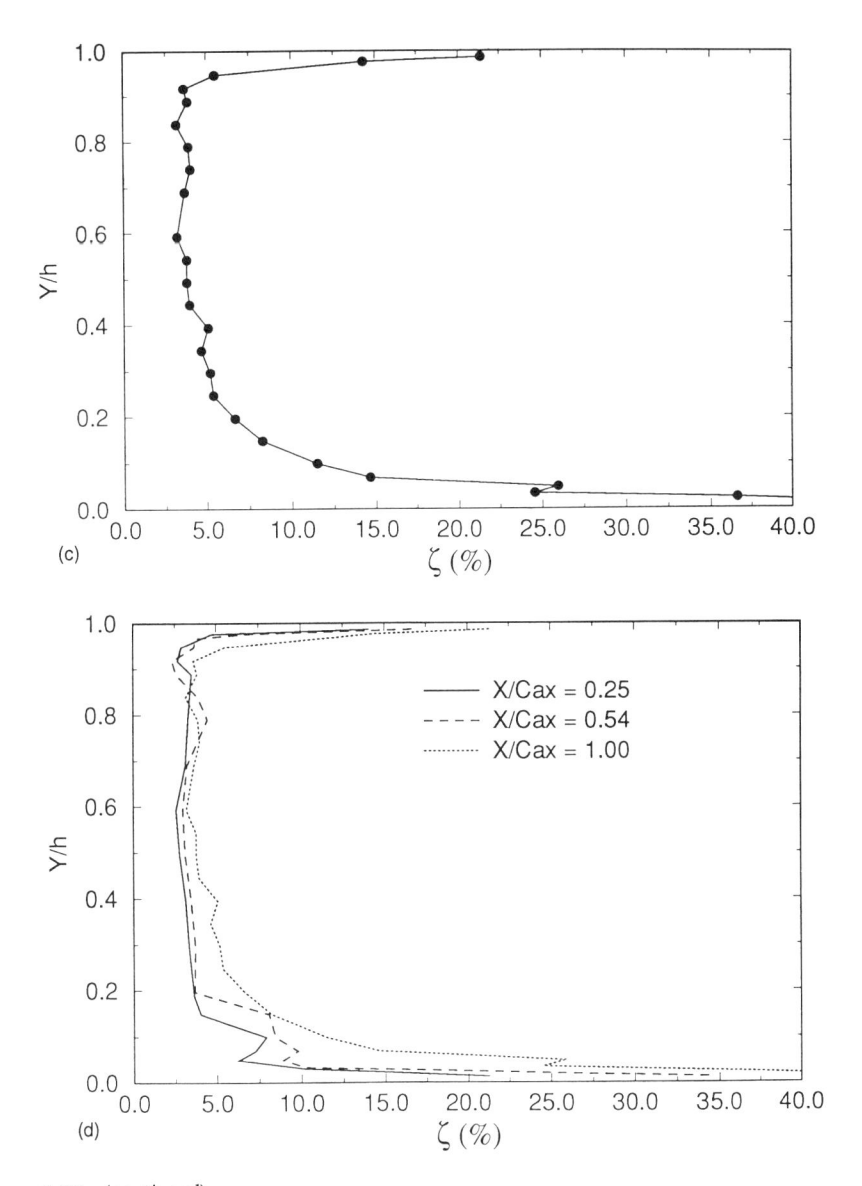

Figure 3.85 (*continued*)

the corner shock and the trailing edge shock. The ejection of coolant will reduce the effective step height. Hence both the corner expansion and the intensity of the reattachment shock will diminish and the blade losses will be reduced (see also Chapter 5).

The lower losses at zero coolant flow ejection at $Y/h = 0.7$ and 0.8 are difficult to understand. The radial migration of low momentum flow in the separated region behind the backwards facing step could have a similar effect as a coolant flow ejection. Differences in secondary flows with and without coolant ejection may be another possibility.

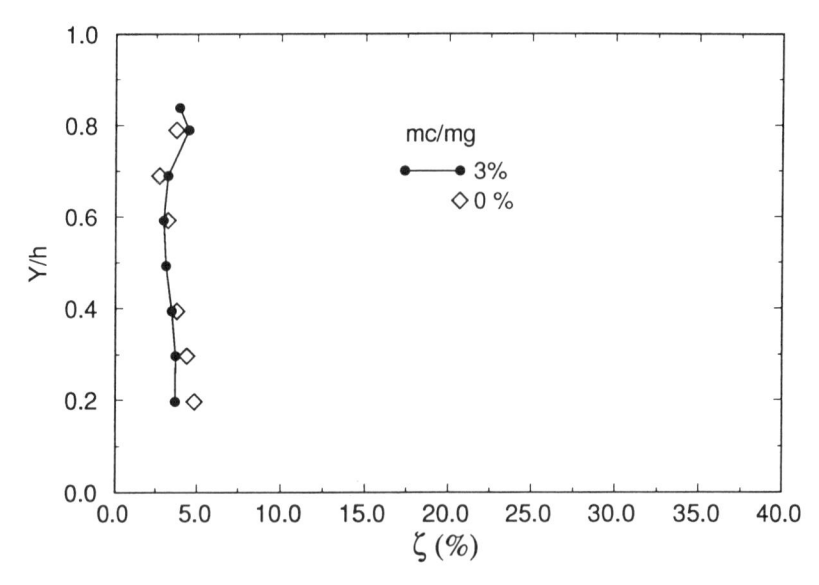

Figure 3.86 Comparison of blade losses at zero and nominal ($c_m = 3\%$) coolant flow rate in plane $X/C_{ax} = 0.54$.

The loss carpet plots for the three downstream measurement planes are presented in Figure 3.87(a–c). Due to problems with the opto-electronic encoder for the angular displacement, the probe position is measured only within an accuracy of ±0.5°. In the axial plane $X/C_{ax} = 0.25$ the wake covers about 45–50% of the blade pitch, except near the end walls. Secondary flows are very little important. The only distinct secondary loss core appears at the hub, with its peak nearly in the centre of the wake. Outside the wake the end wall boundary layer is extremely thin. At the axial distance $X/C_{ax} = 0.54$ the wake has broadened and the hub secondary loss core has taken considerable proportions. The secondary loss core at the tip appears like an island in the wake centre at some distance from the end wall. Right at the end wall a new loss maximum appears. In the following downstream plane at $X/C_{ax} = 1.0$ the upper secondary loss island has migrated further towards mid-span. At the casing the endwall boundary layer appears to be sucked into the wake by the action of the radial pressure gradient. Outside the wake the endwall boundary layer remains thin. In the lower half, the flow field has completely changed. The wake and/or the hub endwall boundary layer have spread over the entire passage. The secondary loss core has entirely merged with the endwall boundary layer. The lateral spreading of the wake mid-span wake profile with increasing axial distance is presented in Figure 3.88.

Wake profile Let us now consider more closely the downstream evolution of the mid-span blade wake. There are different ways of describing the wake profile. In low-speed flow the wake mixing behind cylinders is looked at in terms of the wake-velocity distribution. This is not appropriate for the wake behind blades with transonic and supersonic outlet flow conditions, because

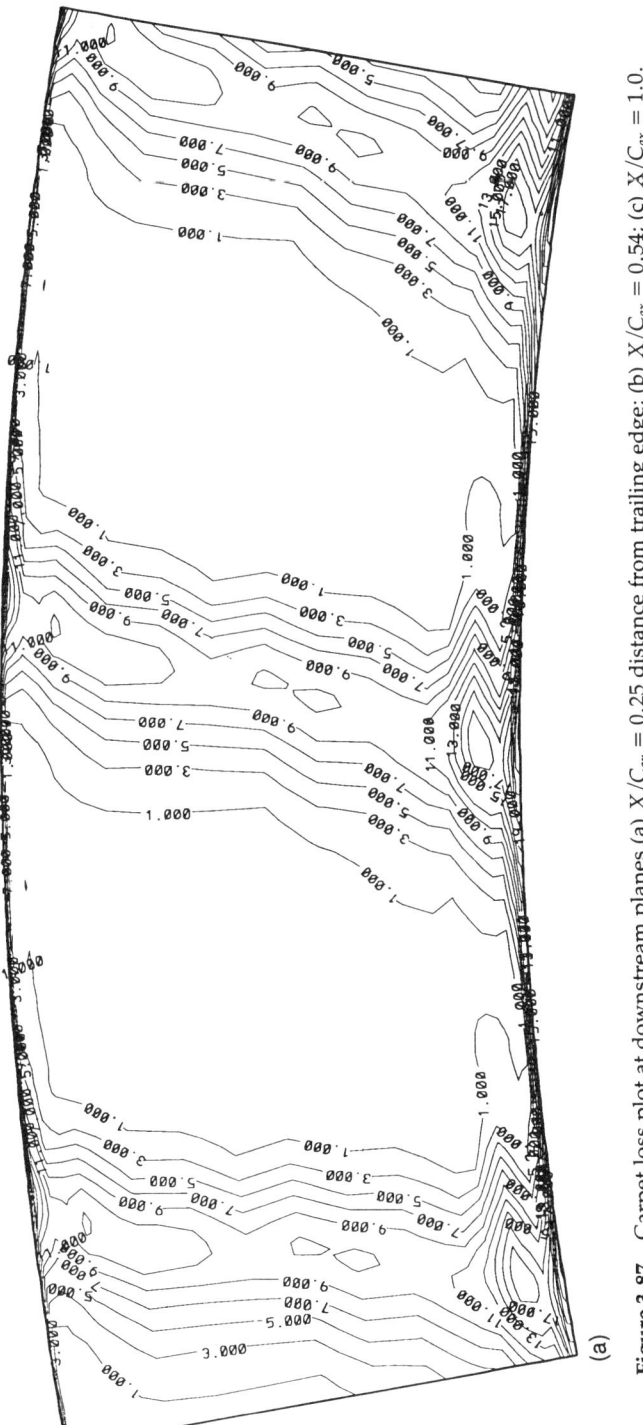

Figure 3.87 Carpet loss plot at downstream planes (a) $X/C_{ax} = 0.25$ distance from trailing edge; (b) $X/C_{ax} = 0.54$; (c) $X/C_{ax} = 1.0$.

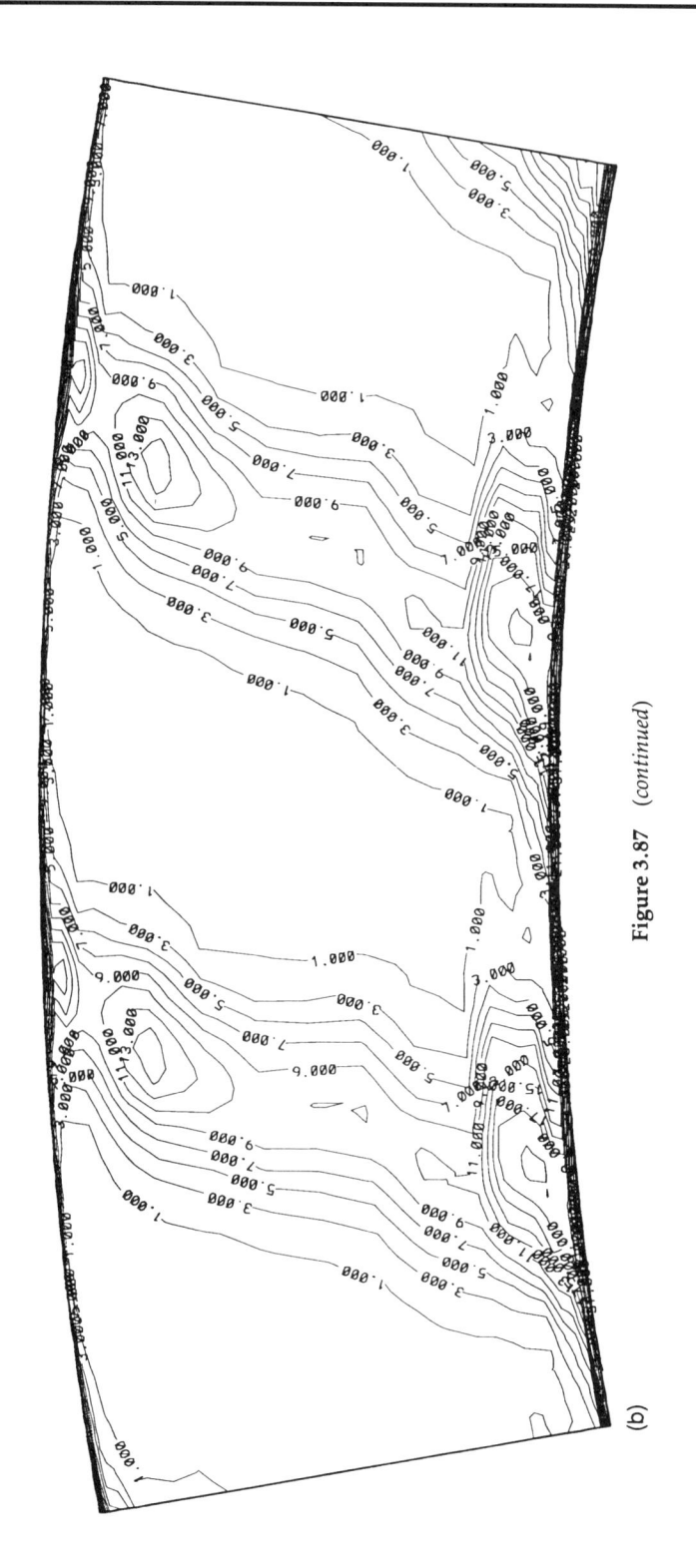

Figure 3.87 *(continued)*

Annular cascade tests

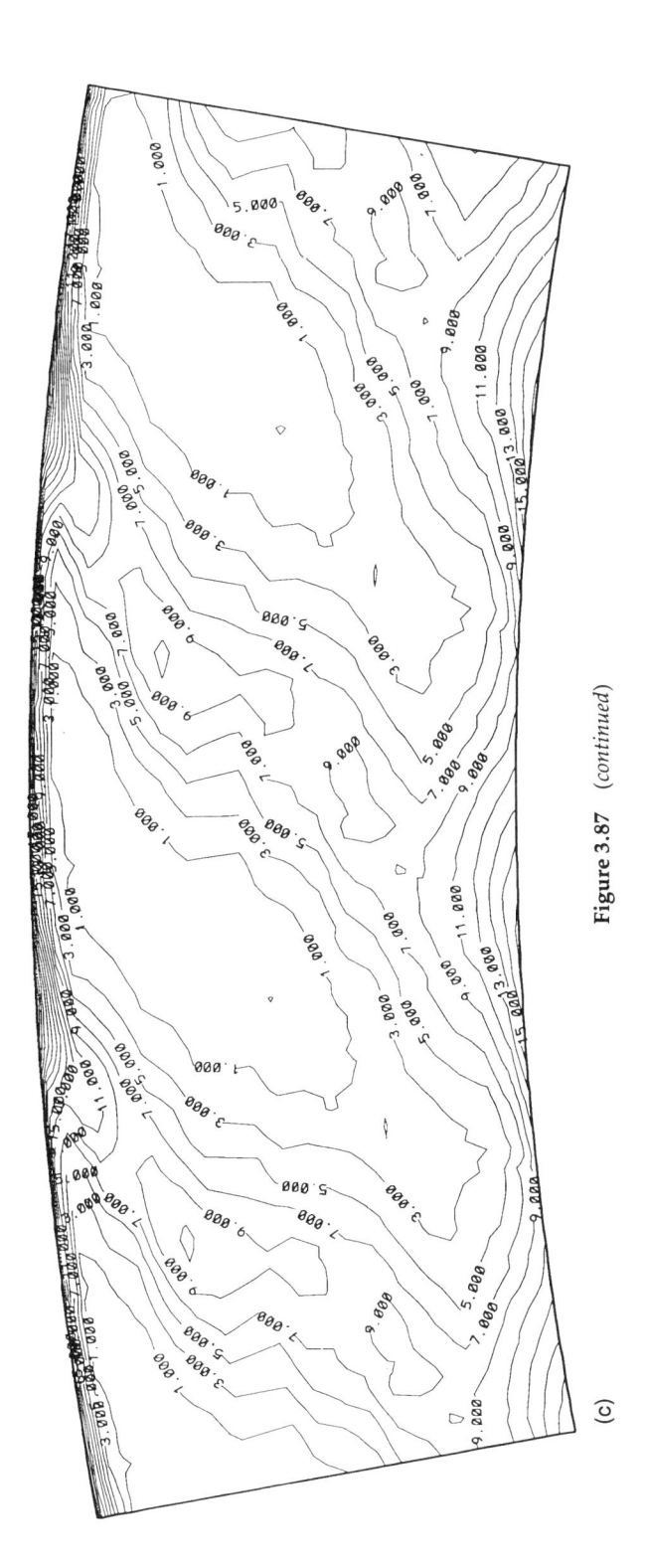

Figure 3.87 (*continued*)

(c)

the static pressure field is strongly non-uniform, with the consequence that the wake crosses a constantly varying pressure field. Hence, the local velocity in the wake depends not only on the wake diffusion process but also on the local static pressure. Of course, the same is to be said when describing the wake by its kinetic energy loss profile. Therefore, the evolution of the wake shape will be studied with regard to its total pressure profile, $1-P_{02}/P_{01}$. Figure 3.88(a) presents these profiles for mid-blade height at the three axial positions $X/C_{ax} = 0.25$, 0.54 and 1.0. The profiles in the first two planes were obtained by averaging over three wakes, the last by averaging over two wakes. The gradual lateral spreading of the wake, accompanied by a progressive reduction of the wake depth, indicates a rather slow mixing process. The decay of the total pressure minimum in the wake centre is presented in Figure 3.88(b).

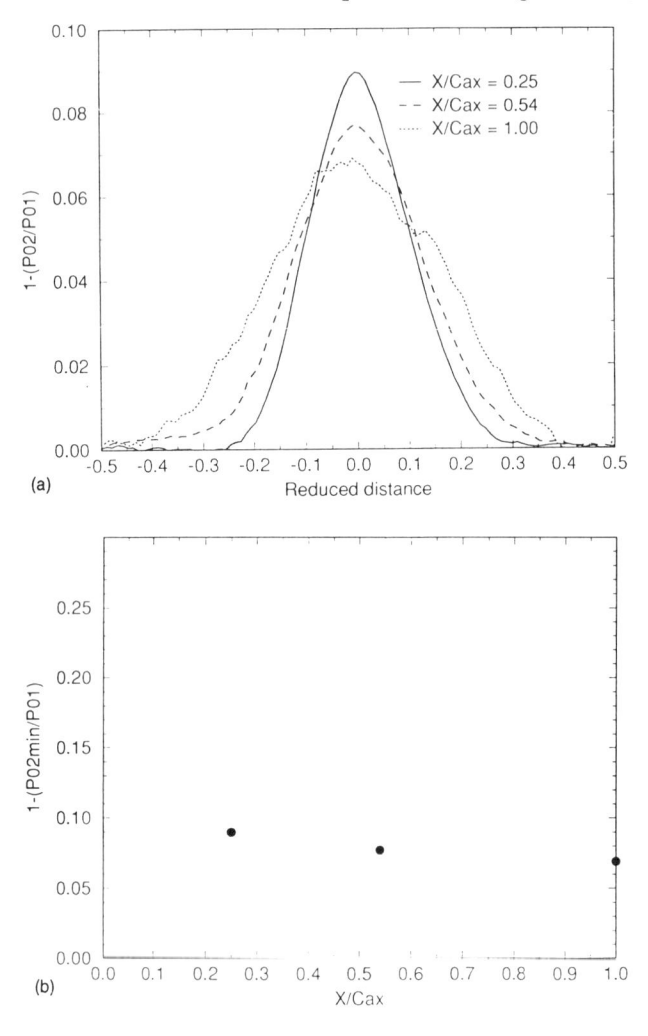

Figure 3.88 (a) Axial evolution of total pressure profiles at mid-blade height; (b) decay of wake total pressure minimum with axial distance; (c) evolution of wake shape parameter σ for Gaussian distribution; (d) comparison of measured and Gaussian wake profiles; (e) self-similarity of total pressure loss profiles.

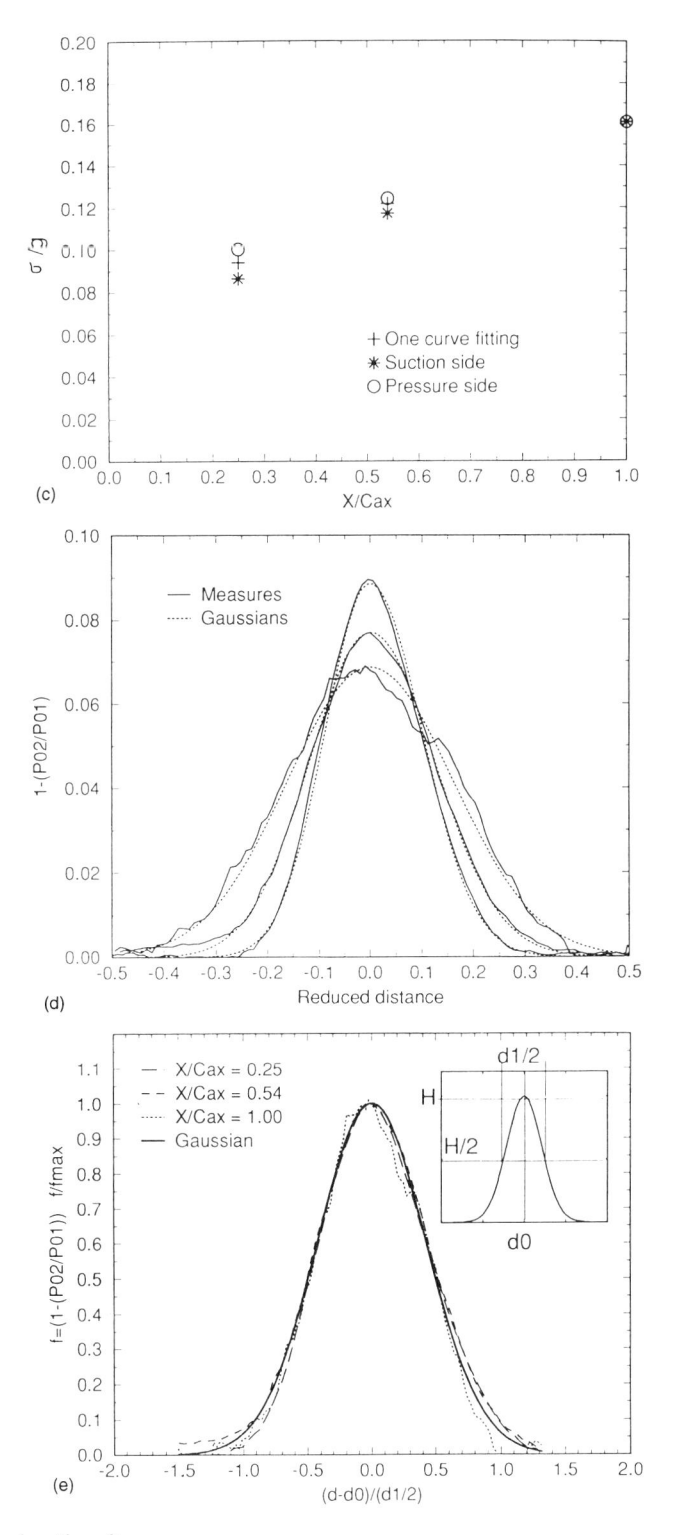

Figure 3.88 (*continued*)

As in Section 5.4.2, the total pressure profiles are approximated by Gaussian distribution functions described by equations (3.31) and (3.32) for the CO_2 concentration profiles. The distribution parameters, σ, characterizing the wake shape were calculated for the two cases of using either a single symmetric distribution curve or an asymmetric distribution with different shape parameters σ for the suction side and pressure side of the wake, Figure 3.88(c). The comparison of the measured and the Gaussian wake profiles show an excellent agreement, Figure 3.88(d).

Using the reduced variable $u = (y-\bar{y})/\sigma$ the distribution law

$$\vartheta(y) = \frac{1}{\sqrt{(2\pi\sigma)}} \exp\left(-\frac{(y-\bar{y})^2}{2\sigma^2}\right)$$

can be written as

$$\vartheta_r(y) = \frac{1}{\sqrt{(2\pi)}} \exp\left(-\frac{u^2}{2}\right)$$

This is the standard normal distribution; the variable has a null mean value and a standard deviation equal to one. This means that choosing an appropriate reduction for a set of Gaussian curves with different σ will lead to collapse them in a unique curve. This is presented in Figure 3.88(e). The figure demonstrates that the total pressure loss curves are similar to each other.

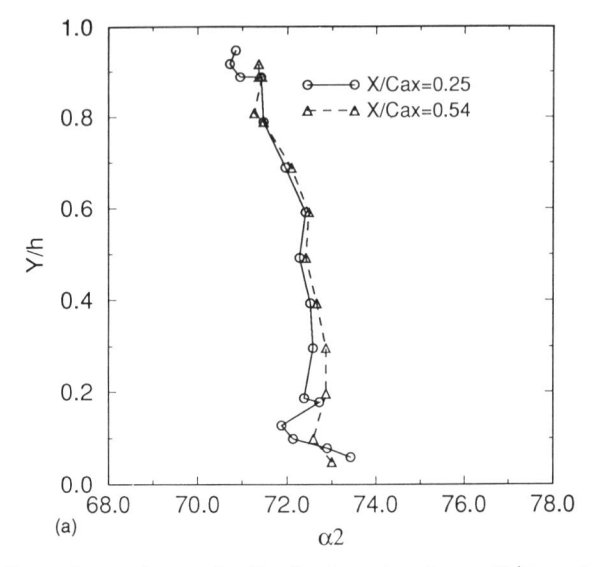

Figure 3.89 (a) Spanwise outlet angle distributions in planes $X/C_{ax} = 0.25$ and 0.54; (b) comparison of outlet angle distribution in plane $X/C_{ax} = 1.0$ with those in planes $X/C_{ax} = 0.25$ and 0.54; (c) spanwise outlet angle distribution at $X/C_{ax} = 1.0$: ─□─ probe measurements, --- derived from position of wake centre with respect to blade trailing edge.

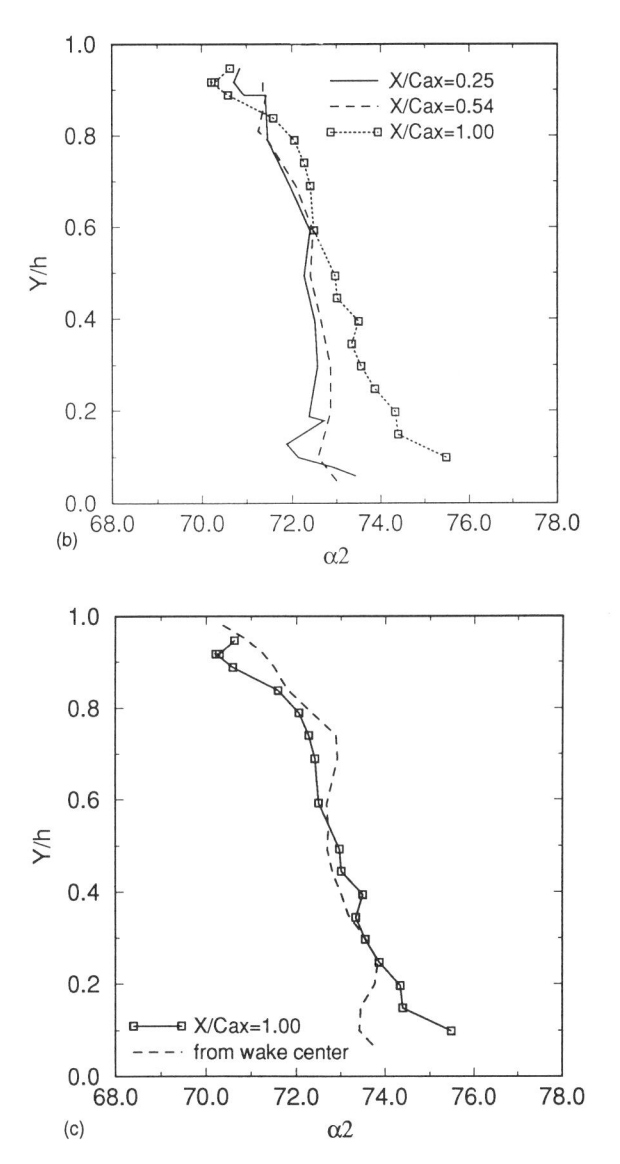

Figure 3.89 (*continued*)

Outlet flow angle The span-wise variation of the outlet flow angle is presented in Figures 3.89(a–c). On p. 245 it was stated that angle measurements for wall distances < 5% were not to be trusted and therefore any data within this zone are omitted from the angle plots. As to the remaining data, the accuracy is estimated to be of the order of ±0.5–0.8°, based on previous experiences with similar probes in two-dimensional transonic flow measurements. An exact error estimation is not possible. The pressure gradient fields are too complex to evaluate the measurements errors related to the probe head size for all conditions.

Figure 3.89(a) compares the outlet flow angle curves for the planes at $X/C_{ax} = 0.25$ and 0.54, which are the most important from the point of view of a turbine design. The overall differences between the two curves are small. Except near hub and tip where maximum local differences of the order of 1° are recorded, the angles differ by less than 0.5° over the major part of the span. the α_2 angle varies from about 71° at the tip to 73° at the hub. The angle variation in the upper half-span is more important than in the low half-span. The nearly constant α_2 in this region is possibly a Mach number effect (supersonic deviation for a convergent blading), which compensates at least partially for the change in the gauging angle $\alpha_2^* = \arccos o/g$ which increases from tip to hub. The typical under- and overturning near the end walls due to the passage vortex is observed only at the hub. The flow-angle distribution at the tip does not allow any conclusions to be drawn as to the action of the tip-passage vortex on the outlet-flow angle distribution. Any such effects appear to be within the data scatter.

Figure 3.89(b) compares the angle variation at 1 axial chord downstream with the two preceding planes. The spanwise angle variation has become much more important. The change is particular important in the lower half-span. This change is naturally related to the strong increase of the hub end-wall boundary layer, put into evidence in Figures 3.85(c) and 3.87(c), which affects more the axial than the circumferential component of velocity. This strong change in the spanwise outlet angle variation between the axial planes $X/C_{ax} = 0.54$ and 1.0 is also evident in the carpet loss plots, Figures 3.87(b,c), which indicate an important inclination of the wake between both planes.

Assuming that the flow direction in the centre of the wake is representative of the mean outlet flow angle α_2, it is possible to evaluate α_2 in a given axial plane from the position of the wake centre with respect to the blade trailing edge. A comparison of this 'wake angle' with the measured flow angle is presented in Figure 3.89(c) for the plane $X/C_{ax} = 1.0$.

Turbulence An attempt was made to record the turbulent fluctuations in the wake with a Kulite pressure probe, Figure 3.90. The assumption is made that

Figure 3.90 Conventional pitot pressure probe combined with fast response Kulite pressure probe.

Annular cascade tests

the wake turbulence is mainly caused by total pressure fluctuations, while static pressure and temperature fluctuations can be considered small. This assumption does not hold in the near wake, but in the first measurement plane at $X/C_{ax} = 0.25$ the streamwise distance from the trailing edge amounts already to 80% of the chord-length.

The turbulence measurements were carried out in the plane $X/C_{ax} = 0.25$ at midspan. The results are presented in Figure 3.91. Both the turbulence and the time-averaged total pressure profile are shown. The data are derived from a continuous midspan traverse over two wakes at $X/C_{ax} = 0.25$. The Kulite data are filtered at 50 kHz. Each point in the figure presents the average of 2000 successive data points. Control measurements with the probe in fixed positions with an averaging over a much longer time agree very well with the results from the continuous traverse.

Maximum values of $Tu = 4.2\%$ are recorded in the wake centre. Considering that due to the finite size of the probe head, 1.6 mm diameter, the pressure fluctuations are averaged over a non-negligible distance, one may expect that the real turbulence levels are slightly higher. In between the wakes the turbulence level drops to $Tu \simeq 1.2\%$. Compared to an inlet turbulence level of $Tu \simeq 1.8\%$, measured with a hot wire probe, a downstream value of 1.2% seems to be slightly high because the magnitude of the velocity fluctuations can be assumed to remain constant from inlet to outlet while the mean velocity increases by a factor four.

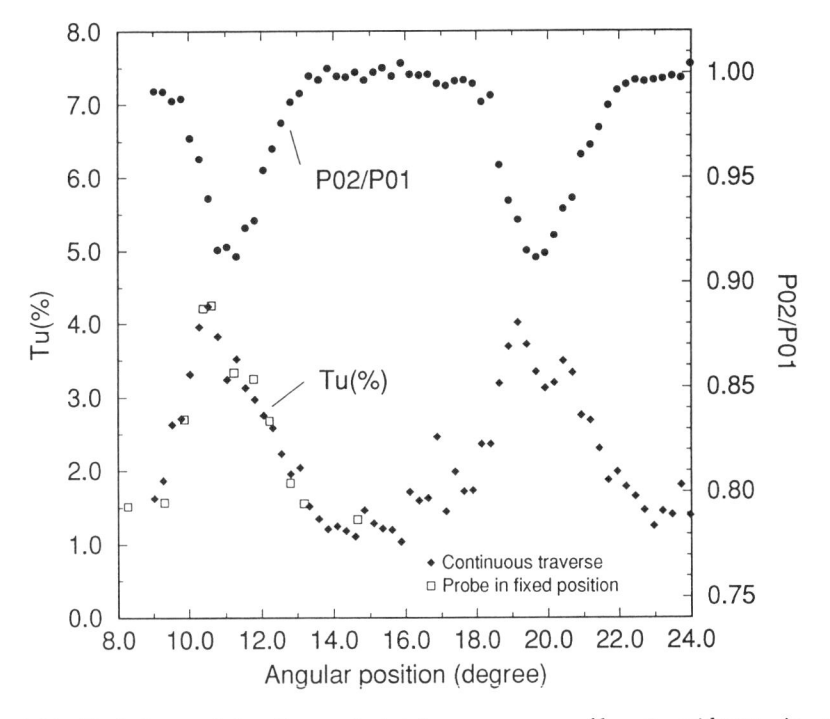

Figure 3.91 Turbulence intensity and total pressure profile at midspan in plane $X/C_{ax} = 0.25$.

It is interesting to notice that the Tu profile is larger than the wake total pressure profile, but a similar observation was made for the inlet end-wall boundary-layer profile in the DLR cascade tunnel.

Another interesting feature is the double peak in the right Tu profile, which may be explained by the effect of the coolant jet. The left Tu profile does not show this peak but the profile exhibits a singular inversion in the same region.

6.3.3 Temperature measurements

Temperature probes There are two families of gas temperature probes for turbomachinery applications: (a) thermocouple-based probes and (b) hot-wire-based probes. Probes of the first category are to date the only ones which are used as standard instruments for temperature measurements. They are used in different configurations with bare, half-shielded and shielded thermocouples. Their main drawback is their slow response time which, even for the smallest thermocouples, hardly exceeds 100 Hz.

Because of the urgent need for fast response probes capable of measuring the phase- and time-resolved temperature fields in turbine stages, the research effort has turned in recent years to the development of hot-wire-based temperature probes. The most promising event over the last years was the development of a double hot-wire aspirating probe, developed by NG at MIT [3.10]; see Figure 3.92. Operated at different overhead ratios, it can be shown that the total temperature and total pressure can be derived from the voltage output of the two wires. Although the working principle is clear, in practice it is very difficult to come forward with reliable, accurate, fast response measurements.

Experiences with double hot wire aspirating probes VKI has been investigating the characteristics of the double hot-wire aspirating probe since 1988 and gave it a high priority over the BRITE project period. The work concentrated on two aspects: the development of appropriate calibration techniques and the miniaturization of the probe. Photographs of the final probe geometry with an

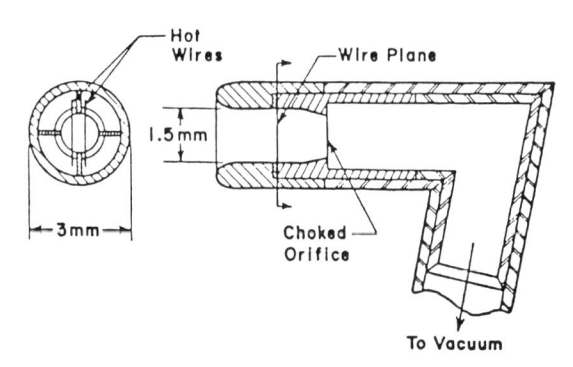

Figure 3.92 NG's double hot wire aspirating probe.

Annular cascade tests

outer diameter of the probe head of 2 mm are shown in Figure 3.93. A typical calibration map, obtained under static conditions, is presented in Figure 3.94, which shows the voltage output of the two wires over a pressure range of 1–2 bars and a temperature range of 25–60°. Measurements behind a fast rotating air jet indicated also a satisfactory response time behaviour, Figure 3.95, but there were clear indications that the response of the probe to short pressure and temperature pulses were quite different from the response to the same pressure and temperature changes under static conditions. The reason for this different behaviour became clear in a test series in which the probe was injected into a hot air jet rather than traversing it. A typical wire response is presented in Figure 3.96. The voltage–time curve $V = f(t)$ is composed of two

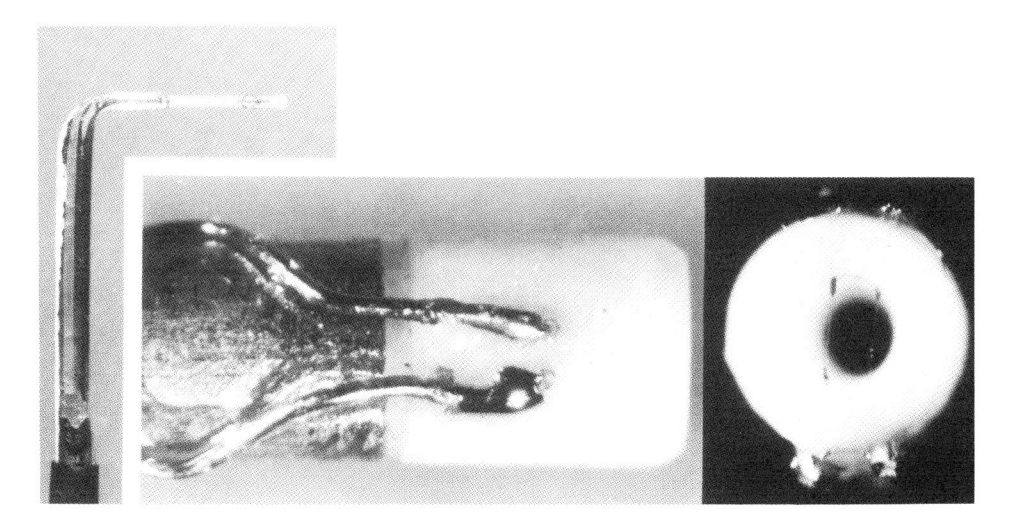

Figure 3.93 Photograph of miniaturized VKI double hot-wire aspirating probe.

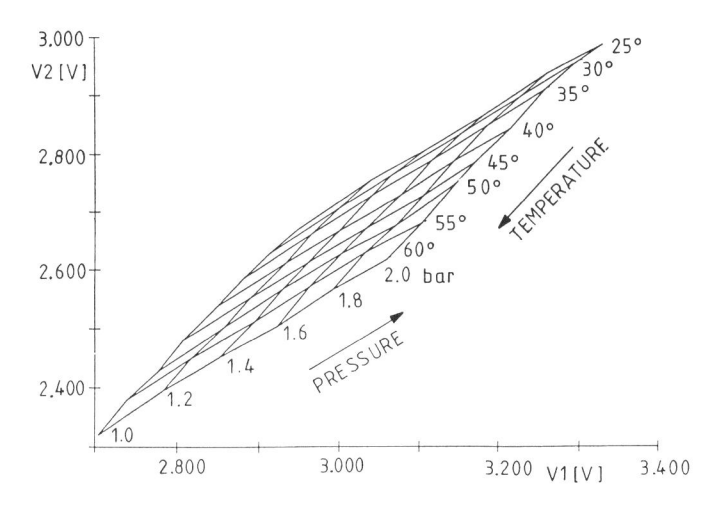

Figure 3.94 Calibration map of double HW probe.

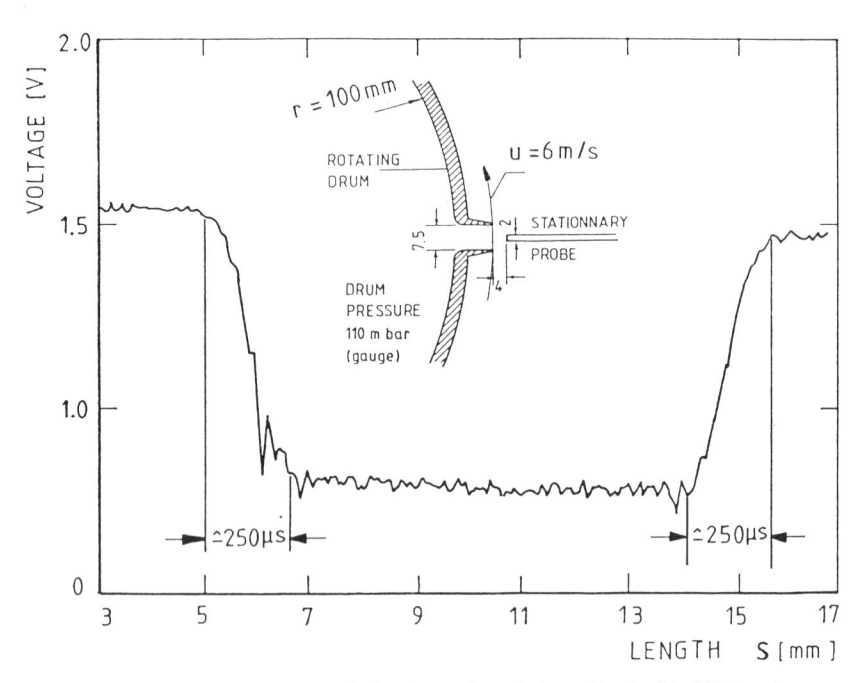

Figure 3.95 Measurements behind rotating air jet with double HW probe.

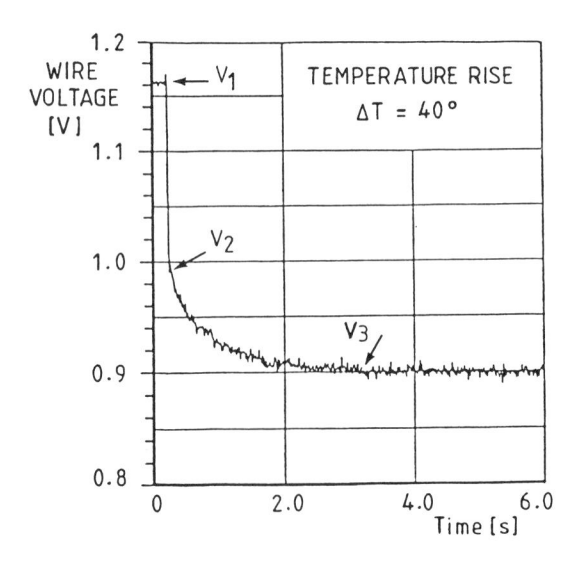

Figure 3.96 Voltage output of double HW probe injected into a hot air jet.

distinct different phases: a sudden voltage drop from V_1 to V_2 upon entering the air jet, followed by a slow asymptotic decrease up to a constant value V_3 for the resting probe. The reason for the voltage decrease from V_2 to V_3 was found to be the gradual heating of the probe head which affects the wire output through heat conduction. The transient from V_2 to V_3 depends on the temperature and pressure jump, but is typically of the order of 1–3 seconds. This is large compared to the running time of the compression tube facility which is of the order of 500 ms only. Hence, wake traverses in this facility would take place right in the early phase of the transient V_2 to V_3, characterized by a strong gradient of the voltage output. Studies are under way to find suitable corrections for this probe head effect, but the project period was too short to reach this objective.

Thermo-couple temperature probe In view of the problems encountered with the double hot-wire aspirating probe, it was decided to carry out the temperature measurements with a thermocouple probe. Figure 3.97 shows a sketch of the probe: two bare thermocouples of 0.1 mm diameter chromel constantan wires mounted side by side with the total pressure probe. Bare thermocouples were chosen to keep the probe head size small. The disadvantage of bare thermocouples is their sensitivity to gas velocity. To reduce this effect measures have to be taken to ensure a low gas velocity at the junction. For the probe in Figure 3.97 the junction velocity depends on its proximity to the pitot probe and on the distance between the junction and the tubes from which the wires emerge. The measuring accuracy of the probe was evaluated by comparing the measurements in the downstream planes, probe positioned between two wakes, with the temperature recorded by a thermocouple probe upstream of the cascade where the Mach number is of the order of $M_1 = 0.15$ only. Through careful adjustment of the position of the thermocouple junctions of the probe it was possible to obtain an agreement within 1%. The measurement accuracy of the upstream probe is estimated to ±0.3%.

Wake temperature profiles The slow response of the thermocouple probe did not allow continuous traversing. Hence a point-by-point procedure had to be adopted. The test program indicated in Table 3.7 was executed.

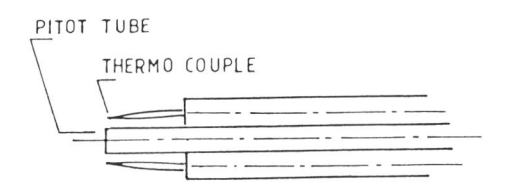

Figure 3.97 Thermocouple temperature probe.

Table 3.7 Point-by-point procedure.

Y/h	X/C_{ax} 0.25	0.54
0.15	×	
0.50	×	×
0.85	×	

Figure 3.98 shows the mid-span temperature and total pressure profiles for the axial planes $X/C_{ax} = 0.25$ and 0.54. For clarity the experimental points of the total pressure measurements have been omitted from these and the following graphs. Two aspects are particularly striking: (a) the lateral extension of the thermal profiles coincides very closely with that of the total pressure profiles and (b) the temperatures in the wake centres differ only by $\approx 4\%$ from the inlet total temperature. Considering the evolution of the temperature ratio from $T_{02}/T_{01} = 0.65$ at the coolant slot exit to 0.955 and 0.96 at the axial distances X/C_{ax} and 0.54, respectively, it appears that the thermal diffusion is very strong in the near-wake region, but very slow in the far-wake region in which the measurement planes are situated.

Figure 3.98(b) shows the total temperature and pressure profiles in the axial plane $X/C_{ax} = 0.25$ for the three spanwise positions: $Y/h = 0.15$, 0.5, and 0.75.

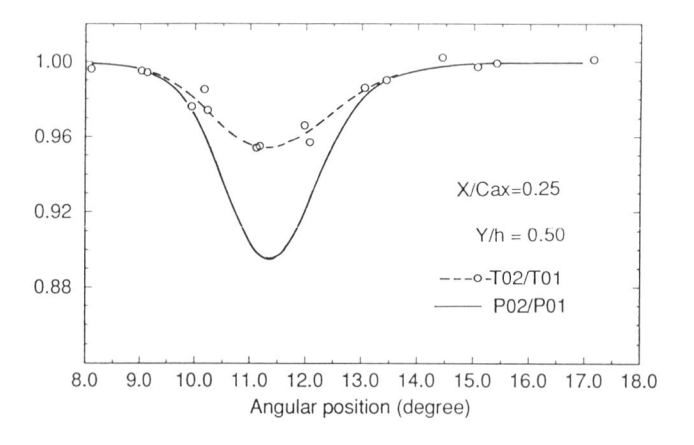

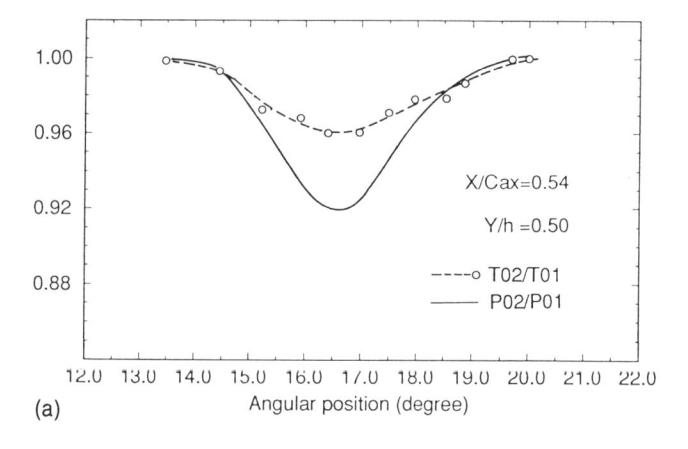

Figure 3.98 (a) Mid-span wake temperature and total pressure profiles at $X/C_{ax} = 0.25$ and 0.54; (b) wake temperature and total pressure profiles at three spanwise positions in plane $X/C_{ax} = 0.25$.

Annular cascade tests

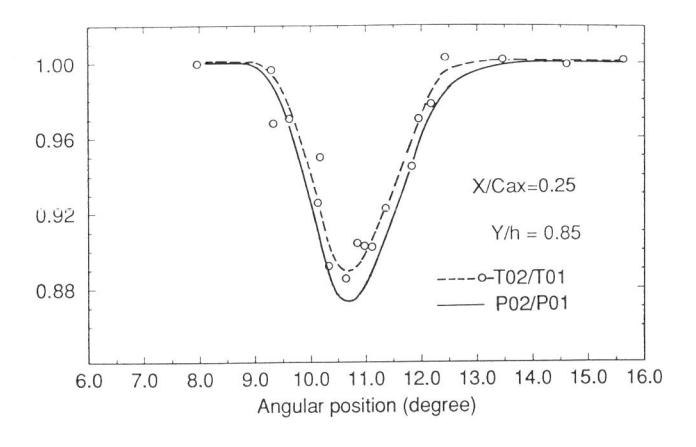

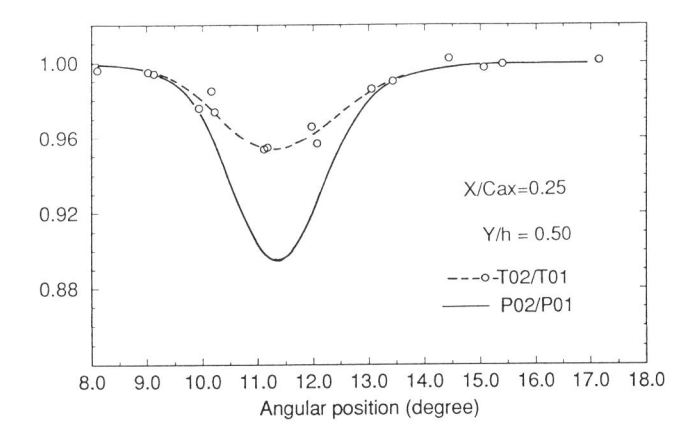

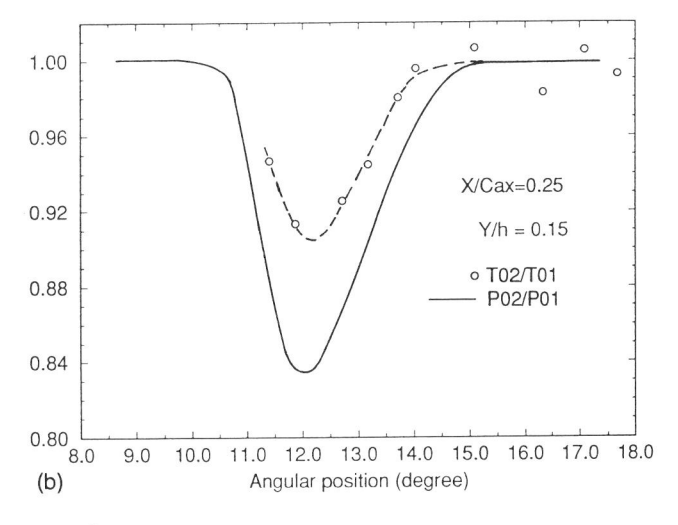

Figure 3.98 (*continued*)

At each graph the temperatures are referred to the inlet temperatures at the same spanwise position (see Figure 3.75). There is a marked difference between the temperature profiles at 15% and 85% blade height to that at 50%. The temperature ratios in the wake centre are respectively $T_{02}/T_{01} = 0.91$ and 0.89 compared to 0.955 at mid-span. These lower values appear to coincide with a thermal wake width which is somewhat smaller than the total pressure profile.

6.4 Measurements in rotation

6.4.1 Motivation and implications

As already indicated in Section 4.2.2, the use of a high-speed rotating probe traversing system was motivated by the idea that it could possibly offer an alternative solution for measuring the performance of transonic turbine guide vanes without increasing the problem of probe blockage effect and thereby open the way to guide vane performance measurements in a turbine stage through probes rotating with the turbine rotor.

The difference of the flow conditions of the BRITE annular cascade as sensed by a fixed and a rotating probe are illustrated in Figure 3.99. At design conditions, the mean isentropic Mach numbers at mean and hub diameters are $M_{2,is} = 1.05$ and 1.12 respectively, but the local Mach numbers across the pitch

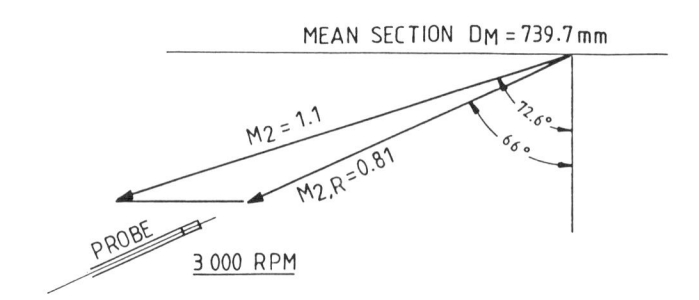

Figure 3.99 Velocity triangle for high-speed rotating probe.

Table 3.8 Relative Mach number and flow angle variation

Hub					Mean				
Absolute			Relative		Absolute			Relative	
M_2	α_2	r.p.m.	$M_{2,R}$	β_2	M_2	α_2	r.p.m.	$M_{2,R}$	β_2
		2000	1.01	72.2			2000	0.90	68.7
1.2	75°	3000	0.92	70.3	1.1	72.6	3000	0.81	66.0
		4000	0.83	68.0			4000	0.72	62.8

Annular cascade tests

are considerably higher, i.e. $M_{2,is_{max}} = 1.1$ and 1.2, respectively. The velocity triangles in Figure 3.99 are drawn for these conditions. Depending on the rotational speed of the probe the relative Mach numbers and flow angles vary as shown in Table 3.8.

With increasing r.p.m. the relative velocity drops rapidly below sonic conditions with the double benefit of gradually reducing the probe blockage effect and avoiding the need of bow-wave shock loss corrections for the total pressure. However, the rotating probe measures a relative total pressure and additional information is needed to allow the calculation of the absolute total presure. One possibility is to measure in addition to the relative total pressure the relative total temperature. The calculation of the guide vane losses for both stationary and rotating measurements is compared below:

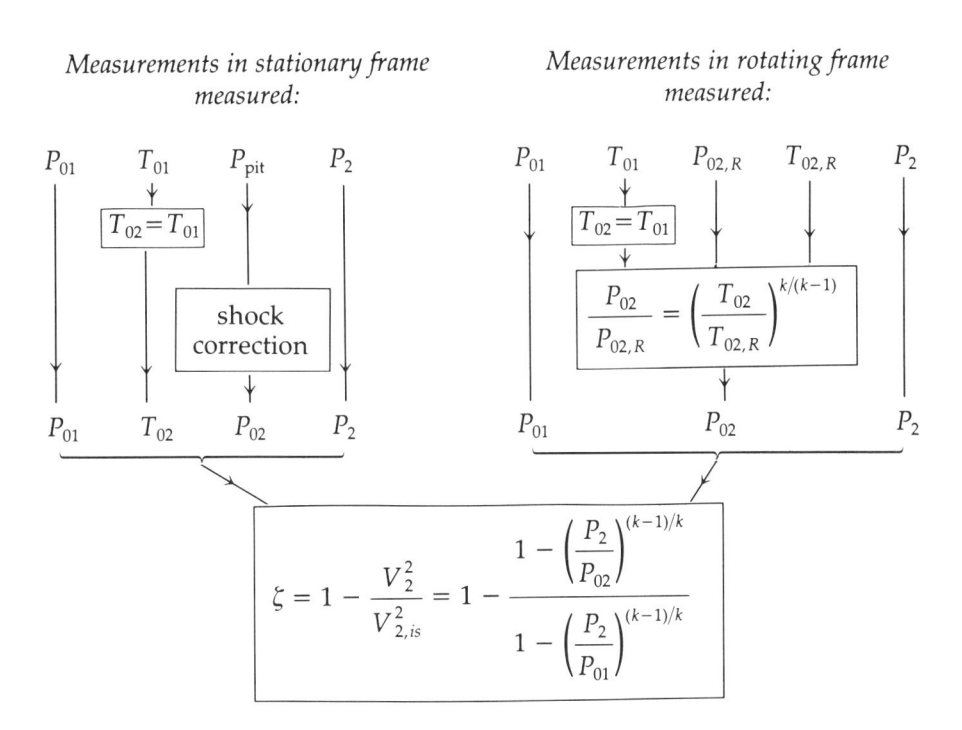

Measurements in stationary frame measured:

P_{01} T_{01} P_{pit} P_2

$T_{02} = T_{01}$

shock correction

P_{01} T_{02} P_{02} P_2

Measurements in rotating frame measured:

P_{01} T_{01} $P_{02,R}$ $T_{02,R}$ P_2

$T_{02} = T_{01}$

$$\frac{P_{02}}{P_{02,R}} = \left(\frac{T_{02}}{T_{02,R}}\right)^{k/(k-1)}$$

P_{01} P_{02} P_2

$$\zeta = 1 - \frac{V_2^2}{V_{2,is}^2} = 1 - \frac{1 - \left(\frac{P_2}{P_{02}}\right)^{(k-1)/k}}{1 - \left(\frac{P_2}{P_{01}}\right)^{(k-1)/k}}$$

As regards the static pressure P_2 it is the same in the stationary and rotating frames. In both cases it is obtained from end-wall pressure tappings. For radii in between the hub and tip radius, it is calculated by linear interpolation between the hub and tip values.

The assumption of $T_{02} = T_{01}$ is reasonable for uncooled blades, see Oldfield et al. [3.9], but does not hold in the case of strong cooling with coolant flow ejection from the blade surface.

The experimental uncertainty of the blade losses depends on the measurement error of each measured value. Figure 3.100 shows the measurement uncertainty for the blade efficiency $\eta = 1 - \zeta$ for the outlet Mach number range $M_2 = 1.0 \rightarrow 1.4$ under the assumption of the following errors:

Measurements in

Stationary frame

$$\frac{\delta P_{01}}{P_{02} - P_{S2}} = 0.005$$

$$\frac{\delta P_{02}^*}{P_{02} - P_{S2}} = 0.01$$

$$\frac{\delta P_{S1}}{P_{02} - P_{S2}} = 0.02$$

Rotating frame

$$\frac{\delta P_{01}}{P_{02} - P_{S2}} = 0.005$$

$$\frac{\delta P_{02,rel}}{P_{02,rel} - P_{S2}} = 0.01$$

$$\frac{\delta P_{S2}}{P_{02,rel} - P_{S2}} = 0.02$$

$$\Delta T_{01} = 0.2 \text{ K}$$

$$\Delta T_{02,rel} = 0.8 \text{ K}$$

Note that the assumed errors for the downstream total pressure and the static pressure are somewhat on the highside and one should be able to do better. Nevertheless, the striking effect for the measurements in the stationary frame is the influence of the static pressure error on η, which grows rapidly with Mach number, Figure 3.100. For the measurements in rotation the uncertainty of η is entirely dominated by the error in the relative total temperature. At low Mach numbers there seems to be a definite advantage for the measurements in the stationary frame, but this changes with increasing M_2.

Besides the uncertainty of η due to measurement errors there is of course the uncertainty due to the probe-blockage effect, which cannot be predicted. Here, the advantage goes of course to the rotating measurement system, provided, of course, that the probe geometry is the same for both types of measurements.

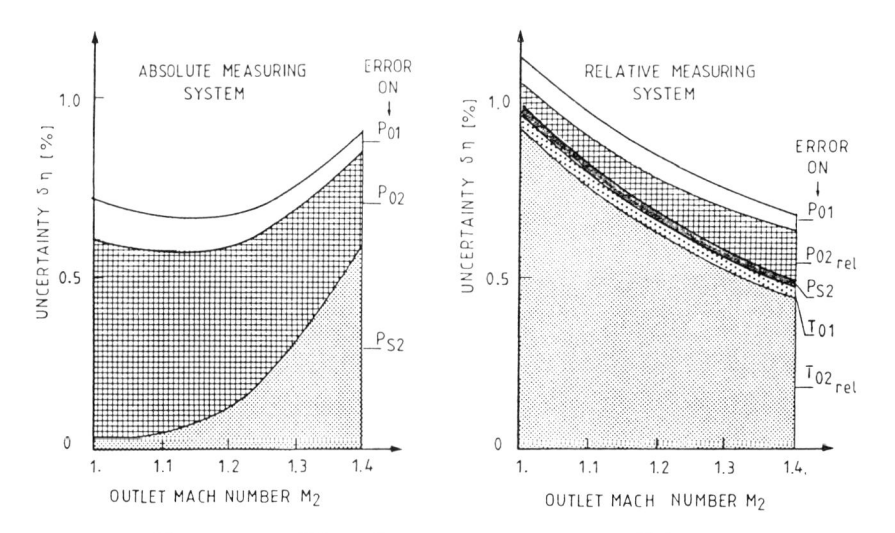

Figure 3.100 Effect of measurement errors on efficiency.

---

*Total pressure behind bow shock.

Annular cascade tests

Due to the lack of an accurate fast response total temperature probe (see Section 6.3.3), it was not possible to calculate the guide vane losses from measurements in rotation. However, the correctness of the relative total pressure measurement can be checked by evaluating the theoretical relative total pressure from the measurements in the stationary frame.

Measured values in stationary frame: T_{02}, P_{02}, P_2, α_2

$$T_2 = T_{02} \left(\frac{P_2}{P_{02}}\right)^{(k-1)/k}$$

$$V_2 = \sqrt{[2C_p(T_{02} - T_2)]}$$

$$W_2^2 = V_2^2 - 2uV_2 \cos \alpha_2 + u^2$$

$$T_{02,R} = T_2 + \frac{W_2^2}{2c_p}$$

$$P_{02,R} = P_2 \left(\frac{T_{02,R}}{T_2}\right)^{\gamma/(\gamma-1)}$$

6.4.2 Data transmission from rotating frame: model tests

Model test rig and simulation of data transmission The main components of the IR data transmission system are already presented in Section 4.2.3 and full details are given in Appendix 3. At present the capabilities of the system will be demonstrated in model tests with well-controlled flow conditions. Figure 3.101 shows the small test rig. A fast response Kulite total pressure probe is

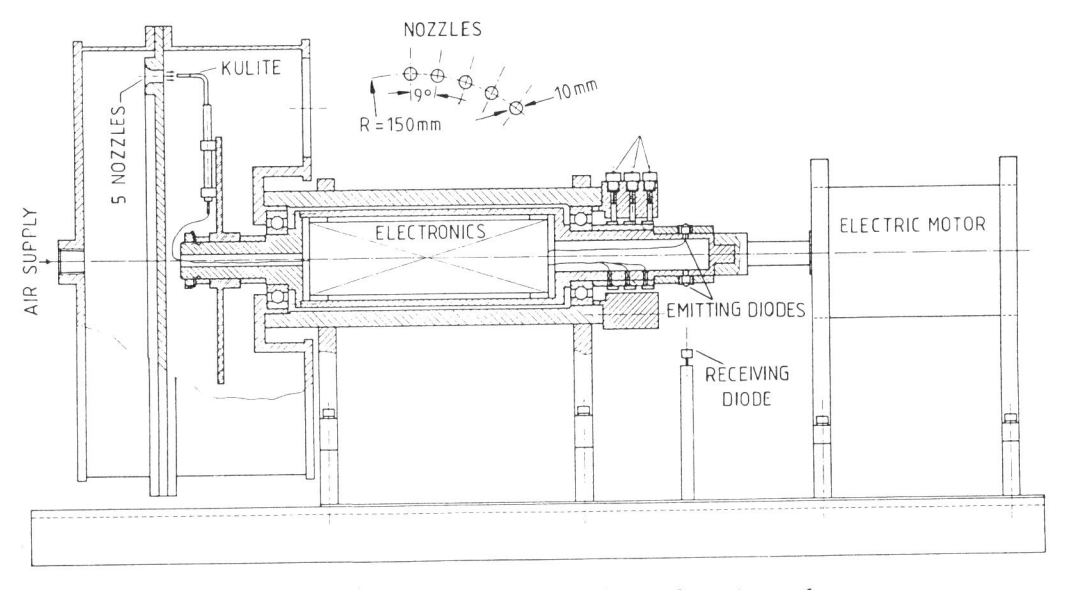

Figure 3.101 Model test rig for high-speed rotating probe.

rotated through air jets exiting from five equidistant closely spaced nozzles arranged circumferentially on the rear wall of a settling chamber. The probe is attached to a disc mounted on a shaft which in its central part houses the electronics for signal amplification and transmission.

The absolute total pressure profile of the air jet was first measured with a small slow-moving pneumatic probe, Figure 3.102. The profile has a nearly trapezoidal shape: a sharp pressure rise, followed by a plateau and an equally sharp pressure drop. The shear layer thicknesses, across which the pressure changes occur, are about 1.2 mm. The measurements with the fast rotating Kulite pressure probe are performed at a peripheral speed of 52 m/s. Under these conditions the pressure rise and fall times amount to $\Delta t = 23$ μs and the air jet passing frequency is 2.2 kHz.

Before carrying out the rotating-probe measurements the transmission system was checked by submitting it to square and trapezoidal waves of 2.5 and 10 kHz.

The oscilloscope traces on the left in Figure 3.103 compare the system output signal with the square-wave input signal. The output signals exhibit at both frequencies a characteristic overshoot at the start and the end of the input signal. On the right side of the figure the output signal is compared to a trapezoidal input signal with a rise time $\Delta t = 20$ μs, which is close to the rise time $\Delta t = 23$ μs experienced by the probe traversing the air jets. Except for a small delay, the output signal follows at 2.5 kHz very closely the input signal. At 10 kHz one observes a very small overshoot.

Kulite pressure probe The probe is equipped with a Kulite XCs-062-5D transducer with a protective screen of type B. The screen is a thin plate with 12 small holes near its outer diameter. The probe was tested over a wide pressure and incidence angle range. The tests showed:

(a) a strong difference between a static calibration (in a pressurized reservoir) and a dynamic calibration (in an air jet);

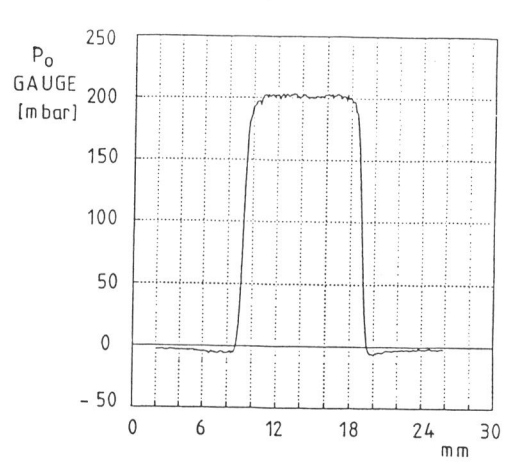

Figure 3.102 Total pressure profile of air jet.

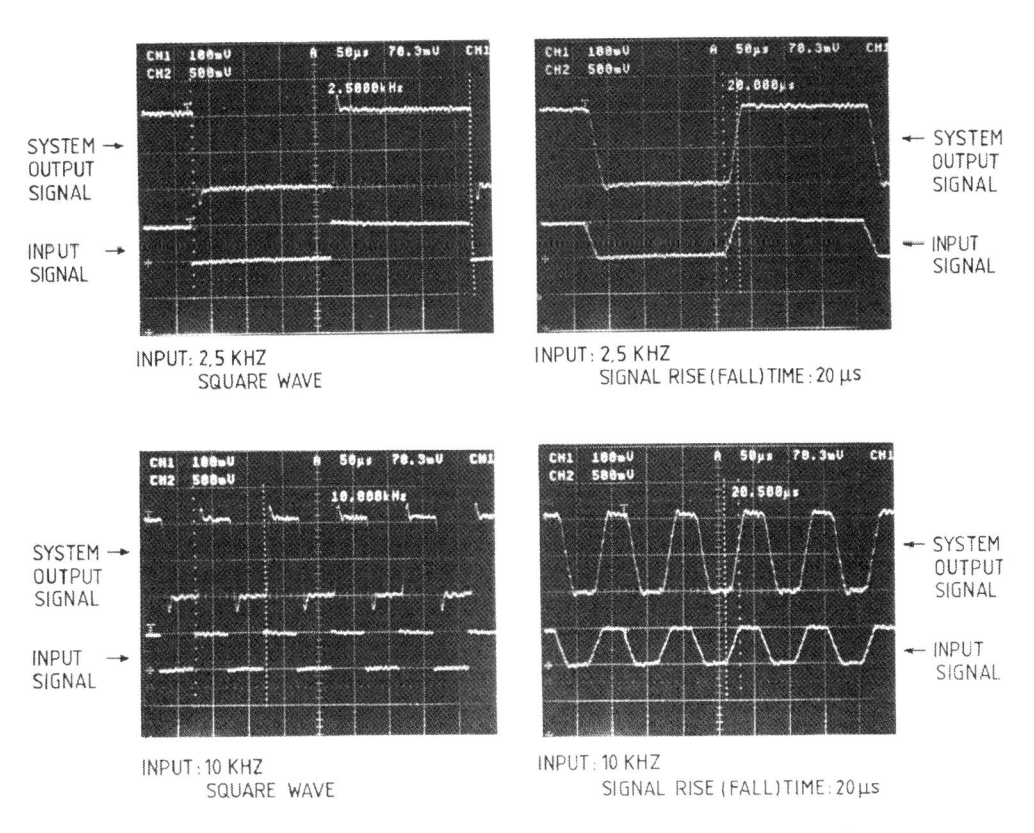

Figure 3.103 Oscilloscope traces for checking opto-electronic data transmission system.

(b) a high sensitivity to incidence angle variations;

(c) a strong asymmetric angle response.

The problems were overcome by putting a thin sleeve over the Kulite transducer, providing thereby a small cavity in front of the screen. The depth of the cavity was systematically varied from 0.1 to 3 mm. At zero cavity the transducer does not sense the total pressure at the air flow but a static pressure corresponding to the air velocity at the position of the holes on the screen, Figure 3.104(a). The overhanging sleeve causes the streamlines on the screen to stagnate near these holes and the underlying pressure ship will sense the corresponding stagnation pressure, Figure 3.104(b). A full total pressure recovery was obtained for a cavity depth of 0.5 mm, Figure 3.105. The effect of this small cavity on the enlargement of the operating range is also impressive. However, the cavity has to be kept minimum, because it affects the frequency response of the transducer.

In rotation the transducer is not only affected by the relative total pressure but also by additional effects like the centrifugal force on the air column in the transducer reference tube (in case of differential transducers) and the direct effect of the centrifugal force on the pressure chip. The influence of both effects

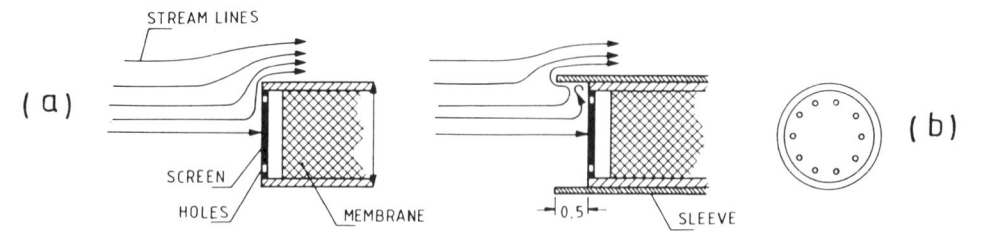

Figure 3.104 Flow around Kulite probe: protective B-screen flush-mounted (left) and with recess (right).

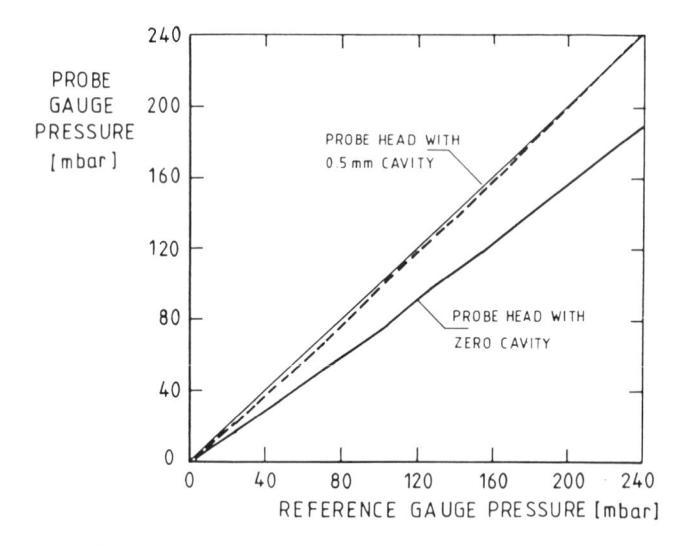

Figure 3.105 Effect of recessed screen (cavity in front of screen) on probe response.

is determined by recording the voltage output of the rotating probe with a cover on the probe head. The pressure shift depends of course strongly on the rotational speed.

This test is also ideally suited to determine the noise of the opto-electronic transmission system including the noise induced by any defects in the bearings. At 3000 r.p.m. the peak-to-peak voltage variations correspond to pressure variations of 4 mbar. This noise is independent of the test conditions (except for the rotational speed) and the noise-to-signal ratio decreases with increasing total pressure in the air jet. In the model tests with an air jet pressure of 200 mbar the noise level is ±1%.

Initially it was intended to build for the rotational tests a probe with the standard VKI probe dimensions, i.e. a probe head length of 35 mm. Calculations demonstrated soon that aerodynamic requirements had to yield to stress requirements. To cope with the bending stresses acting on the probe in the model test rig with rotational speeds up to 4000 r.p.m., the probe head had to be shortened by ≈30%.

Model tests The absolute air jet total pressure was set at $Po = 200$ mbar resulting in an air jet velocity of $V = 169$ m/s. Tests were run from 1000 to 3500 r.p.m. The data are acquisitioned at 1 MHz. The transmission electronics comprise a built-in fifth-order Butterworth filter with a cut-off frequency of 100 kHz. The raw data were numerically filtered at 40 kHz to eliminate the noise related to the eigenfrequency of the Kulite (94 kHz). Figure 3.106 shows the phase-locked averaged relative total pressure profile at 3300 r.p.m. with a peripheral speed of 52 m/s and a jet passing frequency of 2.2 kHz. The profile has been averaged over 100 events. The plateau pressure agrees very closely with the theoretically predicted pressure. The pressure rise and pressure drop can not be accurate because the total probe head diameter, i.e. transducer plus sleeve, is 2 mm compared to a shear layer thickness of 1.2 mm. Also, as long as the velocity in the shear layer of the jet is smaller than the peripheral speed, the incidence angle to the probe is far beyond its operating range.

The transition to the pressure plateau is characterized by a strong pressure overshoot. Since this effect can not be attributed to the transmission system, which has shown to work correctly for pressure rise times characteristic for these tests, see p. 273, it must be assumed that it is due to some pumping effects between the air in the external cavity and the air in the cavity between the screen and the pressure chip.

Outlook on rotating probe measurements in BRITE cascade The above air-jet test is a very severe test case for the rotating-probe measurement system. In fact the pressure gradients and the pressure rise times are much more severe than those encountered during the wake traverses in the BRITE annular cascade. In the first measurement plane at $X/C_{ax} = 0.25$, the wake of the BRITE cascade occupies approximately 50% of the pitch, i.e. with a pitch of 54 mm the

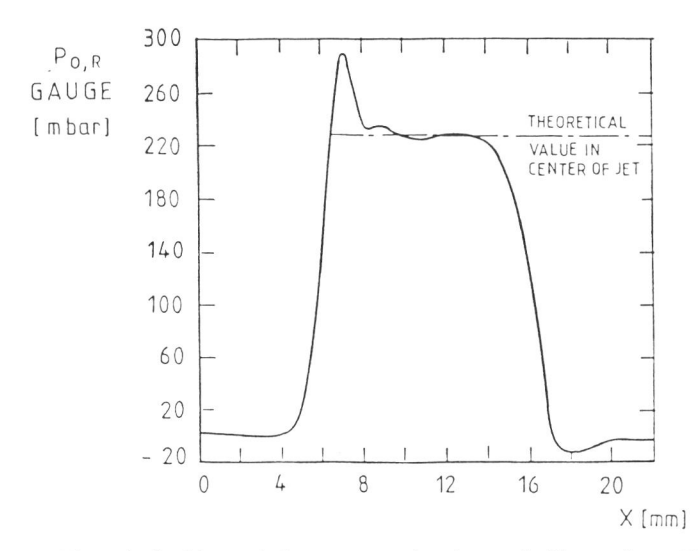

Figure 3.106 Phase-locked-loop air-jet pressure signal recorded by probe at 3300 r.p.m.

total pressure drop and pressure rise through the wake extends each across ≈13.5 mm. For mid-span traverses at a radius of $r = 369.85$ mm the pressure rise time experienced by the probe is then

r.p.m.	2000	3000	4000
$\Delta t (\mu s)$	145	96.8	72.6

i.e. 3–6 times longer than the model test conditions described on p. 275. In addition the probe head is small compared to the wake width and the angle variations through the wake are only of the order of few degrees. Under these conditions, it is reasonable to expect that pressure overshoots like those in Figure 3.106 will not occur.

There is, however, one aspect in the BRITE cascade tests which is much more critical than in the model tests. The relative Mach numbers are of the order of $M_{2,R} = 0.7$ to 0.9 instead of 0.5 and 0.6 in the model tests. Probe blockage effects should therefore be higher.

6.4.3 Preliminary measurements in the CT-3 facility

For the general description of the rotating-probe system the reader is referred to Section 4.2.3. Figure 3.107 shows a photograph of the rotating hub end wall with two Kulite probe probes mounted at 180° distance for balancing reasons. The probes are very similar to those in the model tests. However, since the centrifugal stresses are about 2.5 times higher than in the model tests due to the higher radius of rotation, the probe head had to be stiffened by an additional cross bar as indicated in the figure. The pressure sensors are Kulite XCQ062-50A absolute pressure transducers with B-type screens. Except if otherwise stated, the transducer head is modified by an outer sleeve providing a cavity of 0.5 mm depth in front of the screen (see p. 274).

All tests are performed in the axial plane $X/C_{ax} = 0.25$ at mean blade height. A first test is done with the Kulite probe in a fixed position between two wakes. Figure 3.108 compares the time evolution of the pressure trace of the Kulite with that of the upstream pneumatic probe. The data of the Kulite probe are filtered in this case at 200 Hz. After the first pressure oscillation, the evolution of the pressure traces are nearly identical. For $t > 350$ ms the pressure is nearly constant.

It is well known that Kulite transducers are very sensitive to temperature changes. However, their response time to temperature variations is relatively long. During a test run with the probe rotating at high speed, the probe is exposed (a) to a sudden temperature rise of ≈ 160° at the begin of the test and (b) to much smaller high-frequency temperature variations linked to the blade passing frequency. It is reasonable to assume that the probe will only respond to the overall temperature rise. The comparison of the two pressure traces in Figure 3.108 does provide some information on this effect. Since the pneumatic probe is insensitive to temperature effects, the overall divergence between the curves with time is an indication of the temperature effect on the Kulite

Annular cascade tests

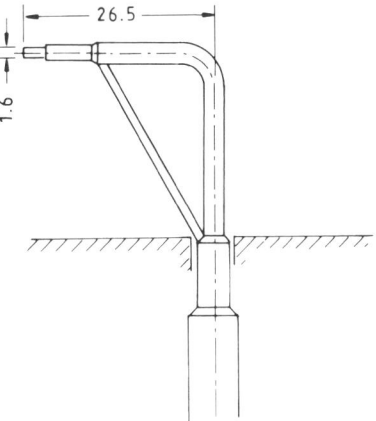

Figure 3.107 Photograph of rotating hub end wall.

pressure probe. The maximum difference in the domain of interest amounts to 0.5% of the total pressure.

The first runs with the rotating probe unit were made to control the dynamic behaviour of the rotating assembly and check the noise level of the data transmission system. The tests were made without flow in the test section at a test section pressure level of ≈ 0.1 bar, which is the typical pressure level prior to a blow-down test. The acceleration tests demonstrated a save operation without any noticeable vibration problems up to the maximum rotational speed of 3500 r.p.m. obtainable with the air motor drive.

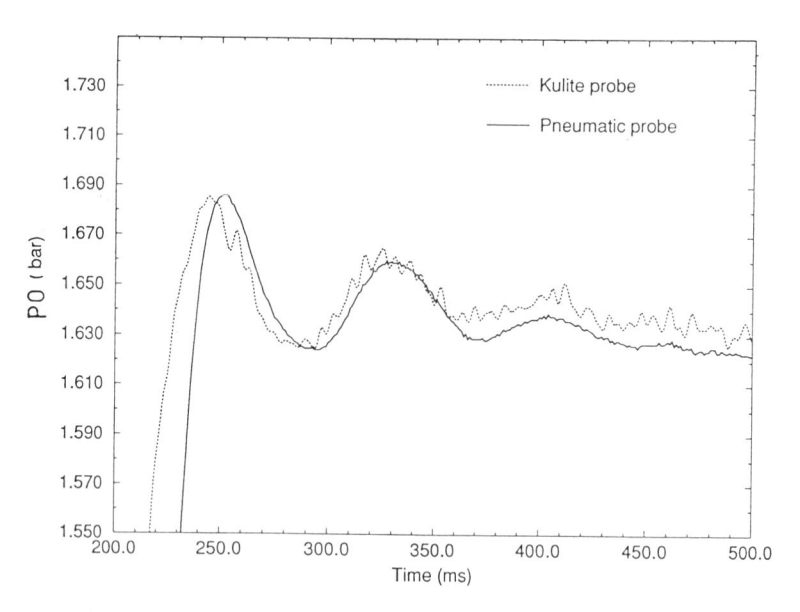

Figure 3.108 Time–pressure traces of downstream Kulite pressure probe (fixed position) and upstream pneumatic probe.

Prior to a test the probe is speeded up to about 80% of the requested r.p.m. During the blow down the probe's rotational speed increases due to frictional forces acting on the rotating end wall, as demonstrated in Figure 3.109. The curve presents the variation of the rotational speed over 300 ms from time $t = 200$ ms to $t = 500$ ms. The opening of the shutter valve for releasing the flow from the compression tube occurs at $t = 160$ ms at a rotational speed of 2500 r.p.m. The high speed data acquisition system for recording the relative

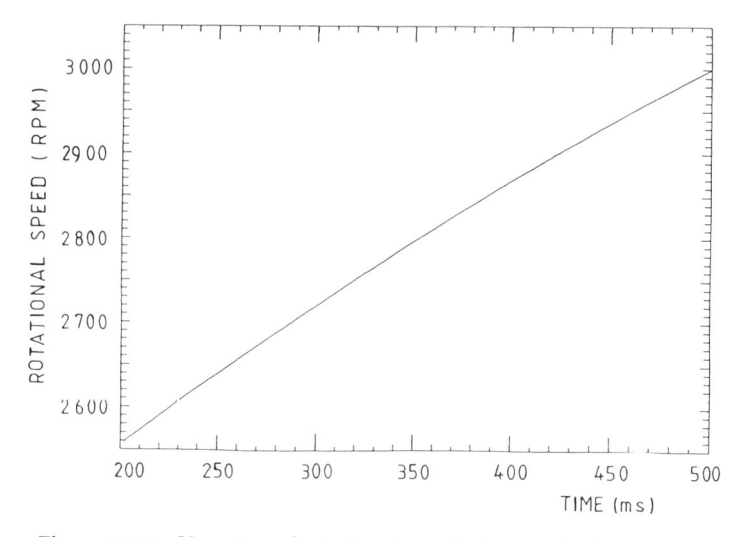

Figure 3.109 Variation of rotational speed of probe during test run.

total pressure would start typically at $t = 350$ ms, i.e. after the initial strong pressure fluctuations, Figure 3.108. The ensemble averaging is done in general between time $t = 420$ and 500 ms. In the present case the rotational speed has reached 2900 r.p.m. at $t = 420$ ms and the total variation in r.p.m. over the averaging period is $\approx 3.5\%$.

Similar as in the model tests the noise of the data transmission system is controlled by recording the signal output of the rotating probe with its probe head covered up. At a rotational speed of 2500 r.p.m. the noise band is 6.5 mbar. The noise does change little with r.p.m. in the domain of interest i.e. from 2000 to 3500 r.p.m. Compared to a total relative pressure of the order of 1.2 bar during a blow down test, the noise-signal ratio is $\approx 0.5\%$.

The above test also provides information on the effect of the centrifugal forces on the transducer. This information is used to correct the relative total pressure measurements in the following tests.

6.4.4 Relative total pressure wake profiles

The relative total pressure measurements described hereafter correspond to the rotational speed curve of Figure 3.109. Figure 3.110 shows a typical raw data string belonging to a total data acquisition over 150 ms, i.e. between time $t = 350$ ms and $t = 500$ ms. At a mean rotational speed of 2900 r.p.m. over this period and a total of 43 blades the blade passing frequency is ≈ 2.1 kHz. With a data aquisition frequency of 400 kHz the distribution of the relative total pressure over 1 pitch is defined by 192 points and the total number of periodic events over this time is 312. The determination of the exact blade passing time

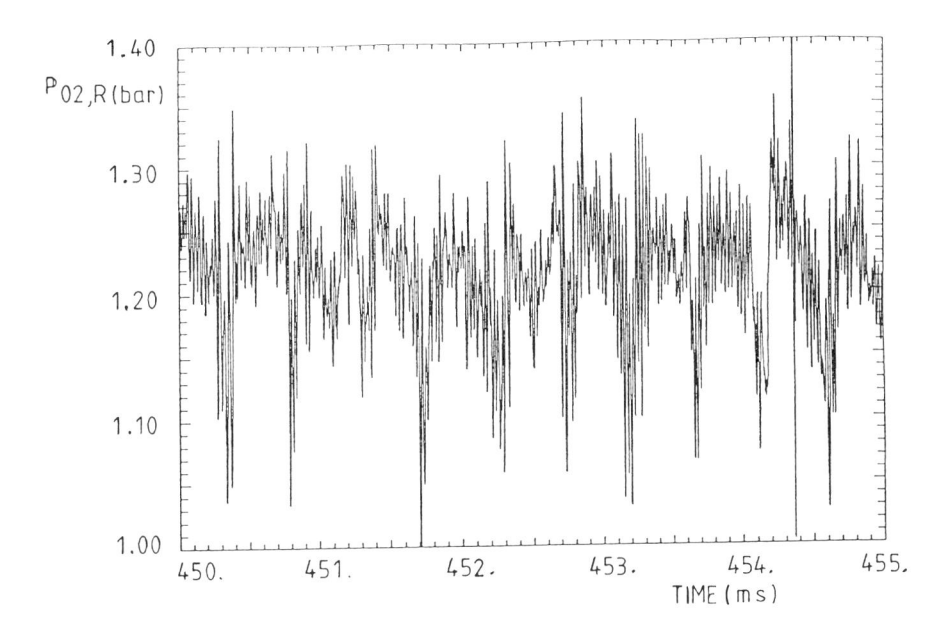

Figure 3.110 Raw data string recorded by Kulite pressure probe.

for the ensemble averaging is based on the accurate measurement of the rotational speed.

The averaging process is done using directly the row data or after filtering them numerically at 40 or 20 kHz. The signals were averaged over 50, 150 and 300 events. Figure 3.111 shows a comparison of the ensemble averaged signal over 300 events for both the raw signal and the signal filtered at 20 kHz. Except for small wiggles in the curve derived from the raw data, there are no significant differences between the curves. The averaging over a large number of periodic events, however, has the disadvantage that the rotational speed changes by a significant amount, 3.5% in the present case. This affects of course the relative flow angle and the relative total pressure via the change of the relative velocity. To reduce these variations the averaging should be done with a minimum, but sufficient, number of events. Figure 3.112 compares the relative total pressure profiles averaged over 50 and 150 events after a filtering of the raw data at 20 kHz. The small differences between both traces clearly indicate that 150 events are largely sufficient for the averaging procedure. The relative total pressure, non-dimensionalized with the upstream total pressure, is plotted in Figure 3.113.

6.4.5 Comparison with traverse results from stationary measurements

As indicated in Section 6.4.1, the only way to ascertain the accuracy of the relative total pressure measurements is by comparing them with the relative total pressure profile derived from the measurements in the stationary frame.

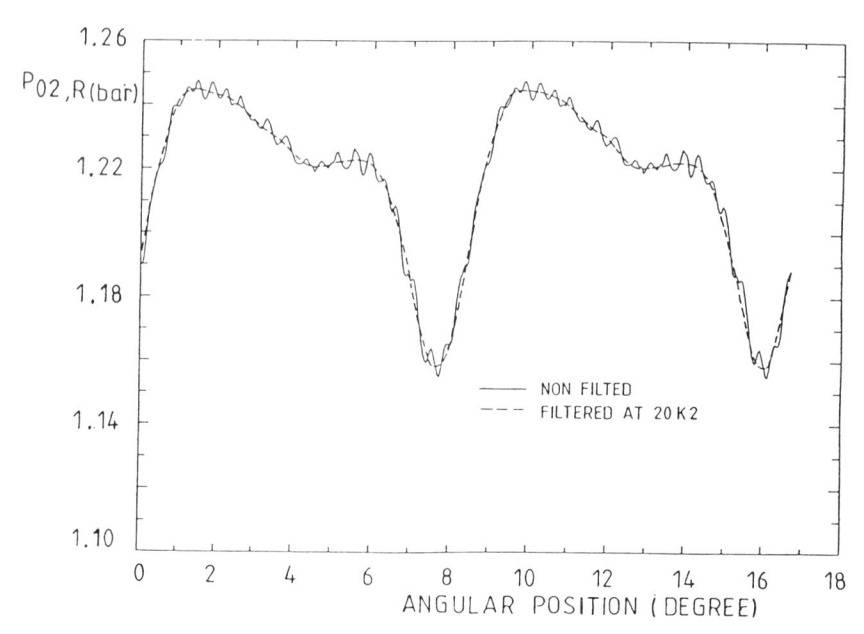

Figure 3.111 Ensemble-averaged relative total pressure profile over 300 pitches for raw and filtered (20 kHz) data.

Annular cascade tests

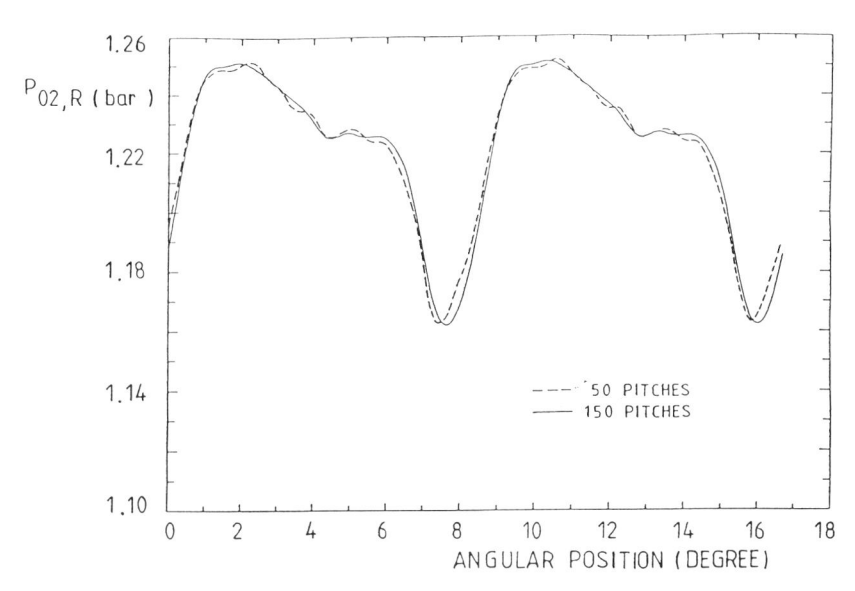

Figure 3.112 Ensemble-averaged relative total pressure signal for 50 and 150 pitches.

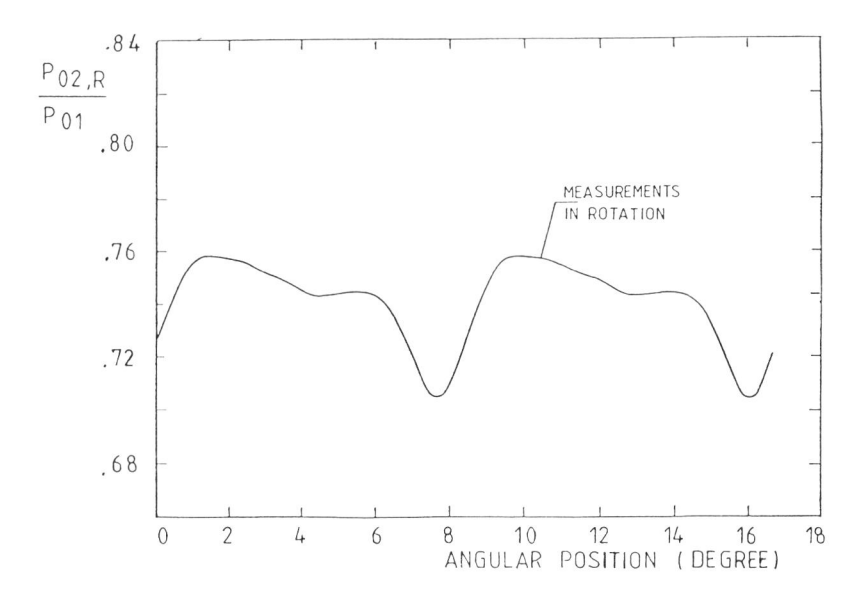

Figure 3.113 Non-dimensionalized relative total pressure profile.

Figure 3.114 presents the stationary flow data P_{02}/P_{01} and P_2/P_{01} and the derived relative total pressure ratio $P_{02,R}/P_{01}$. Figure 3.115 compares the measured relative total pressure with that derived from stationary measurements. The Kulite probe measurements with the rotating probe system have been corrected for temperature effects as described in Section 6.4.3.

The differences between the two curves in Figure 3.115 are surprisingly large

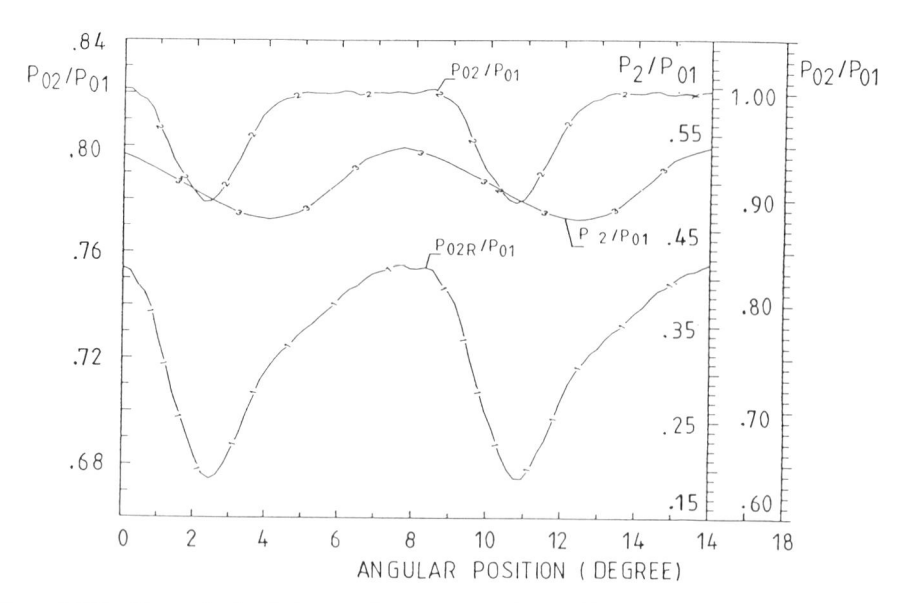

Figure 3.114 Theoretical relative total pressure ratio profile derived from stationary flow measurements.

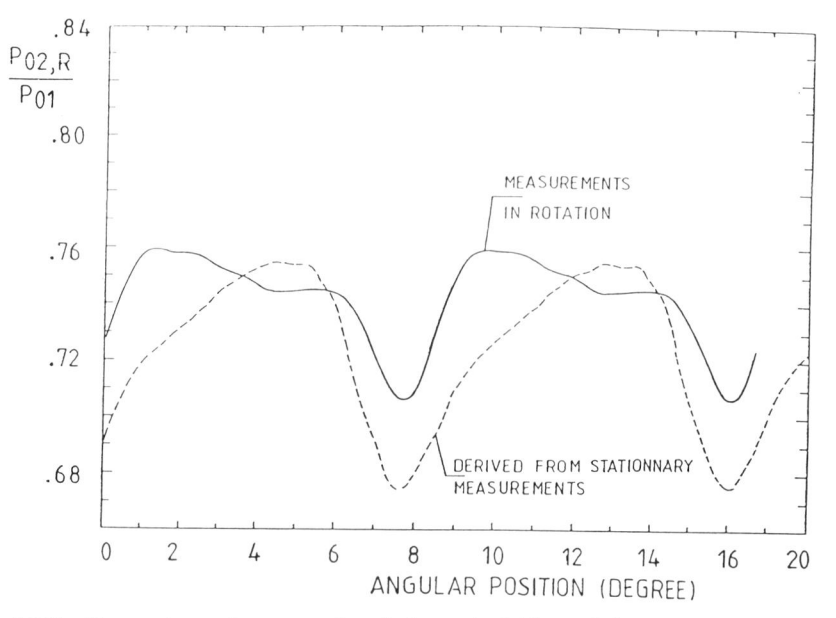

Figure 3.115 Comparison of measured and theoretical (derived from stationary measurements) relative total pressure ratio profiles.

and at first sight unexplainable. In analysing the possible reasons, it appears that the static pressure is the key problem, and this in spite of the expected reduced blockage effect resulting from the reduced Mach number with respect to the rotating probe. In fact, a comparison of the probe blockage effect between a stationary and a rotating probe is only meaningful for the same probe geometry and so was the initial intention. However, in the course of the project this aspect has changed profoundly. The stationary probe design was completely reviewed, resulting in probes with extremely long heads and drastically reduced blockage effects. On the other hand, the probe head of the rotating probe had to be shortened and stiffened to support the high centrifugal forces, thereby increasing the blockage effect. Obviously the differences in the probe geometry and the associated differences in the probe-blockage effects outweigh the advantage of the reduced Mach number for the rotating probe.

The interpretation of the relative total pressure measurements is not an easy task. In the absolute flow the measured total pressure contains only information on the total pressure loss through the blading. In the corresponding relative flow, the measured relative total pressure depends not only on the total pressure loss but also on the static pressure and the resultant relative velocity. A change in static pressure, induced by the probe stem, will therefore have a direct impact on the relative total pressure, but affects the absolute outlet pressure only indirectly via the dependence of the total pressure losses on the exit Mach number. The sensitivity of the relative total pressure to a change in the static pressure is also demonstrated in Figure 3.116. Without changing the average downstream static pressure for the mid-span flow conditions, the local static pressure distribution is modified by shifting it circumferentially by 3.65° with respect to the absolute total pressure profile. The effect on the calculated relative total pressure is surprisingly big. The calculated value $P_{o2,R}/P_{o1}$ has at present nearly the same shape as the measured

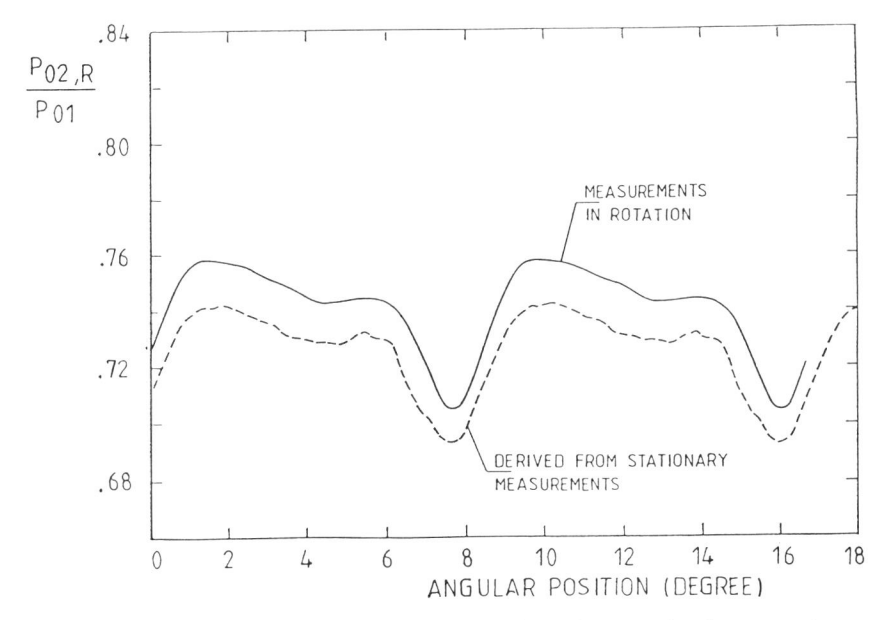

Figure 3.116 Comparison of measured and modified theoretical relative total pressure profiles (modification obtained through circumferential shift of static pressure profile in Figure 3.114).

value $P_{o2,R}/P_{01}$ except for a general shift between the two curves by $\approx 2\%$. The strong similarity between the two curves is of course purely incidental and a different shift of the static pressure distribution results in a different relative total pressure profile.

This leads us of course to question the opportunity of relative total pressure measurements in a turbine rotor, because of the difficulties in the interpretation of the results. The upstream effect of the rotor blades on the guide vane exit static pressure field will indeed be even stronger than that exerted by the probe stem.

Finally, it is worth while mentioning that the rotor inlet relative total pressure profile depends strongly on the axial distance between stator and rotor. The total pressure wake of the guide vane is convected downstream following the main flow direction while the static pressure gradients propagate in a different direction, in transonic flow often almost normal to the flow direction. Hence the wake traverses an ever changing static pressure field and thereby the relative total pressure profile will vary much more strongly with axial distance than the absolute total pressure profile.

7 Viscous flow calculations

7.1 Basic equations

The three-dimensional steady compressible viscous flow is governed by the well-known set of Navier–Stokes equations. The system of partial differential equations is solved by a time-marching procedure. Accordingly the equations appear in the following unsteady form:

$$\frac{\partial f}{\partial t} + \frac{\partial F_k}{\partial X_k} = \frac{\partial F_{vk}}{\partial X_k} \tag{3.34}$$

where f is the vector of unkonwns (density, ρ, velocity components, V_i and total internal energy, E), F_k are the flux vectors of the convective terms, F_{vk} are the flux vectors of the diffusive terms, and the X_k are Cartesian coordinates. The equations are discretized by a second-order accurate finite-volume approach for the convective terms while centred finite differences are used for the diffusive terms. The relative flux vectors are

$$f = \begin{bmatrix} \rho \\ \rho V_i \\ \rho E \end{bmatrix} \quad F_k = \begin{bmatrix} \rho V_k \\ \rho V_i V_k + p\delta_{ik} \\ (\rho E + p)V_k \end{bmatrix} \quad F_{vk} = \frac{1}{Re}\begin{bmatrix} 0 \\ \tau_{ik} \\ \tau_{ik}v_k - q_k \end{bmatrix} \tag{3.35}$$

The equation of state for a perfect gas completes the set of relations:

$$p = (\gamma - 1)e\rho = (\gamma - 1)\left(E - \frac{\langle V_i^2 \rangle}{2}\right)\rho$$

Turbulent flows are modelled according to the Favre-averaging technique, which allows accounting for the velocity and density fluctuations. The Boussinesq approximation is introduced to compute the Reynolds stresses and turbulent heat fluxes that are respectively set proportional to the effective viscosity μ_{eff} and effective conductivity k_{eff}. The simulation codes devloped for the computations account for the molecular viscosity so that μ_{eff} and k_{eff} are defined as

$$\mu_{eff} = \mu_l + \mu_t, \qquad k_{eff} = k_l + k_t$$

Accordingly, the stress tensor, τ, and heat flux, q, are written as follows:

$$q_k = -\frac{\gamma k}{Pr} \cdot \frac{\partial e}{\partial x_k}, \qquad \tau_{ik} = \mu\left[\left(\frac{\partial V_i}{\partial x_k} + \partial V_k \partial x_i\right) - \frac{2}{3}\left(\frac{\partial V_j}{\partial x_j}\right)\delta_{ik}\right]$$

All quantities are non-dimensionalized by the reference upstream velocity $\sqrt{(P_0/\rho_0)}$ and total temperature T_0, whereas the axial chord is the reference length. All other reference values are directly derived from these.

The physical domain is limited by inlet, exit, solid and periodic boundaries. The three boundary conditions (total pressure, total temperature, pitchwise flow angle β_1 for two-dimensional flows) are imposed at inlet, whereas only the static pressure p is imposed at exit, since the axial velocity component is subsonic in all the current set of calculations. In three-dimensional flows the spanwise flow angle is specified upstream, while in the exit section the static pressure is assigned at the hub and computed thereon by using the radial equilibrium equation. Along the solid boundaries a no-slip wall condition is imposed, while for temperature adiabatic condition is imposed.

7.2 Numerical algorithms

7.2.1 Two-dimensional algorithms

Two different algorithms have been used to approach the two-dimensional flow because of the need to implement the k–ε turbulence model to estimate the turbulence behaviour downstream of the cascade.

Implicit BL code A The numerical procedure to solve the system of equations is based on the approximate factorization scheme. To make the implicit coupled time-dependent algorithm less computationally costly, Pulliam [3.14] proposed a scalar form, adopted for the present set of calculations, that retains the intrinsic robustness of the original algorithm even for internal viscous flows (Michelassi et al. [3.15]). The governing equations are still that of system (3.34), but a general coordinate frame (ξ, η) is adopted in place of the Cartesian one. The convective, F_k, and diffusive, F_{kv}, Jacobians in the directions ξ, η, are now written in terms of contravariant velocity components.

The form of the standard approximate factorization algorithm in two dimensions can be written as follows:

$$[I + \Theta \Delta t(\delta_\xi F_1 - \delta_\xi^2 F_{1v})][I + \Theta \Delta t(\delta_\eta F_2 - \delta_\eta^2 F_{2v})]\Delta f = RHS \quad (3.36)$$

in which I is the identity matrix, θ is a parameter that allows weighting of the explicit–implicit nature of the space operator, δ is a centred difference operator, Δt is the time step, $\Delta f = f^{n+1} - f^n$, and RHS represents the convective and diffusive fluxes. The introduction of a set of eigenvalues, Λ, and eigenvectors, T, of the convective Jacobians allows writing $F_1 = T_\xi \Lambda_\xi T_\xi^{-1}$, $F_2 = T_\eta \Lambda_\eta T_\eta^{-1}$ that can be plugged into equation (3.36) to obtain the final form of the algorithm:

$$T_\xi[I + \Theta \Delta t(\delta_\xi \Lambda_\xi)]N[I + \Theta \Delta t(\delta_\eta \Lambda_\eta)]T_\eta^{-1}\Delta f = RHS \quad (3.37)$$

in which the matrix $N = T_\xi^{-1} T_\eta$ is solution independent. The solution process requires only scalar pentadiagonal matrix inversion with a reduction of nearly 50% of the operations required by the standard algorithm in which a block matrix inversion is required. Since the steady-state solution is sought, a local time-stepping formula is introduced. Second and fourth-order dissipative terms are implemented in the implicit part as well as in the explicit one ([3.14]–[3.16]). The fourth-order implicit damping requires a pentadiagonal scalar matrix inversion that largely enhanced the algorithm stability with respect to the standard three-diagonal matrix inversion with implicit second-order damping. The CFL number ranges between 5 and 10. The basic turbulence model implemented in this code is based on the algebraic Baldwin–Lomax (BL) formulation [3.42].

Explicit code B The second algorithm is based on an explicit pseudo-physical time-dependent technique (Martelli and Boretti, [3.18], [3.19]). The equations, written in Cartesian coordinates like the system (3.34) together with the k–ε transport equations, are discretized in time using an explicit, dissipative, four-stage Runge–Kutta algorithm. The equations are discretized in space by using a cell-centred, finite-volume formulation with hexagonal elements. The time derivatives and the source terms are evaluated at the cell-centre grid point. The spatial differences are evaluated as net fluxes across the faces of the cell, computed at the cell faces, and divided by the cell area. Global conservation is ensured by evaluating the flux vectors on the faces of the boundary-conforming mesh cells.

The scheme is second-order accurate in space on a regular grid. When the previous time-dependent equations (3.34) are written in the form

$$\frac{\partial f}{\partial t} = T_i(f) + T_v(f),$$

where T_i and T_v are inviscid and viscous operators respectively, the solution at the time $(m + 1)t \Delta t$ is expressed as a function of quantities evaluated at the time $(m)\Delta t$ according to the following recursive scheme:

$$f^{(0)} = f^{(m)}$$

$$f^{(k)} = f^{(0)} - C_k \Delta t[T_i(f^{(k-1)}) + T_v(f^{(0)})] + DISP, \qquad k = 1, \ldots, 4 \qquad (3.38)$$

$$f^{(m+1)} = f^{(k)}$$

where $DISP = [AV^2 D^2(f^{(0)}) - AV^4 D^4(f^{(0)})]$. D^2 and D^4 approximate second- and fourth-order differences respectively, AV^2 and AV^4 are numerical viscosity coefficients needed to stabilize the iterative process. The maximum CFL number resulting from the four-stage algorithm is 3.60, but a smaller value is adopted because of the full coupling of the k–ε conservation equations with the flow variables conservation equations, the neglected contribution of viscous effects to the definition of the maximum time step, and the strong mesh-size variations throughout the computational domain. The increase in robustness of the code has allowed the use of the k–ε model, which is really hard to treat because of the strong non-linearity of the source terms and, further, of the damping parts of the low Reynolds form implemented.

7.2.2 Three-dimensional algorithm

The two-dimensional implicit algorithm (code A) has been extended to the three-dimensional flows, retaining the same formulation. The general coordinate system (ξ, η, ζ) is adopted. The form of the standard algorithm in three dimensions can be written as follows:

$$T_\xi * [I + \Theta\Delta t(\delta_\xi \Lambda_\xi)] * N * [I + \Theta\Delta t(\delta_\eta \lambda_\eta)] * P * [I + \Theta\Delta t(\delta_\zeta \Lambda_\zeta)] T_\zeta^{-1} \Delta f = RHS \qquad (3.39)$$

The matrices $N = T_\xi^1 T_\eta$ and $P = T_\eta^1 T_\zeta$ are still solution independent. Since the steady-state solution is sought, a local time-stepping formula is introduced. The remaining part of the two-dimensional algorithm is substantially unchanged. The BL turbulence model is retained with some modifications to treat multiple boundaries.

7.3 Coolant flow ejection models

7.3.1 State of the art

A thorough review of the approaches proposed in the literature has been carried out, and about 25 papers have been analysed [3.20]–[3.41]. A great amount of publications deals with the experimental aspect of a general interaction between a main flow and a cross-flow jet without specifically referring to a cascade geometry. Most papers dealing with the numerical study are based on an incompressible flow model, only few treat compressible flow. Most authors use a $k-\varepsilon$ turbulence model with a high-Re formulation and wall functions after Launder and Spalding [3.29]. Leschziner and Rodi [3.30] introduce some modifications in the model to take into account surface curvature and strong accelerating flows. The conditions imposed at the jet are usually very simple: constant cross-flow velocity and zero value for the other components. Some authors [3.27], [3.28], [3.36] assume constant total pressure (incompressible flow) and others specify a parabolic distribution of the cross-flow velocity. Kar and Poulose [3.37] specify the velocity distribution through a second-order polynomial for the components in the plane normal to the jet. Most authors assume constant turbulence level and length scale at jet inlet. Only Viegas *et al.* [3.35] give some conclusions on the use of the BL model, but little information on the way they treat the jet zone is given in the paper. Unfortunately no in-depth analysis of the behaviour of the different numerical models for that mixing area is presented. The global approach proposed by Kollen and Koschel [3.41] is suitable to evaluate the mixing process but can not be implemented in an NS code.

On the basis of this investigation the code was modified following three different guidelines:

- mesh improvement and point clustering in the mixing region;
- improvement of the turbulence model in the jet mixing area;
- new boundary conditions in the area of the jet exit.

7.3.2 Modification of the code

The implicit two-dimensional code A with the algebraic BL model has been extended to the flow ejection case. To model the mixing process new boundary conditions have to be imposed in the jet area. Further, an appropriate turbulence modelling is stated for the viscosity value in the region close to the jet exit, accounting for interaction of the two streams and the wall distance. According to the previous literature analysis three different formulations have been considered.

(a) The two velocity components and the temperature are specified on the blade surface at the position of the jet. Accordingly, the usual balances are performed on the control volume and the pressure is computed. With this procedure the jet mass flow can not be fixed and is a result of the calculation. This procedure was originally tested in the code, exhibiting stability problems. Difficulties have been found also in the iterative procedure set up to enforce the required mass flow rate. This approach has been abandoned.

(b) The jet mass flow has been fixed according to the data and distributed along the jet slot. The flow angle and total temperature are fixed. The total pressure at the jet exit is a result of the calculation and no guess has to be provided; the true experimental jet mass flow rate and total temperature are automatically enforced. In fact the jet flow rate might be expressed as

$$(\rho v)_c = \frac{P_{0c}}{\sqrt{T_{0c}}} \cdot \sqrt{\left[\frac{2\gamma}{\gamma - 1}\left(1 - \left(\frac{P}{P_{0c}}\right)^\alpha\right)\right]\left(\frac{P}{P_{0c}}\right)^\beta}, \qquad \alpha = \frac{\gamma - 1}{\gamma}, \quad \beta = \frac{1}{\gamma}$$
(3.40)

The static pressure P at the jet exit might be computed by the NS solver, while P_{0j} is the jet total pressure necessary to give the required mass flow rate. Observing that

$$(\rho u)_c = (\rho u)_{\text{inlet}} C_m \frac{\text{pitch}}{\text{jet slot}}$$

in which C_m is 0.0, 0.02, 0.03, 0.04, and

$$(\rho u)_c^2 T_{0c} \frac{\gamma - 1}{2\gamma} = \text{constant}$$

it can be proved that equation (3.40) has a monotonic behaviour with P_{0j}. The only remaining unknown in equation (3.40) is P_{0j}, which can be determined by an iterative process to match the local total conditions to the desired mass flow rate at every NS iteration. Observe that in this way the static pressure computed by the NS solver in the jet slot at every iteration is correctly affected by the presence of non-zero velocities on the blade. This modification affects only the boundary conditions and leaves the rest of the code totally unaltered. This procedure has proved to be stable without introducing significant reduction in the convergence rate.

(c) The experimental total pressure, total density, and flow direction at the jet exit have been assumed as data. This appeared more physical and related to the actual experimental conditions. Unfortunately the correct value of the total pressure is not always known so that it needs to be guessed to match the correct mass flow rate. The procedure results in a modification of the previous method (b) since the local mass flow rate at the jet slot is computed through an isentropic expansion from the total and the local static pressure conditions. This feature introduces quite large oscillations because the pressure is not fixed; it is computed by the energy equation. This approach requires a more complex procedure compared with method (b) and does not ensure a better stability or quality of results.

Note that all approaches assume the same fluid for jet and main stream (i.e.

the same γ and k). All calculations presented hereafter have been carried out by using form (b). Regarding the evaluation of the viscosity in the jet region we have considred several possibilities, according to the suggestions found in the literature. Viegas *et al.* [3.35] assumed the jet behaving as two boundary layers separated by an inviscid core and consequently computed the turbulent viscosities in the jet region. This interpretation is consistent with a set of experiments mentioned in [3.35]. In the top surface boundary layer at the lip of the jet exit, the eddy viscosities were computed as part of the overall computation process. To account for the 'history' of the jet development, the eddy viscosity at the jet exit station can be blended with the local eddy viscosity given by the BL model through a linear weighting function. According to the previous studies two formulations have been investigated:

(a) The BL model is used for the inner layer with an adapted evaluation of the distance from the wall.

(b) The eddy viscosity at the jet exit station is blended with the local eddy viscosity given by the BL model through a linear weighting function that gives full weight to the upstream values at the jet exit and full weight to the BL model at the trailing edge; this approach is more complex and it requires a large number of mesh points in the jet area. The tests carried out showed no significant modification to the results, and the simpler approach (a) was adopted.

7.4 Mesh generation

The accuracy needed in the jet-mixing region requires a refined mesh generator. A flexible interactive grid generation program has been designed and used in connection with the viscous solver. The geometry encountered in nearly all the blade cascades may be conveniently mapped by using H-, C- or O-type grids. Among these three, the H-grid with a periodic grid point distribution in the flow field upstream and downstream of the blade has the simplest structure. Normally the standard H-grid results in very skewed cells downstream from the trailing edge with a detrimental effect on the quality of the calculations. The C-grid allows a better resolution of the blade nose, especially for turbine blades, but normally has the same problems as the H-grid downstream of the blade. The O-type grids seem to fulfill the requirement of a proper mesh resolution in both the leading- and trailing-edge regions, but it is more suited for isolated aerofoils than for blade cascades. While retaining the general H-grid structure, the grid generator is based on a mixed elliptic and algebraic method that is used to compute an initial coarse computational grid with constant point spacing. The method allows the identification of several regions in which elliptic or algebraic variable-order polynomials (of the form $x_i = \sum_{k=1,n} \alpha_k x_j^{k-1}$ in which x,y are the coordinates and n is the order of the polynomial) are used. On the coarse mesh, Figure 3.117, a refined grid is

Viscous flow calculations

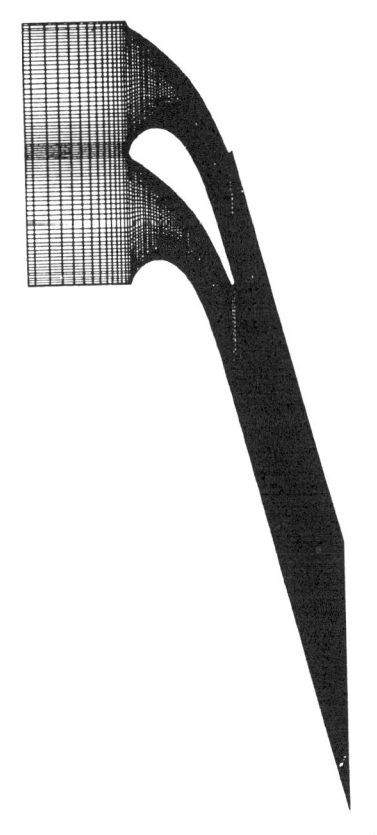

Figure 3.117 Coarse point periodic grid.

algebraically interpolated and clustered along the initial coarse grid lines, Figure 3.118. The elliptic grid generation is particularly suited for leading- and trailing-edge regions. This technique generally yields very fine grids quite similar to elliptic meshes, with a dramatic decrease in computational time. To avoid the grid skewness in the flow field upstream and downstream of the blade a special type of H-grid was selected in which the point distribution periodicity on the lower and upper boundaries is abandoned (Figure 3.118). The shape of the periodic boundaries is given by second- or third-order polynomials. The shape of the upper and lower sides is clearly periodic, but the point distribution on these sides is not. This grid is evidently much more regular than the one shown in Figure 3.117. The grid cells are not distorted in the flow region behind the blade, thereby allowing a more accurate discretization of the differential equations. Experience has shown that curved periodic boundaries downstream of the blade with non-periodic points distribution can introduce mass errors because of the imperfect correspondence of the upper and lower discretized boundaries (Figure 3.118). Therefore in the calculations that boundary has always been assumed straight.

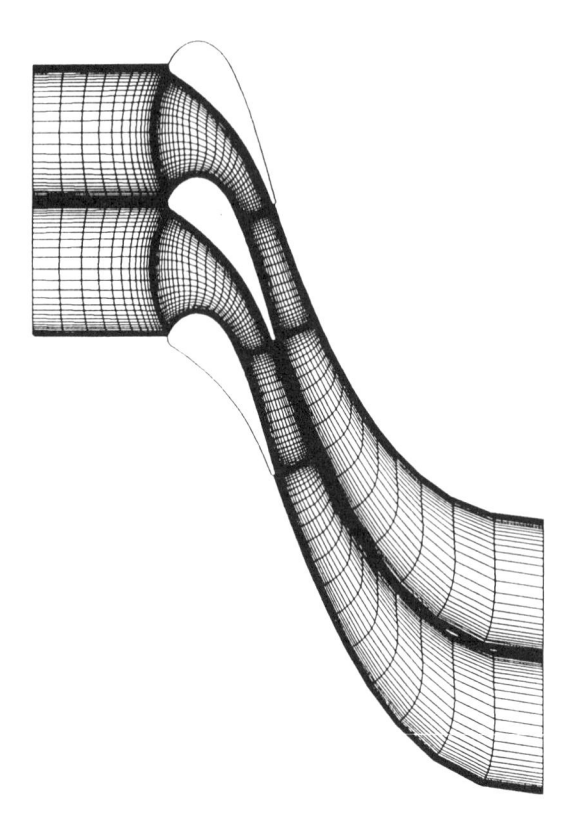

Figure 3.118 Refined not-point-periodic grid.

7.5 Turbulence models

7.5.1 Algebraic model

The algebraic turbulence model (implemented in the code 2D-A and 3D) is based on the well-known Baldwin–Lomax (BL) formulation. The BL model divides the flow into two zones, the first of which, close to the wall, is governed by the van Driest damping expression for the mixing length, which increases monotonically with the wall distance. The thickness of this region is approximately determined by the location of the point where the product of the distance from the surface and the absolute value of the local mean vorticity first becomes a maximum. In the evaluation of the law-of-the-wall coordinate y^+, the velocity gradient at the wall appearing in the wall friction velocity has been replaced by the maximum value of the vorticity along the normal to the wall in order to avoid ill-posed conditions if separation occurs. While referring to the original paper [3.42] for further details on the model, special care was taken to average the various boundaries contributions to the turbulent viscosity in the flow core. Michelassi *et al.* [3.15] found it convenient to weight the walls contributions according to the following formula:

$$\mu_{t\,core} = \frac{1}{\sum_{i=1,nw} \frac{1}{W_i}} \sum_{i=1,nw} \left(\frac{\mu_{t\,core}}{W}\right)_i \qquad (3.41)$$

in which nw is the number of boundaries delimiting a cross-section, and W is the Van Driest damping function. This expression seems to ensure the necessary blending of contributions from the boundaries. Note that equation (3.41) is applied only in the flow core, while μ_t in the inner layer in proximity of the walls is computed without averaging. The BL model optionally includes a simple laminar/turbulent transition criterion according to which μ_t is set equal to zero if the ratio between the maximum turbulent to molecular viscosity in a cross-section is smaller than 14.

7.5.2 Two-equation model

The problem of algebraic turbulence models is the absence of any significant transport mechanism; the effect of vorticity transport alone cannot simulate the turbulent kinetic energy transport. This transport mechanism is relevant in mixing processes as we are dealing with. The experience carried out in a duct flow with a k–ε model [3.17] suggested the introduction of two equation models for cascade flow. The implementation of this class of models is not straightforward because of the presence of stagnation points, but work is in progress for a first set of IGV geometries.

Because of the need to treat the region close to the ejection hole in which a boundary layer interacts with a nearly inviscid jet it has been necessary to use a low-Re form of a k–ε model [3.43] that accounts for the molecular viscosity of the fluid. In the basic equations (3.34) two new unknowns, ρk and $\rho \varepsilon$, are added; F and G are extended with the classic transport terms for the turbulent quantities and a source term S is added on the right-hand side where compressibility effects are also taken into account following the suggestions of Wilcox and Rubesin [3.46]. The corresponding set of terms and low-Re functions can be found in [3.19].

New boundary conditions are added. The turbulence level and length scale are specified at the inlet. Along the solid boundary turbulence kinetic energy and dissipation rate vanish. At the outflow boundary the unknowns are extrapolated from the internal cells. Observe that no transition criteria is specified and the boundary layer is assumed completely turbulent.

7.6 Two-dimensional flow calculations

The final set of calculations for the two-dimensional linear cascade refers to the most significant experimental results of the DLR tests. The test conditions are described by the exit isentropic Mach number M_{is}, and by the jet mass flow rate expressed in terms of percentage of the IGV mass flow rate. The calculations, summarized in Table 3.9, have been performed for both the trailing-edge ejection blade (TE blade), and the pressure-side ejection blade (PS blade). The total temperature ratio between the jet and the mainstream (T_{ojet}/T_0) used in the two-dimensional calculations has been set at 0.65 in order to reproduce the same density ratio of the experimental tests at DLR, in which the fluid ejected was CO_2.

Two typical computational grids for the trailing edge and pressure-side ejection geometries are shown in Figure 3.119. The grids have 175 × 73 points. The trailing-edge region is reasonably well refined and allows the most important features of the flow patterns to be captured. Downstream of the trailing edge the streamwise point distribution is kept as regular as possible to allow the description of the wake decay. The grids are obtained by using the elliptic approach in the trailing-edge area, whereas the rest of the grid is computed by using the algebraic option. For the two-dimensional implicit code A the CFL number used in the time-step formula [3.15] ranges from 5 to 10. The artificial second- and fourth-order damping weights were minimized to ensure wiggle-free solutions and still maintain an acceptable convergence rate. In fact, due to the implicit treatment of both second- and fourth-order damping, the weights of the artificial terms might affect the convergence rate. Typical weights for second-order damping range between $\frac{1}{8}$ and $\frac{1}{2}$ (the lower value refers to shock-free flows), while the fourth-order ranges between $\frac{1}{64}$ and $\frac{1}{16}$. For the two-dimensional explicit code B the time step was limited to 30–70% of the maximum explicit value. The artificial damping terms have weights that are typically double those adopted in the implicit code A. The presence of the two-equation turbulence model often necessitated raising the second-order damping weight up to unity. The runs were performed on an IBM Risc 6000 super workstation with 32-bit accuracy and on a CRAY-YMP with 64-bit accuracy. No appreciable differences were found in the converged results. Typical computational times range from 1 to 4 hours on the IBM for both the codes, while approximately 400–800 s are needed on the CRAY supercomputer for the implicit code A. These computer times are required to reach overall averaged residuals of the order of 10^{-6}–10^{-7}.

Table 3.9 Summary of computations. (The isentropic Mach number has an indicative value. For real computations M_{is} matches exactly the experimental value)

M_{is}	$C_m(\%)$	0.0	2.0	3.0	4.0
0.70					
0.80		✓	✓	✓	✓
0.90					
1.00					
1.05		✓	✓	✓	✓
1.10					
1.20		✓	✓	✓	✓

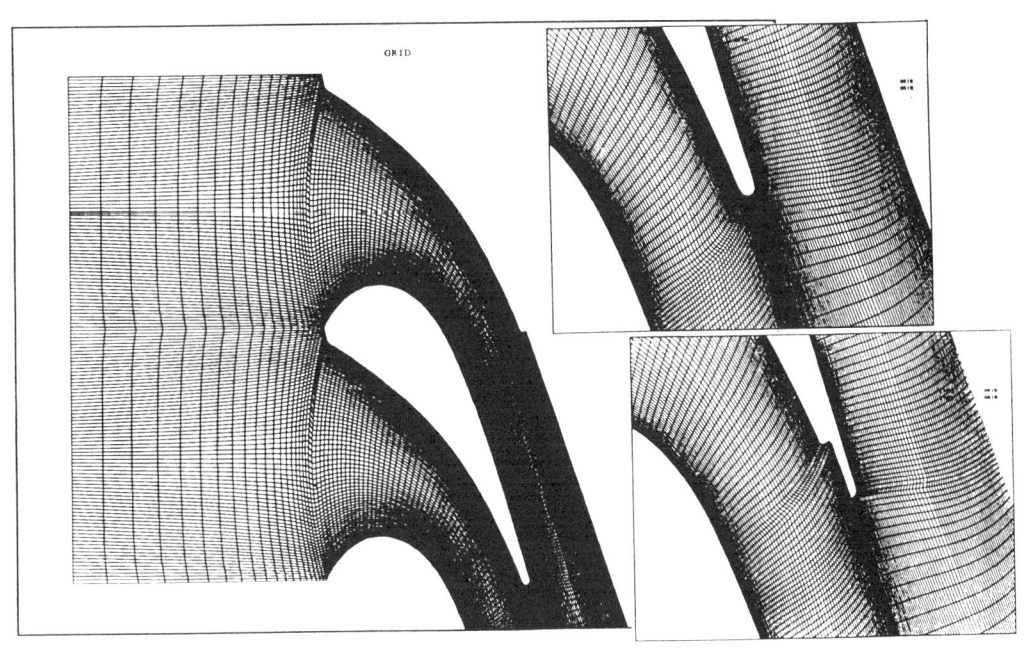

Figure 3.119 Computational grids.

7.6.1 Blade-velocity distribution (isentropic Mach number)

Figure 3.120 shows the blade isentropic Mach number distributions computed by the explicit solver with the two-equation turbulence model. At transonic (nominal) outlet Mach number one observes a very weak shock located at approximately 40% of the chord in both the TE and PS ejection cases. The computations by the implicit solver show a constant velocity downstream of the shock for the TE blade at $C_m = 0\%$, whereas for the same conditions the explicit solver shows an evident deceleration together with a pressure recovery that is maximum at $M_{is} = 1.20$. A more complete set of calculations was performed with the BL turbulence model and implicit code A. Figure 3.121 shows the computed M_{is} distributions for some of the Mach number–ejection flow rate combinations summarized in Table 3.9. The results obtained with the refined grid might be considered grid independent, since further grid refinement did not affect the isentropic Mach number profiles.

The subsonic solutions are practically independent of the coolant flow since the jet mainly influences the shock strength. In the supersonic flow regime the shock position is unchanged in the presence or absence of the coolant jet. For the TE blade the coolant ejection does not produce big changes in the computed flow pattern, and the position of the shock remains substantially unchanged, but the strength is largely reduced by the presence of the jet, regardless of the mass flow rate.

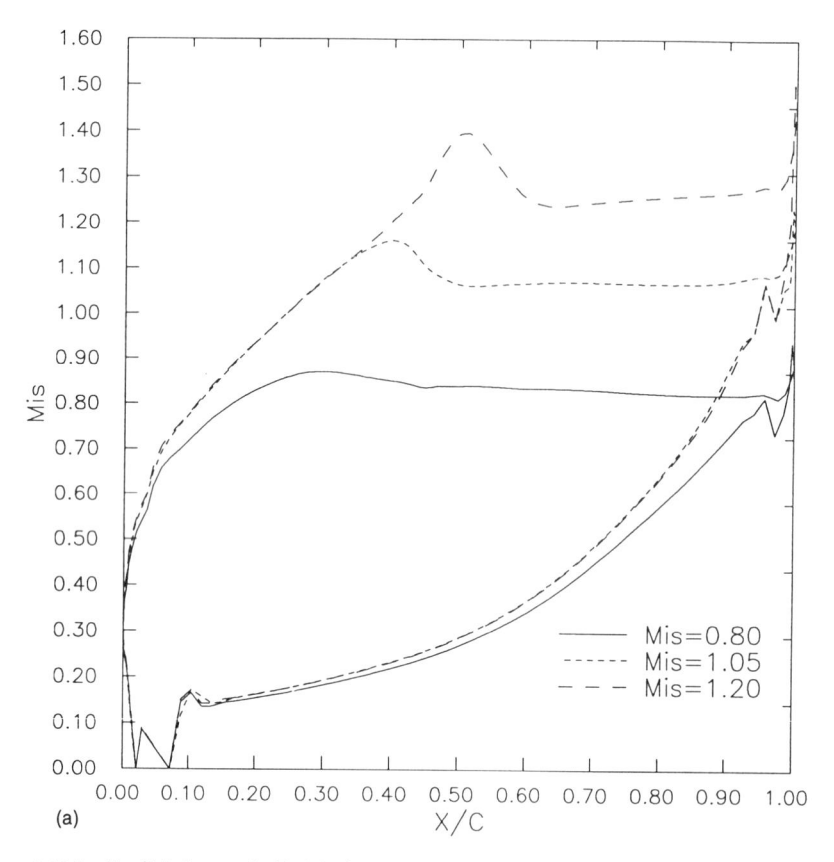

Figure 3.120 Explicit $k-\varepsilon$ code B: M_{is} for (a) TE ejection, $C_m = 0\%$; (b) PS ejection, $C_m = 0\%$.

Figure 3.121(c) shows the base pressure coefficient $C_b = (P_b - P_2)/P_{(01)}$ computed for the TE ejection blade. Not all the computed values are reported, since, due to the presence of pressure wiggles in the proximity of the trailing edge, some computed values were out of scale. The general trend shows an increase of C_b as the jet flow is switched on. It is interesting to observe that C_b increases with C_m and M_{is} up to $M_{is} = 1.05$, then decreases again due to the change in the shock system from normal to oblique.

Figure 3.122 refers to the PS blade. Comparing the TE-ejection blade Mach number distributions at $M_{is} = 0.8$ with and without jet flow it is possible to observe that the computed throat section acceleration remains unchanged for the two-blade geometries. For this set of calculations it was necessary to switch on the transition criteria used in the BL model. The plots refer to the same conditions as for the TE blade and show a quite different behaviour of the blade at the supersonic flow regime. In fact at $M_{is} = 1.20$ it is possible to detect two shocks at approximately 30 and 50% of the blade chord, the first of which is generated by the sharp corner at the jet, the second at the blade trailing edge. For this geometry the shock is entirely smeared by the coolant ejection at the nominal Mach number $M_{2,is} = 1.05$, whereas only the first shock is smeared

Viscous flow calculations

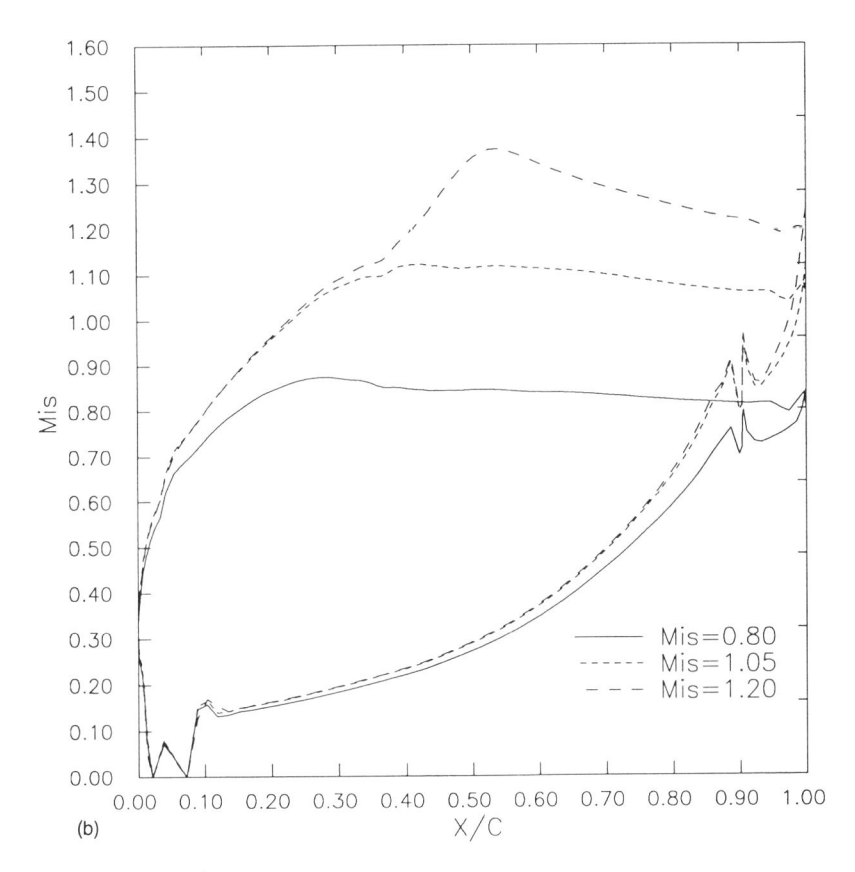

Figure 3.120 (*continued*)

when injecting at supersonic conditions, i.e. at $M_{2,is} = 1.2$. The heavy influence of the jet on the flow pattern for the PS blade is partly due to the coolant slot exit positioned considerably upstream with respect to the TE. For this geometry the computations reveal the heavy influence of the transition model. Figure 3.123 shows the effect of the introduction of the transition criterion on the isentropic Mach number. Without transition the computations predict a nearly constant pressure and velocity, whereas, when fixing the transition at $\mu_t/\mu = 14$ (which corresponds approximately with the shock impingment on the suction side) the flow accelerates from sonic to supersonic. This large difference in the predictions was found only at the nominal exit Mach number for the PS blade, while it produced virtually no changes in all the other configurations where the boundary layer was always considered completely turbulent.

Figure 3.123(c,d) shows the base pressure coefficients. Actually, the PS blade needs the computation of two base pressures, one located at the coolant slot exit on the pressure side (C_{b1}), the other at the trailing edge (C_{b2}). C_{b1} grows with M_{is} because of the increased difference between the base pressure and the pressure upstream of the trailing edge shock, which has the same effect as the coolant flow ejection. This behaviour is contrary to that of C_{b2}, where the coolant ejection produces a decrease in the computed base pressure. This might be explained by observing that the PS ejection accelerates the flow field near the trailing edge with large changes in the flow pattern.

Effect of blade geometry The blade geometry in the trailing-edge region affects the development of the flow field because of the different shock-expansion wave patterns at transonic and supersonic flows, whereas the flow fields for the TE and PS blades look much the same for the subsonic cases.

The TE blade has a unique expansion-shock waves pattern departing from the trailing edge. In fact the computed isentropic blade Mach number distribution for $C_m = 0$ shows only one velocity peak and one shock on the suction side, which is evidently generated at the pressrue side of the trailing edge. Conversely the PS blade exhibits a much more complex flow field since the backward-facing step on the rear pressure side and the blade trailing edge cause two separate expansion-shock waves, which can be seen for the $C_m = 0$ configuration (Figure 3.122(a)). In fact the M_{is} distribution on the suction side of the blade shows an up–down behaviour that might be interpreted by the

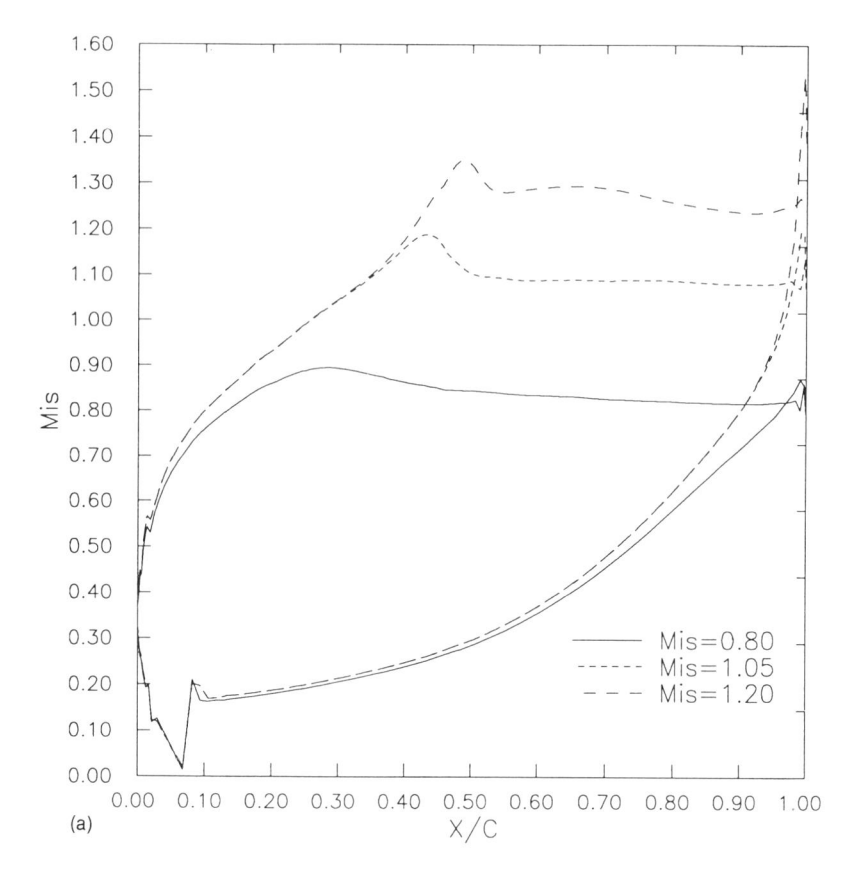

Figure 3.121 (a) Implicit BL code A: M_{is} for (a) TE ejection, $C_m = 0\%$; (b) TE ejection, $C_m = 3\%$; (c) implicit BL code A: base pressures for TE ejection.

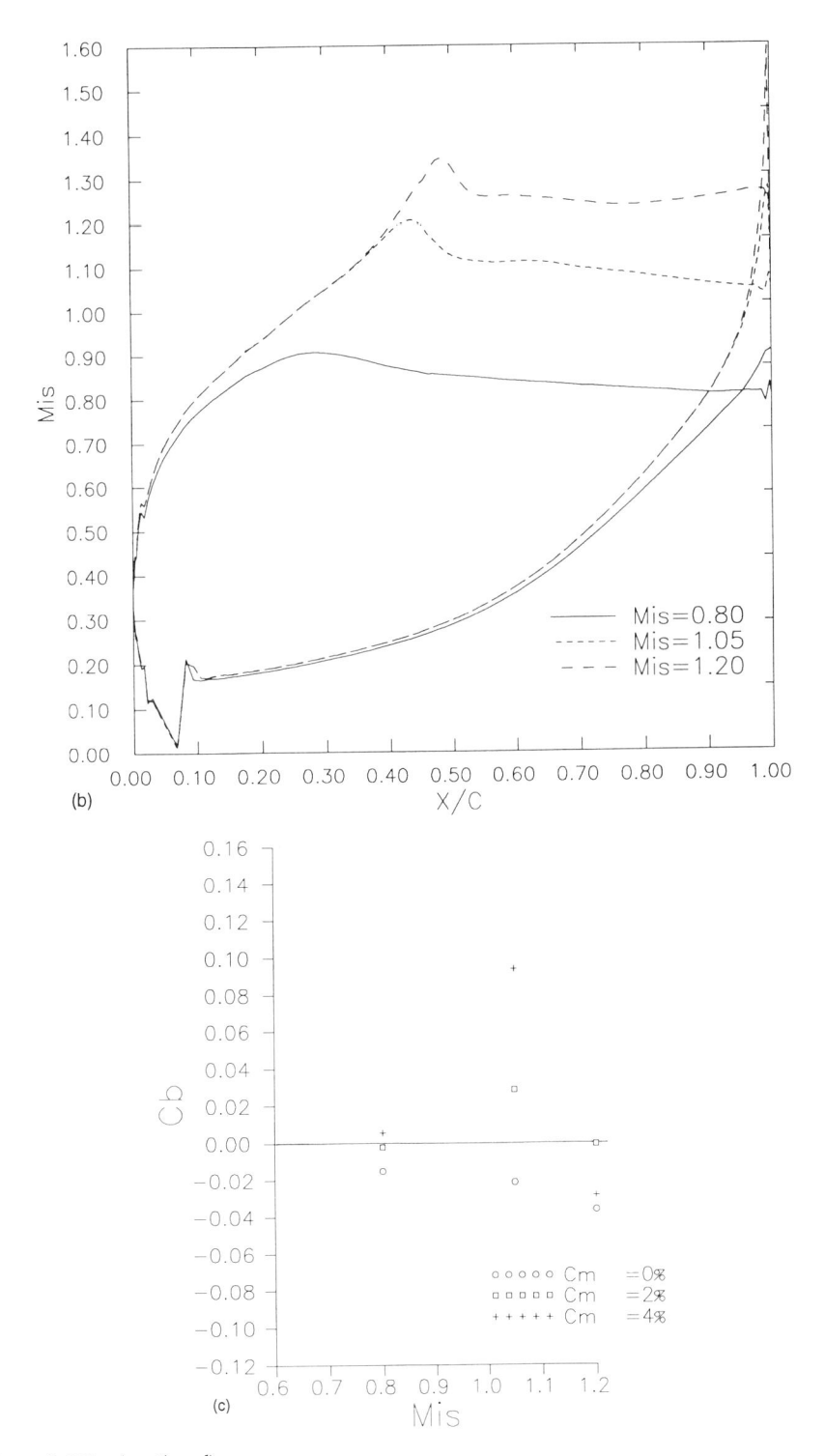

Figure 3.121 (*continued*)

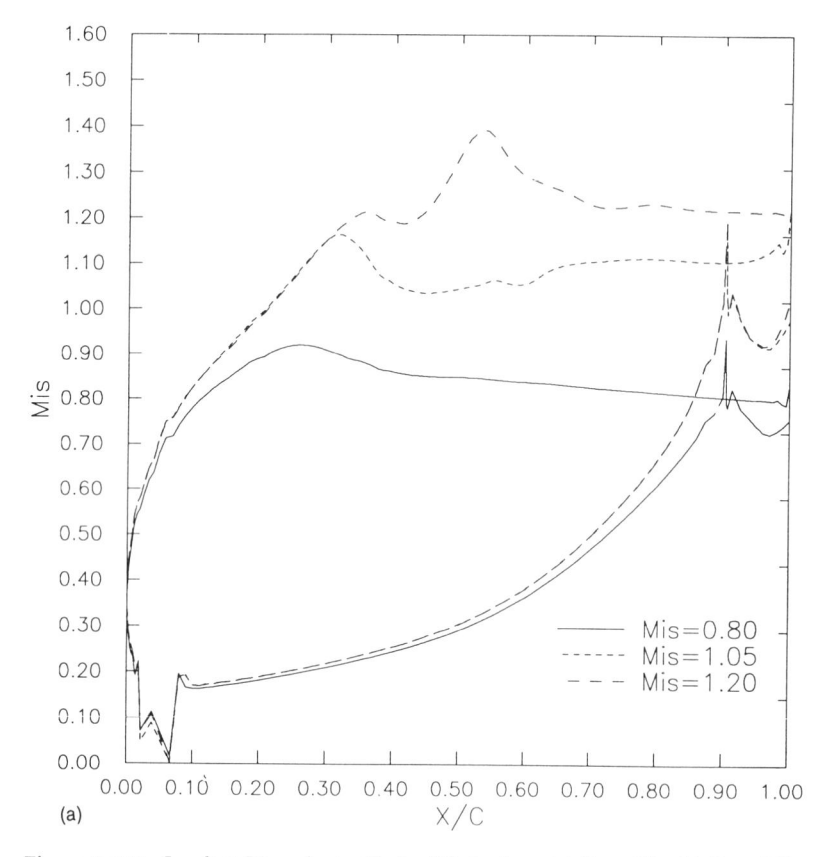

Figure 3.122 Implicit BL code A: M_{is} for PS ejection: (a) $C_m = 0\%$; (b) $C_m = 3\%$.

presence of an expansion wave originating at the pressure-side step. The expansion wave is followed by a first weak shock departing from the reattachment point between the backward-facing step and the trailing edge. This pattern is followed by the usual expansion-shock waves at the blade trailing edge, where the two streams from the suction and the pressure sides merge together, as can be seen in Figure 3.124, which shows the isolines of the function S, defined as $S = M_{local}(\text{grad } P)|(\text{grad } P)|^{-1}$, which is particularly suited for shock visualization.

The flow pattern in the vicinity of the trailing edge is also of great interest. The TE ejection blade has a quite thick trailing edge, which is necessary for a mechanically and thermally resistant blade, despite the coolant ejection slot. Figure 3.125 shows the computed flow pattern for the two blade geometries. For $C_m = 0$ the TE ejection blade shows two quite long counterrotating vortices, whereas the PS ejection blade exhibits in addition to the trailing edge vortices a large recirculation region behind the backwards facing step. This produces a thicker wake for the second blade as compared to the first in case of no coolant ejection.

Viscous flow calculations

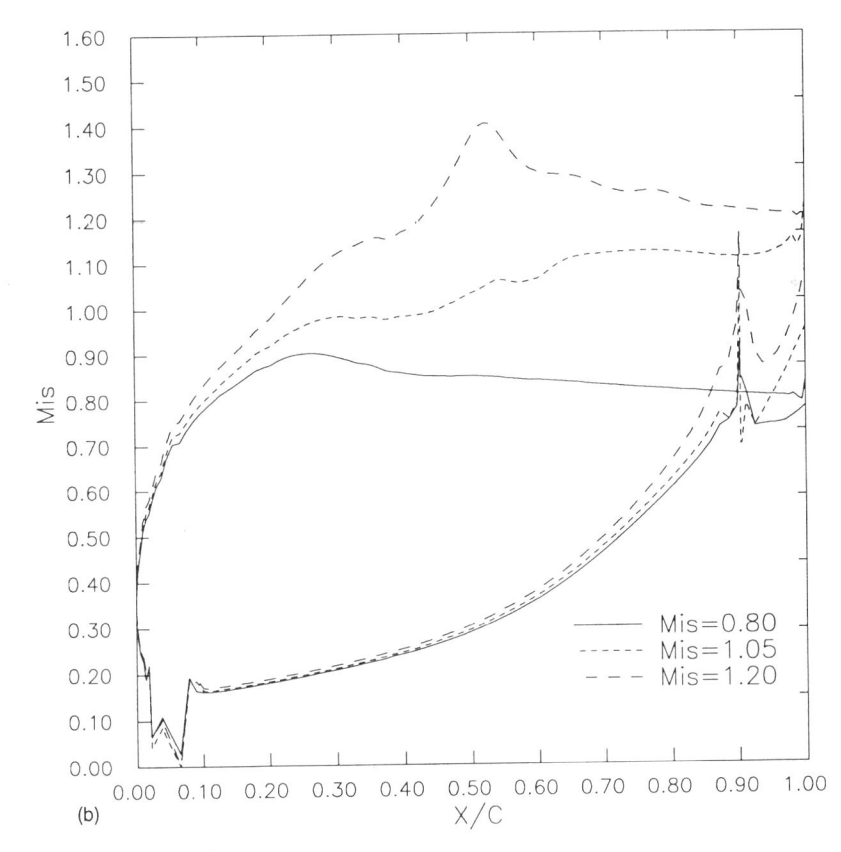

Figure 3.122 (*continued*)

Effect of coolant flow ejection A much deeper understanding of the blade flow fields is obtained by studying the effect of coolant flow ejection. Figures 3.120–123 indicate the absence of shocks in the transonic flow regime when switching on the coolant ejection. The computations refer to the set of measurements in which the fluid ejected was cold air.

Figure 3.124 shows for the PS ejection blade the shock patterns for $C_m = 0$ and 3%. While the differences in the far downstream region are small, the expansion wave followed by the shock departing from the step on the rear pressure side is evident for $C_m = 0\%$ whereas it has nearly disappeared for $C_m = 3\%$ and also the downstream shocks appear somewhat smeared. This explains the large differences found in the blade Mach number distributions.

The flow visualizations given in Figure 3.125 show the dramatic changes induced by the coolant ejection in the trailing-edge region. The TE ejection destroys the two large trailing edge vortices. Only a small vortex is found on the suction side of the ejection, but this small structure is not steady and tends to shed during the iterations. Since the flow is subject to a very smooth turning on the pressure side due to the coolant injection the expansion and shock

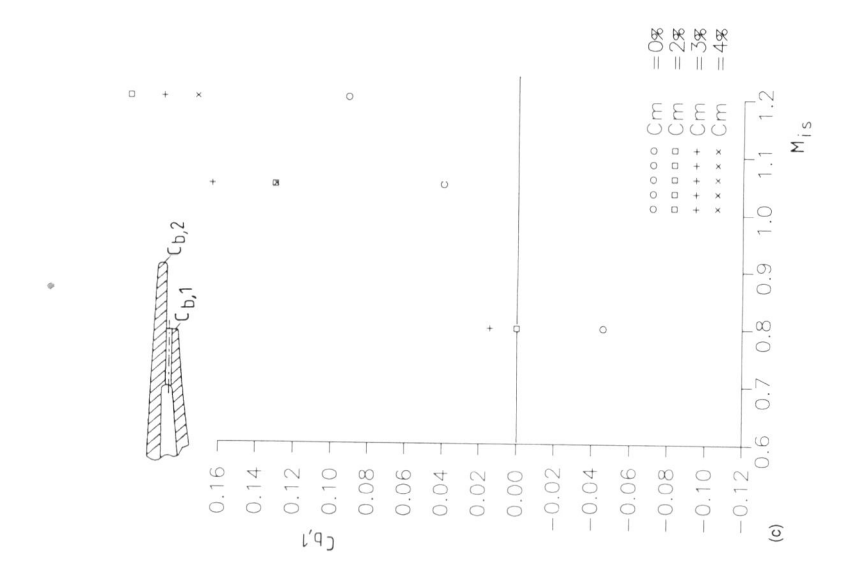

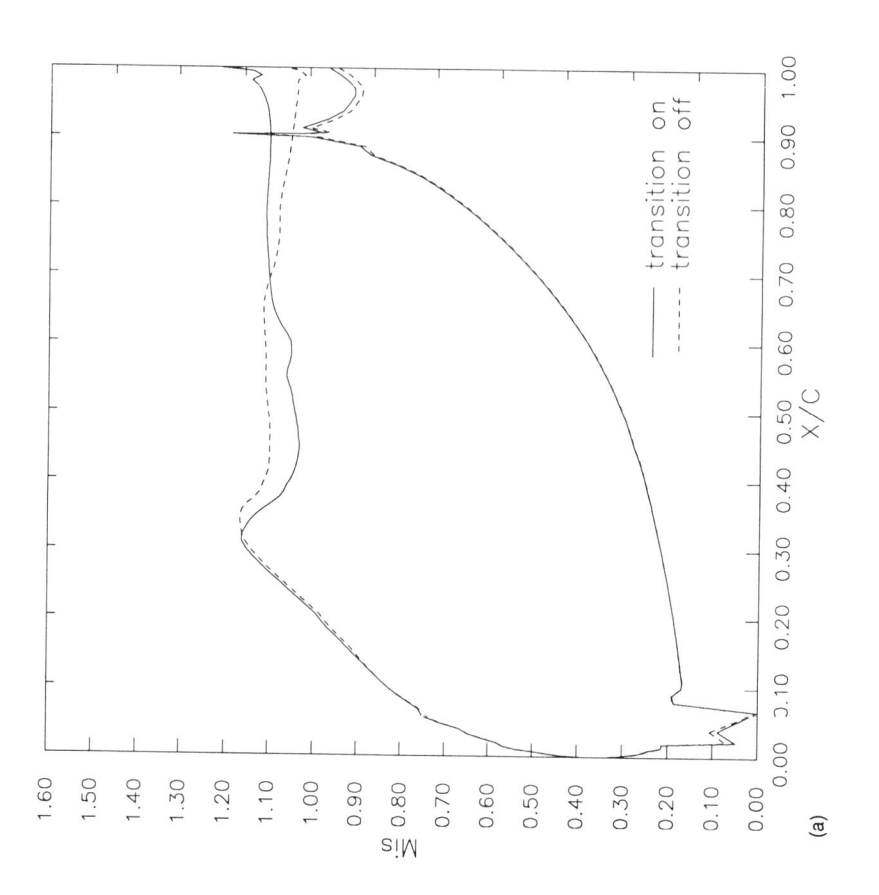

Viscous flow calculations

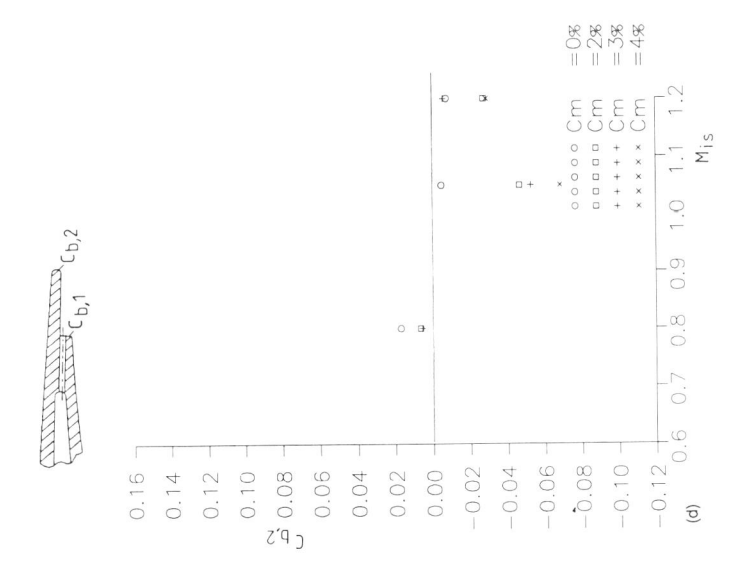

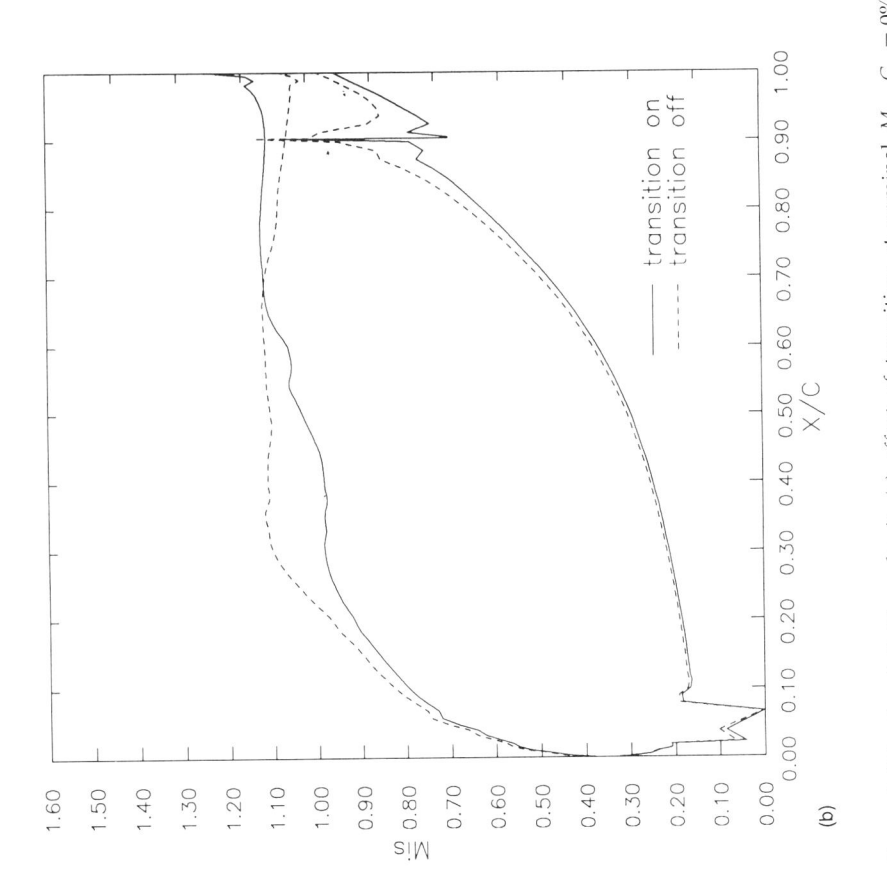

Figure 3.123 Implicit BL code A: (a) effect of transition at nominal M_{is}, $C_m = 0\%$; (b) effect of transition at nominal M_{is}, $C_m = 3\%$; (c) base pressures C_{b1} for PS ejection; (d) base pressures C_{b2} for PS ejection.

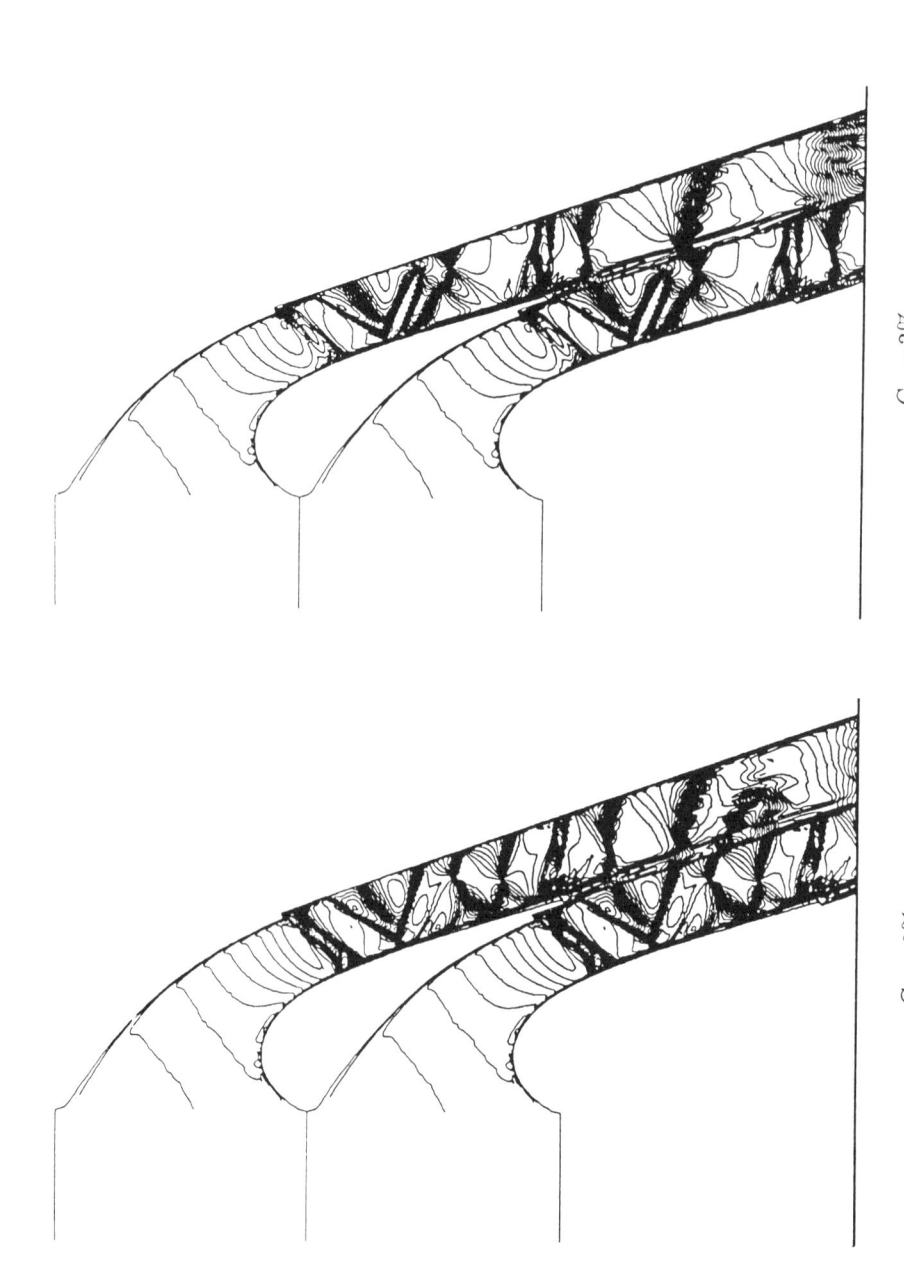

Figure 3.124 Computed shock pattern, $M_{is} = 1.2$.

$C_m = 3\%$

$C_m = 0\%$

Viscous flow calculations

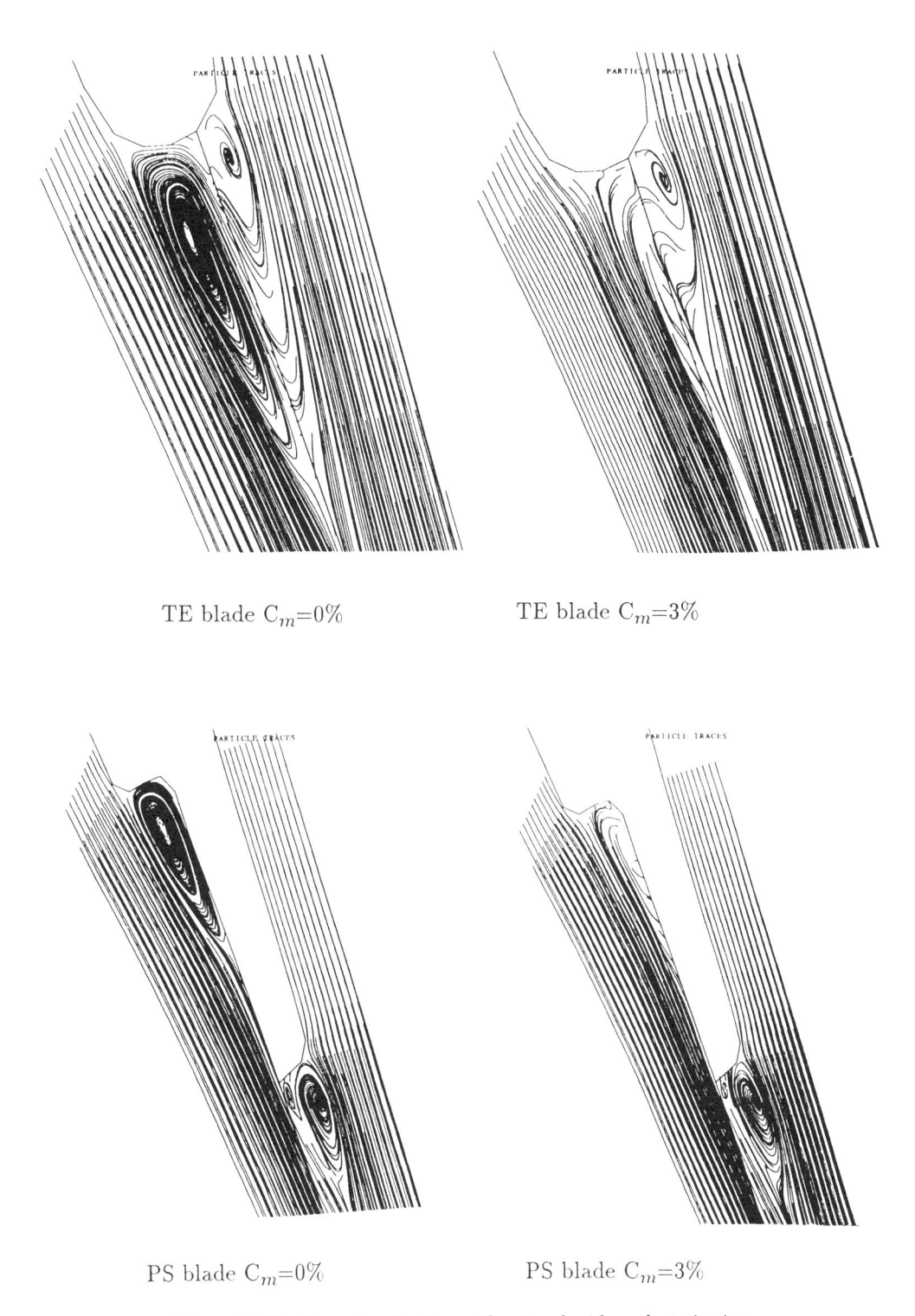

TE blade C_m=0%

TE blade C_m=3%

PS blade C_m=0%

PS blade C_m=3%

Figure 3.125 Flow visualization without and with coolant ejection.

waves cannot have the same strength as found for the $C_m = 0$ case. The PS ejection has not the same beneficial effect. Figure 3.125 shows that the PS ejection kills the recirculation bubble downstream of the pressure side step, but the trailing edge recirculations survive having the same size as for the $C_m = 0$ case, with a detrimental effect on the general performances of the blade. The pressure isolines in Figure 3.126 show the large differences between a subsonic shock free and a supersonic flow field.

7.6.2 Blade performance calculations

Losses might be computed by the following standard formula:

$$\zeta = \frac{(V_{is}^2 - V^2)}{V_{is}^2}$$

where V_{is} and V ar the isentropic and real velocity far downstream. The ζ can be evaluated through the mass averaged mean total pressure $\overline{P_0}$ to obtain

$$\zeta = \frac{[\overline{P_0}^{(\gamma-1)/\gamma} - 1]}{[(\gamma-1)/2]M_{is}^2} \tag{3.42}$$

where ρ is the density and M_{is} is the isentropic Mach number. Recall that P_0 is non-dimensionalized with respect to the inlet value $P_{(0\,inl)}$. Because of the

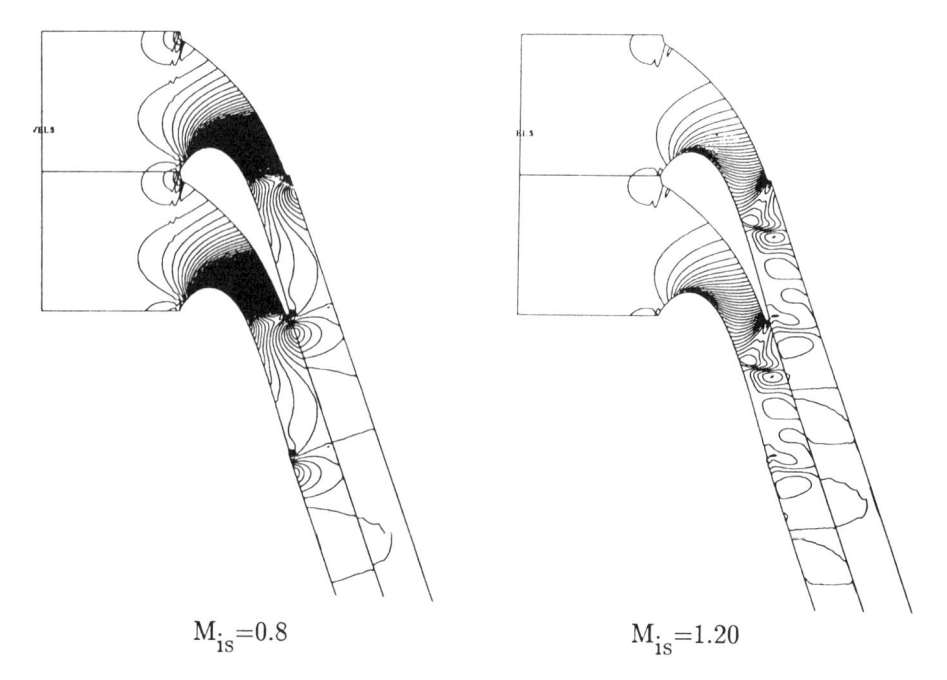

$M_{is}=0.8$ $M_{is}=1.20$

Figure 3.126 Pressure isolines (TE).

presence of two streams with different total pressure (P_0) and temperature (T_0) a new loss coefficient can be evaluated:

$$\zeta = 1 - \left(\frac{V}{V_{is}}\right)^2 (1 + C_m) \bigg/ \left[1 + C_m \left(\frac{V_{is\,c}}{V_{is}}\right)^2\right] \quad (3.43)$$

in which C_m is the coolant jet mass flow rate, $(V/V_{is})^2$ and $(V_{is\,c}/V_{is})^2$ can be computed from mj. Because of the small coolant mass flow rates C, equation (3.43) gives practically the same results as equation (3.42).

A new approach to the computation of ζ has been developed to reduce the error related to the averaging of non-uniform total pressure. In place of averaging the total pressure in the exit section of the computational domain, a set of conservation equations for mass, axial and tangential momentums, and energy, is applied to a control volume between the trailing edge and the mesh boundary far downstream. The equations are then solved by an iterative procedure between a control plane close to the trailing edge, where data are provided by the Navier–Stokes solver, and the plane far downstream. This technique is theoretically correct and avoids the effects of numerical dissipation and diffusion related to the mesh structure and the Navier–Stokes solver algorithm downstream of the trailing edge. This method also provides results in better agreement with experiments than those coming from the total pressure-averaging procedure, and will be used in the following section.

Loss calculation The results of the calculations are presented in Tables 3.10–13. Tables 3.10 and 3.11 show the results obtained with the implicit code A and the BL turbulence model, whereas Tables 3.12 and 3.13 report the coefficients obtained with explicit code B and the k–ε model. The introduction of the two-equation turbulence model generally produces a decrease in the computed loss coefficients.

Table 3.10 Loss coefficients for TE ejection blade (Implicit B-L Code A)

C_m (%)	0	2	3	4
$M_{is} = 0.8$	0.07		0.091	
$M_{is} = 1.05$	0.052	0.071	0.075	0.068
$M_{is} = 1.22$	0.052		0.06	

Table 3.11 Loss coefficients for PS ejection blade (Implicit B-L Code A)

C_m (%)	0	2	3	4
$M_{is} = 0.8$	0.078		0.085	
$M_{is} = 1.05$	0.057	0.062	0.064	0.061
$M_{is} = 1.22$	0.055		0.064	

Table 3.12 Loss coefficients for TE ejection blade (Explicit Code B, k–ε model)

C_m (%)	0	2	3	4
$M_{is} = 0.8$	0.045			
$M_{is} = 1.05$	0.03		0.044	
$M_{is} = 1.22$	0.039			0.037

Table 3.13 Loss coefficients for PS ejection blade (Explicit Code B, k–ε model)

C_m (%)	0	2	3	4
$M_{is} = 0.8$	0.037			
$M_{is} = 1.05$	0.029		0.036	
$M_{is} = 1.22$	0.056			0.072

The formulation with a two-equation turbulence model provides quite good results since the computed loss coefficients ζ are typically 0.01 smaller than those given by the BL model. Further they show a similar trend to the experiments. The wake computed by the two-equation model appears less spread than the one computed by the algebraic model; this could be the reason for the smaller losses.

The computed loss coefficients show that the coolant ejection from the trailing edge has a beneficial effect on the blade performances, though the PS ejection shows higher ζ than those of the TE blade at any coolant flow rate. This might be explained by the fact that the TE ejection kills, or drastically reduces, the trailing edge vortices, thereby avoiding one of the main sources of wake losses (see Figure 3.125). Although the PS ejection succeeds in convecting away the first recirculation bubble located downstream of the step (see Figure 3.125), the trailing-edge vortices are still present at any coolant flow rate. These two counterrotating vortices survive the coolant ejection and represent a fixed amount of vortex-induced losses. So, on top of the extra expansion-shock waves induced by the step on the pressure side of the PS blade with respect to the TE blade, the presence of the extra vortices plays a significant role in the determination of the overall loss coefficient.

The calculations do not identify any clear effect of the C_m percentage changes on the loss coefficient. In fact when C_m ranges between 2 and 4% no significant changes are found on ζ, and, moreover, the increase or decrease of C_m does not always have the same effect on the losses.

7.6.3 Evolution of wake profile

The wake profiles computed at seven positions downstream of the blade trailing edge (axial coordinate $x/C_{ax} = 2, 4, 8, 15, 25, 45, 80\%$) trace the evolution of the wake decay and the extent of the coolant jet influcence downstream of the cascade. All the following plots refer to the computations carried out with the implicit code A and the BL model. The profiles computed with the explicit code B are only marginally different and will not be shown.

Figures 3.127 and 3.128 show the evolution of the total pressure and total temperature profiles at the nominal Mach number for both blades, with and without a 3% coolant ejection.

The analysis of the numerical results shows that in the case of the TE ejection blade the wake and coolant jet mix quite soon. The comparison of $C_m = 0\%$ and 3% demonstrates that the structure of the total pressure wake remains substantially unaltered by the presence of the coolant jet, apart from a small kink that can be seen at the stations $x_{axial} = 1$ and 2 mm ($X/C_{ax} = 0.023$ and 0.046). It is also necessary to observe that the ratio $P_{o(c)}/P_o$(inlet) is around 0.6–0.8 in the wake and the mixing of the two streams is small and acts in a quite narrow region. The tests showed that the increase in the coolant flow rate slightly modifies the wake profiles, but not the general trend. This actually confirms what was found about the kinetic loss coefficient. At $x_{axial} = 35$ mm ($X/C_{ax} = 0.813$) the total temperature mixing process is always completed since the temperature profile closely resembles a flat line, but the total pressure wake is still quite deep and a non-uniform P_o profile exists at one chord downstream of the trailing edge. The Mach number seems not to influence the general wake decay process, although, as expected, the loss of total pressure and the wake thickness increase with Mach number because of shock–boundary layer interactions for choked conditions. For the PS ejection blade the general behaviour is similar; the wake mixing starts early so that no significant bimodal total pressure plots appear in the first section (1 mm downstream of the trailing edge). This might be explained with the presence of large trailing edge vortices that undeniably enhance the wake mixing. Moreover, the coolant ejection from the PS ejection blade takes place further upstream with respect to the TE ejection blade. Small differences appear in the total temperature plots. Several test with different grids seem to demonstrate that the numerical diffusion has been kept under control.

7.7 Three-dimensional flow calculations

In accordance with the experimental program, three-dimensional calculations were performed only for the PS ejection blade at $C_m = 0\%$ and 3%. The isentropic exit Mach number, computed by the inlet total pressure and the exit hub pressure, is 1.12 for the nominal test conditions. In order to match as close as possible the experimental inlet conditions the inlet total pressure profile was set to give a boundary layer thickness of 0.01% of the blade height. This produces an almost flat P_0 profile in agreement with the experiments where it was not possible to find any appreciable non-uniformity in the inlet total pressure. The experimental inlet total temperature, Figure 3.75, was also specified for the calculations. The flow quantities are made non-dimensional

310 Investigation of the wake mixing process

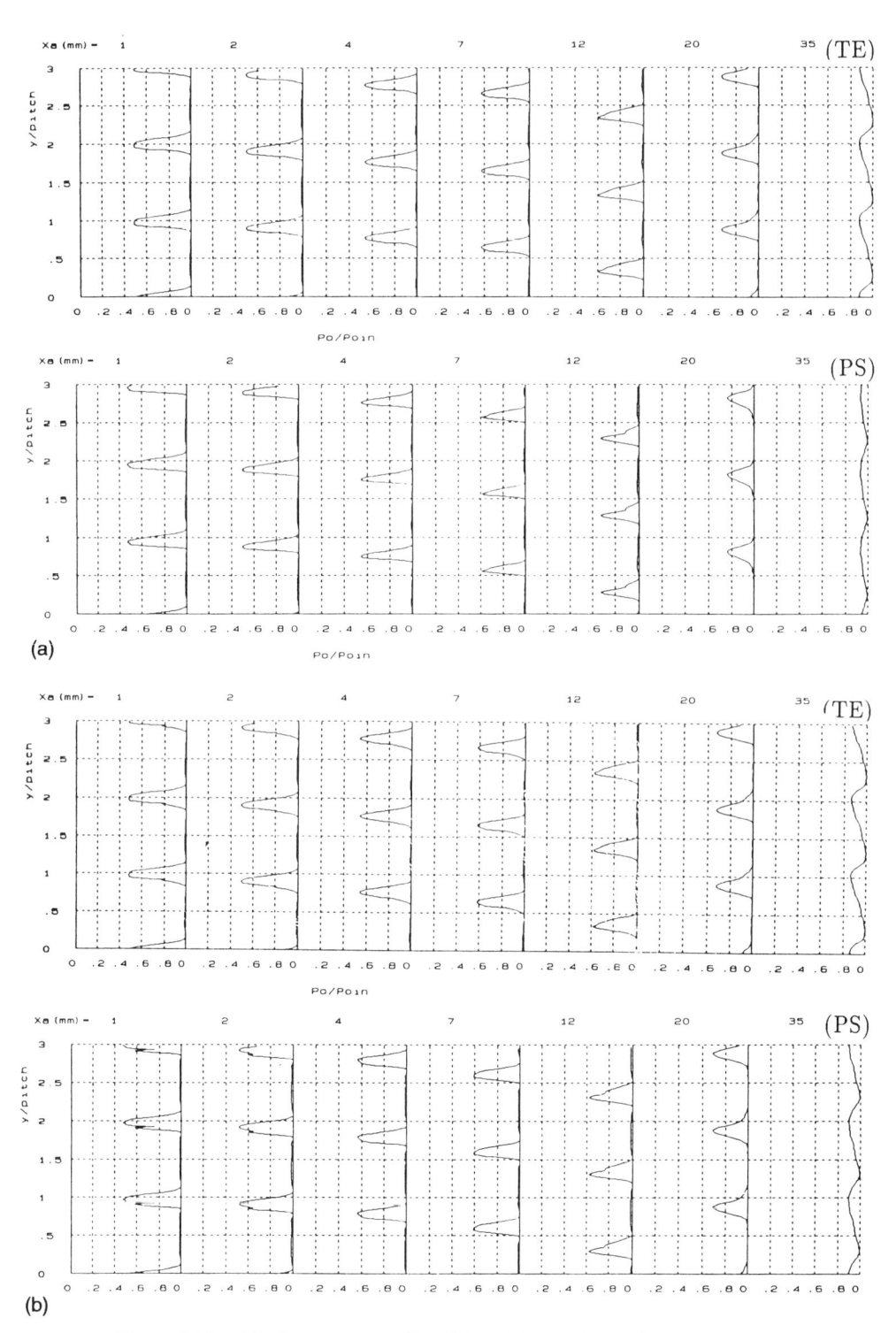

Figure 3.127 Total pressure profiles: $M_{is} = 1.05$: (a) $C_m = 0\%$; (b) $C_m = 3\%$.

Viscous flow calculations

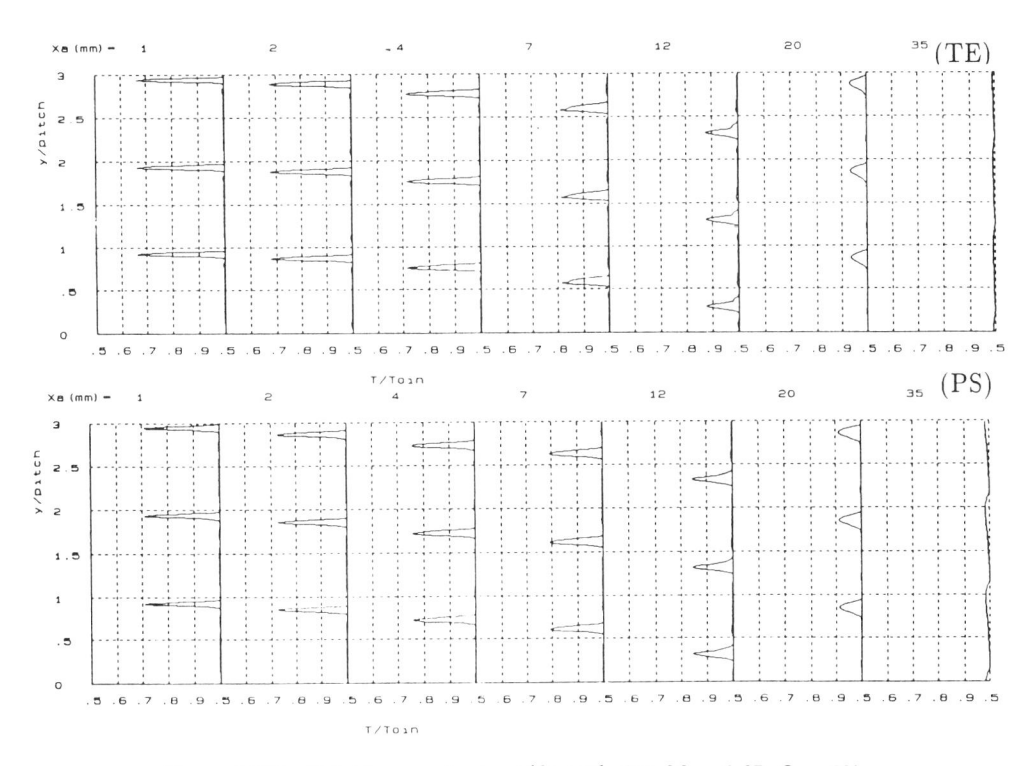

Figure 3.128 Total temperature profile evolution $M_{is} = 1.05$, $C_m = 3\%$.

with respect to the inlet maximum total pressure and temperature, whereas lengths are non-dimensionalized with respect to the blade chord.

Flow simulations were performed using three grids, $101 \times 35 \times 35$, $175 \times 41 \times 45$, and $175 \times 45 \times 65$ with more than a half-million grid points. Figure 3.129 shows a perspective view of the coarse $101 \times 35 \times 35$ grid for two consecutive blade passages. The tentative grid allowed a first qualitative evaluation of the flow pattern. The two refined grids have the same point distribution in the streamwise direction as adopted in the two-dimensional calculations, whereas the point distribution in the cross-section was decreased to maintain the grid points around a half-million. The final computational grid has 65 points in the radial direction with a strong point clustering in the hub and shroud regions since the previous $175 \times 41 \times 45$ grid showed an excessive growth of the hub boundary layer. The three-dimensional solver was run with CFL numbers ranging from 5 to 10. The second- and fourth-order damping weights range between $\frac{1}{8}$ to $\frac{1}{2}$ and $\frac{1}{32}$ to $\frac{1}{16}$ respectively. A special boundary treatment of the artificial dissipation terms [3.15] reduced the artificial fluxes on the boundaries, thereby producing very low mass flows errors (typically 0.05–0.1%) despite the relatively coarse grids. The final grid runs were performed on a CRAY-YMP supercomputer where approximately 5000–6000 seconds were needed to reach an overall residual of the order of 10^{-5}.

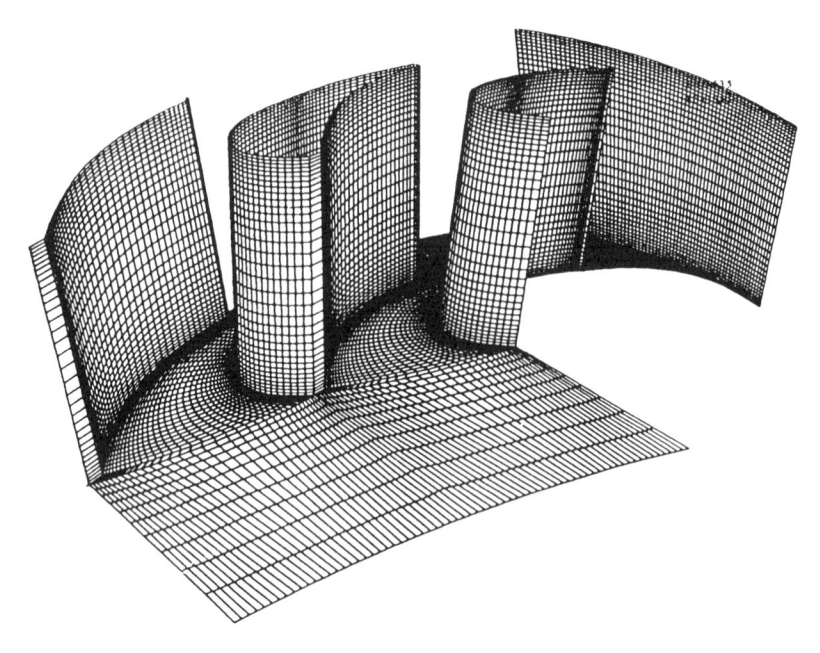

Figure 3.129 175 × 41 × 45 computational grid.

7.7.1 Blade velocity distribution and cascade flow fields

The blade velocity distributions are presented along the blade chord. Figure 3.130(a) refers to the refined grid calculations and shows five sections located at the hub of the radial IGV (0% of total blade height), 10%, midspan, 90% and at the shroud section (100%). The $C_m = 0\%$ plot shows basically what was found for the two-dimensional test cases at M_{is} ranging from 1.05 to 1.2. In fact the hub section has a stronger contraction ratio with respect to the shroud section because the blade grows radially from the hub with a constant section. A shock might well be located at approximately 35% of the blade chord at a radial distance up to 50% of the blade height. The additional two sections are basically subsonic and the flow stays at $M = 0.9$–1.0 from 30% of the blade chord on. This means that the flow is shock free for approximately 50% of the blade height. The blade velocity distribution exhibits only one shock whereas the two-dimensional calculations performed for the PS ejection blade showed two consecutive shocks departing from the two corners on the pressure side. This might be explained by observing that the two shocks in the two-dimensional calculations were quite evident only at $M_{is} = 1.21$, while the present run refers to $M_{is} = 1.12$. Moreover, the smaller grid point number in the pitchwise direction adopted in the three-dimensional calculation does not ensure the same good description of the blade boundary layer.

Figure 3.130(b) shows the blade velocity distribution for $C_m = 3\%$ on the five sections described above. The effect of the coolant ejection is quite similar to what was observed for the two-dimensional test cases, where the shock was

Viscous flow calculations

(a)

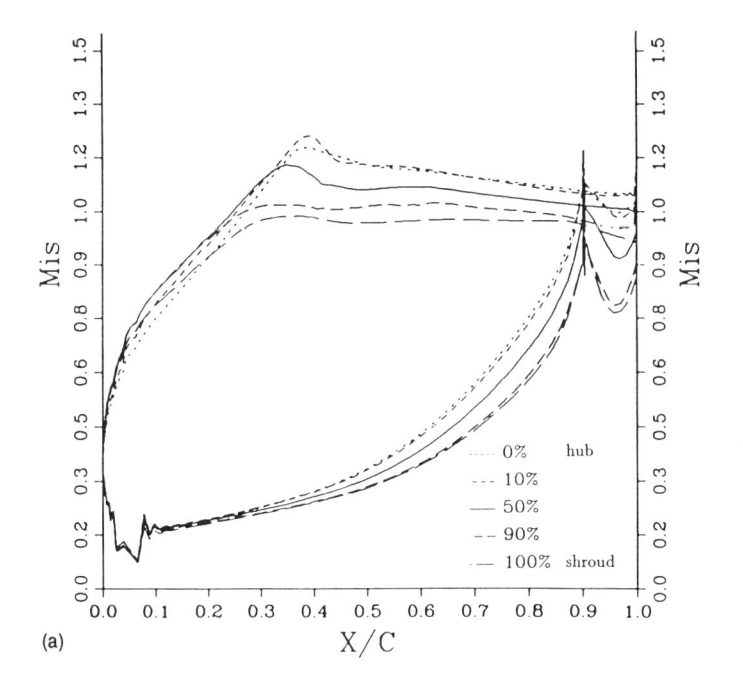

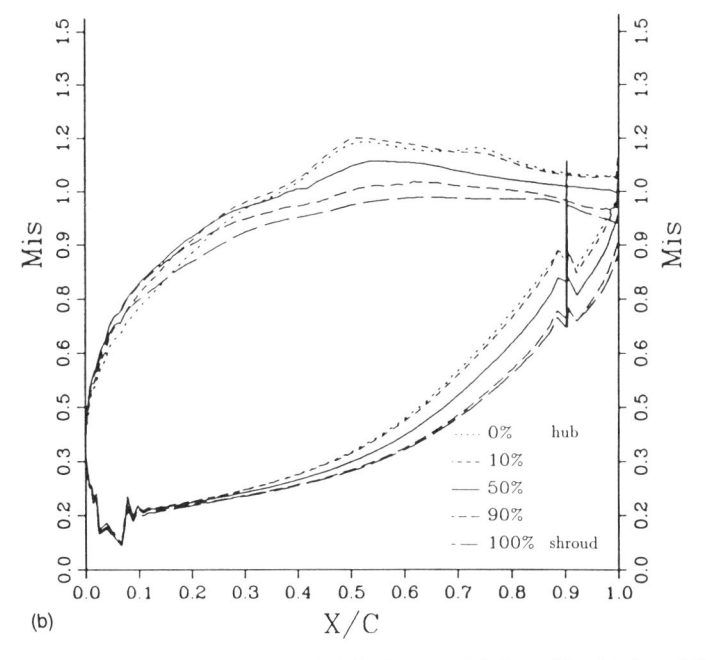

(b)

Figure 3.130 M_{is} for $M_{hub} = 1.12$, PS ejection: (a) $C_m = 0\%$; (b) $C_m = 3\%$.

smeared out. Still the three-dimensional calculations show that the coolant ejection produces a marginally supersonic region on the suction side that extends from 50% to 100% of the chord. The flow feels the presence of the coolant ejection all along the blade height. This is partly due to the computation of the coolant flow conditions. In fact in the three-dimensional calculations the jet conditions were specified differently at every blade height depending on the local static pressure. Because of the radial nature of the flow, the static pressure tends to grow in the blade-tip region to produce the radial pressure gradient necessary to bend the flow. In order to have $C_m = 3\%$ of the inlet mass flow rate at every blade section, the jet total pressure also grows towards the blade tip. The converged results show a uniform coolant flow rate with a variable total pressure and velocity along the slot height, with a maximum difference between base and tip of approximately 20%. It is now debatable whether the experimental coolant jet conditions really give this non-uniformity. Unfortunately no measurements are available in the jet slot, so that any direct verification of the coolant jet modelling strategy is impossible.

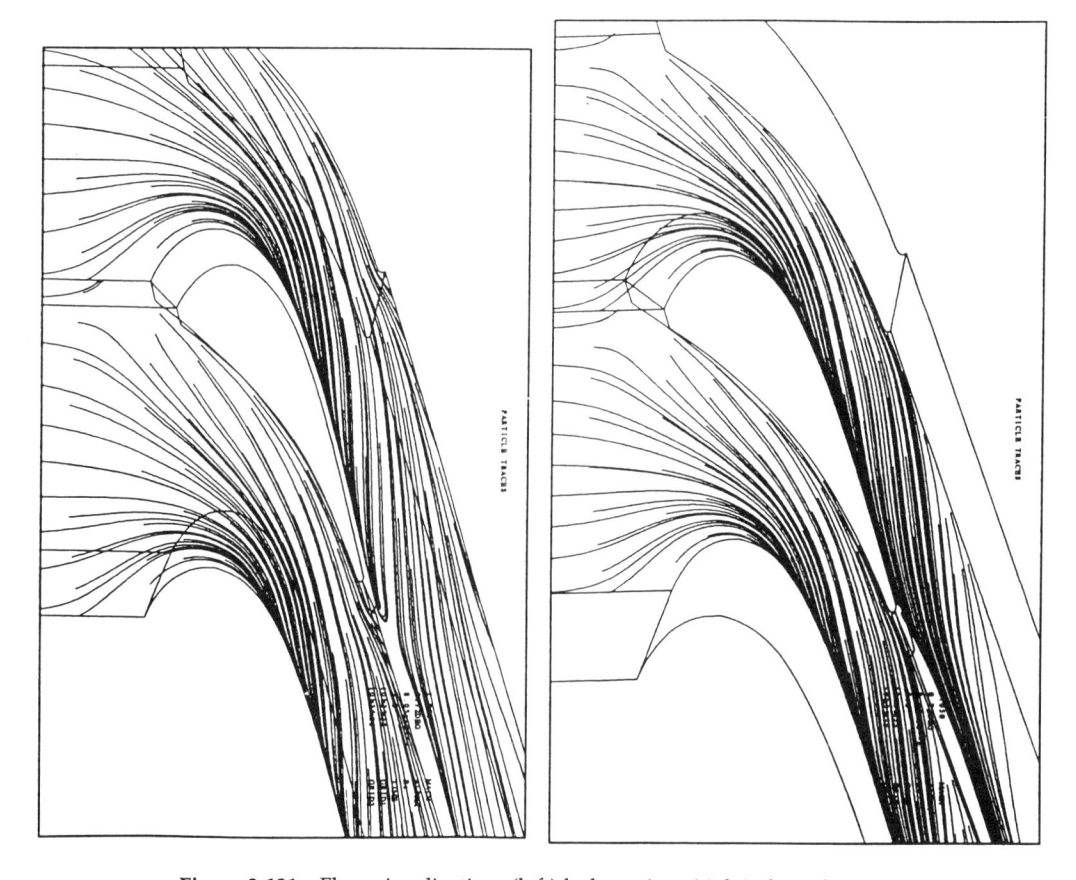

Figure 3.131 Flow visualization: (left) hub section; (right) shroud section.

Viscous flow calculations

7.7.2 Flow visualization

The three-dimensional qualitative flow visualization is performed for the $175 \times 41 \times 45$ computational grid for the TE ejection blade. The visualization is obtained by injecting a number of particles into the flow and following their trajectories, or traces, so as to have an image of the mean flow path. Figure 3.131 shows the visualization of the hub and shroud sections flow. The two sections behave almost identical. This indicates that the two passage vortices are nearly symmetric and of comparable strength. The hub section also shows a small recirculation departing from the trailing edge. The leading edge region close the hub section (Figure 3.132) shows a saddle point ahead of the blade from which the two branches of the horseshoe vortex depart.

The visualization of the flow on the suction and pressure sides of the blade is of great interest to understand the effect of the coolant ejection. Figure 3.133 is obtained by injecting particles in the computational plane closest to the blade suction and pressure sides. Particles are also injected upstream and downstream of the blade so as to demonstrate the effect on the blade on the approaching flow and deep inside the wake. In Figure 3.133(a) the flow structure on the suction side shows the effect of the growth of the two counter-rotating passage vortices. The interaction with the radial pressure gradient strongly bends the low momentum stream towards the hub section. This effect is particularly evident for $X/c_{ax} > 60\text{--}70\%$, where the boundary

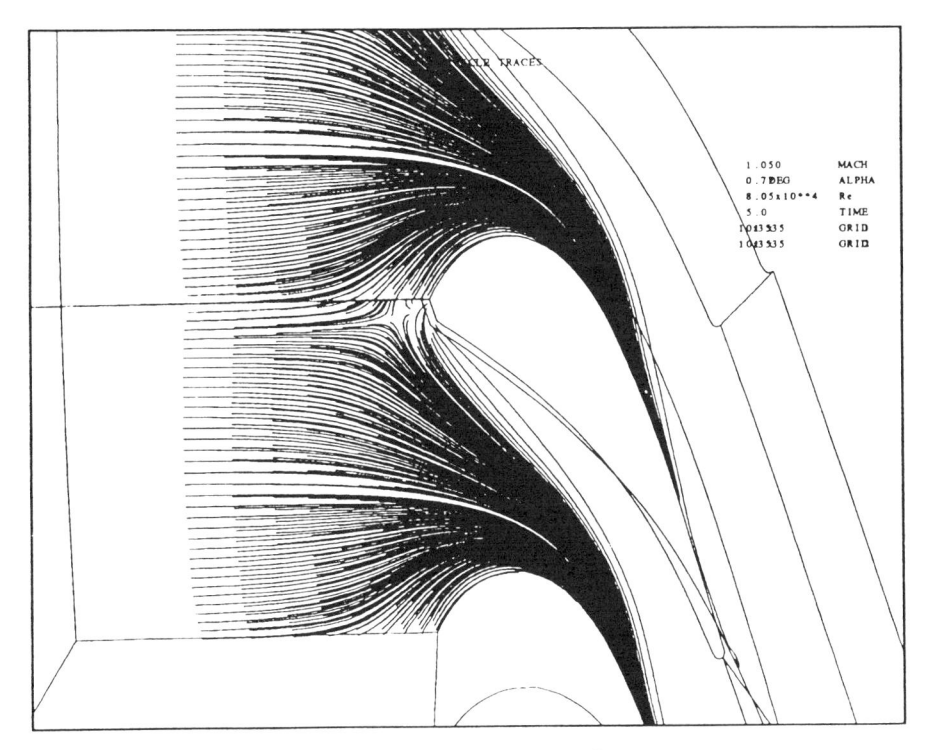

Figure 3.132 Saddlepoint visualization.

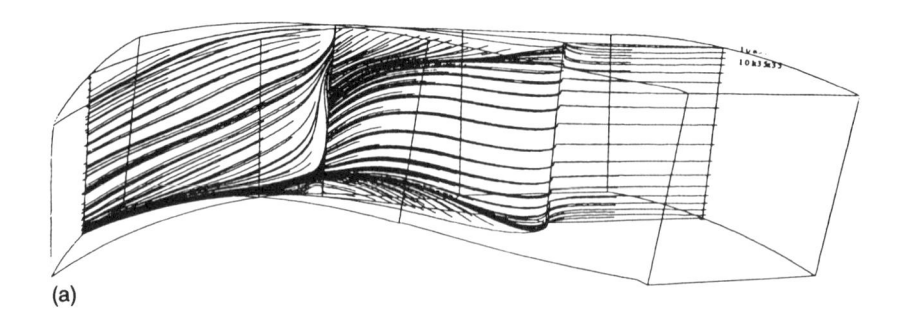

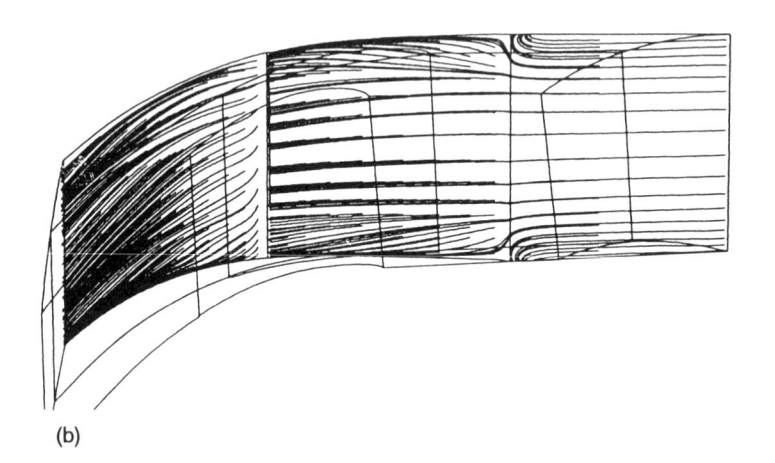

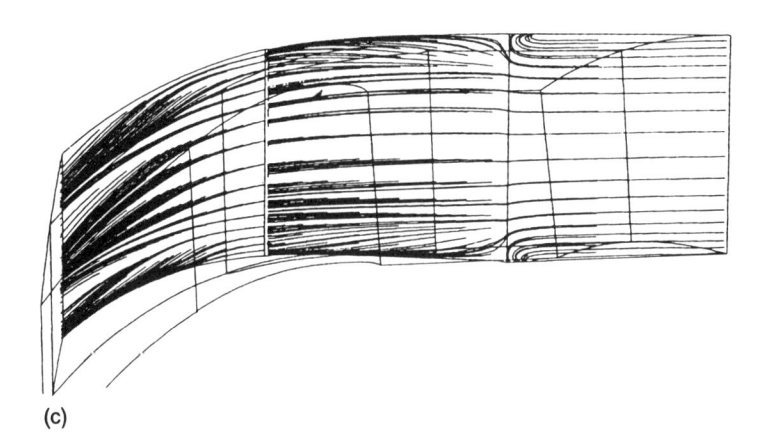

Figure 3.133 Flow visualization: (a) of the blade suction side ($C_m = 0$ or 3%); (b) of the blade pressure side ($C_m = 0$%); (c) of the blade pressure side ($C_m = 3$%).

layer grows because of the shock–boundary-layer interaction. From this location on, the flow particles lose their momentum while the radial pressure gradient remains practically unchanged. In the very low momentum area located immediately downstream of the trailing edge the flow goes in the radial direction and the particles are directly transported towards the hub. This is not surprising because of the quite thick trailing edge of the blade under investigation. Observe the presence of a small hub recirculation close to the trailing edge. Figure 3.133(a) is virtually the same for either $C_m = 0\%$ or $C_m = 3\%$ coolant flow. Figure 3.133(b) reports the same visualization for the pressure side at $C_m = 0\%$. The traces of the horseshoe vortices in front of the blade are evident upstream of the leading edge. The particles are strongly bent towards the casing, as could be expected from the saddle-point visualization of Figure 3.132. Observe that the wake structure of the pressure-side visualization closely resembles that of the suction side when $C_m = 0\%$ and again the flow particles are bent towards the hub, however, at a much smaller rate than on the suction side. The effect of the coolant flow ejection ($C_m = 3\%$) from the trailing edge is evident in Figure 3.133(c). While the flow paths ahead of the blade and in the blade passage remain substantially unchanged, the coolant ejection clearly straightens the particle traces in the wake region. The flow is strongly energized by the coolant jet in the region where the two-dimensional calculations showed the presence of the two counterrotating trailing-edge vortices. These vortices are killed by the coolant ejection, with a beneficial effect on the wake. The two-dimensional total pressure profiles downstream of the blade showed that the effect of the coolant ejection was felt only very close to the trailing edge. This analysis is confirmed by the three-dimensional computations where the wake flow path is straightened only immediately downstream of the trailing edge. After a few millimetres the flow is bent again strongly towards the hub.

7.7.3 Loss and angle distribution

The kinetic loss coefficent is defined as

$$\zeta = 1 - \frac{1 - (P_l/P_{0,l})^m}{1 - (P_l/P_{0,1})^m}$$

in which $m = (\gamma-1)/\gamma$, P_l and $P_{0,l}$ are the local static and total pressures and $P_{0,1}$ is the total pressure at midspan upstream. The local values of the exit flow angle and total and static pressure are area-averaged in the tangential direction so as to give a unique spanwise average profile at $x/C_{ax} = 0.25$ and 0.54.

Figure 3.134 shows the computed profiles for $C_m = 0\%$ and 3%. The computations show a quite thick boundary layer in the hub region in which the kinetic loss coefficient tends rapidly to values of the order of 10%. Observe that the computed kinetic loss coefficients seems to decrease in the flow core with increasing downstream distance. This is due to the area-averaging technique,

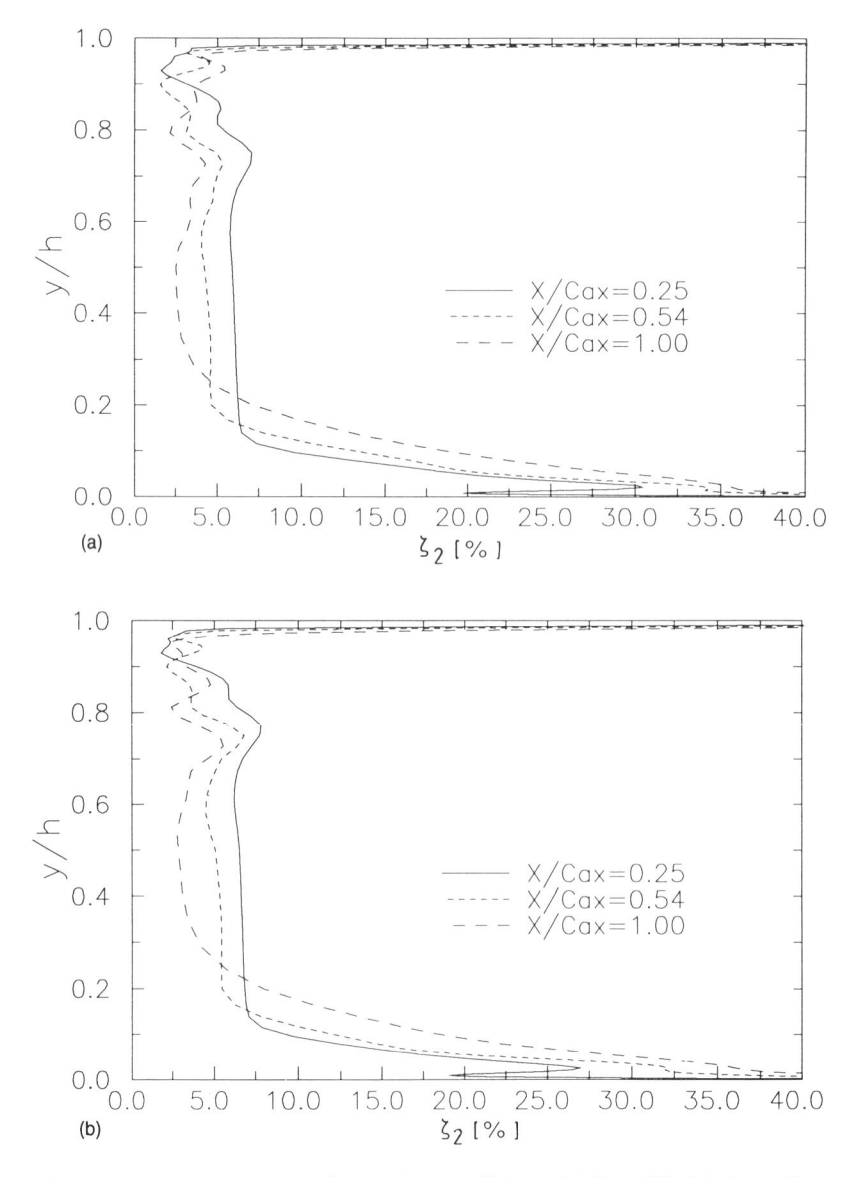

Figure 3.134 Area averaged kinetic loss coefficient: (a) $C_m = 0\%$; (b) $C_m = 3\%$.

used in conformity with experiments, which should be replaced by a mass-averaging in the pitchwise direction.

The computed profiles for $C_m = 3\%$ do not differ much from those for $C_m = 0\%$, except for some differences close the blade tip. This feature is in total agreement with the flow visualization and the two-dimensional cross-sections downstream of the blade, in which the influence of the coolant ejection was no longer felt for $x/C_{ax} > 0.1$. The kinetic loss is small in the blade-tip region, while in the hub region the growth of the boundary layer is strong and

Viscous flow calculations

probably related to an insufficient number of points. The area-averaged exit flow angles are reported in Figure 3.135. The flow patterns are essentially the same at the three downstream positions. The passage and horseshoe vortices are responsible for the characteristic under and overturning by approximately 5° with respect to the core flow, which has an exit flow angle of approximately 72°.

Figure 3.136 shows the pitch-averaged isentropic Mach number along the blade height. The flow clearly shows a strong radial gradient of pressure which tends to the experimental value of $M_{is} = 1.12$ at the hub.

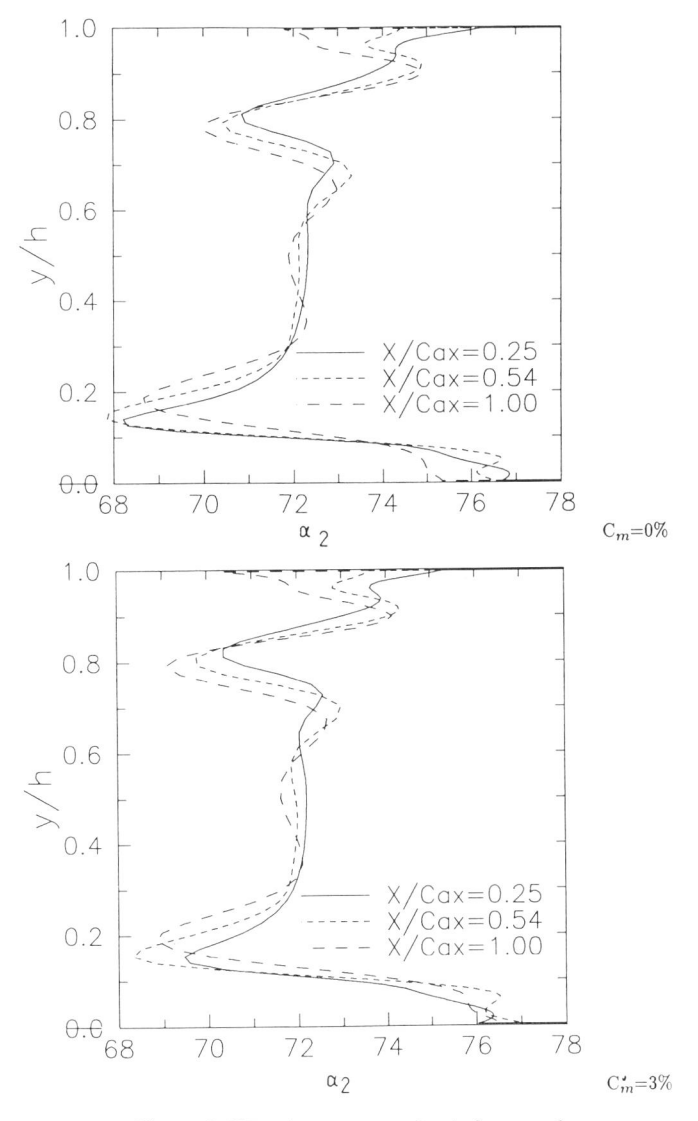

Figure 3.135 Area averaged exit flow angle.

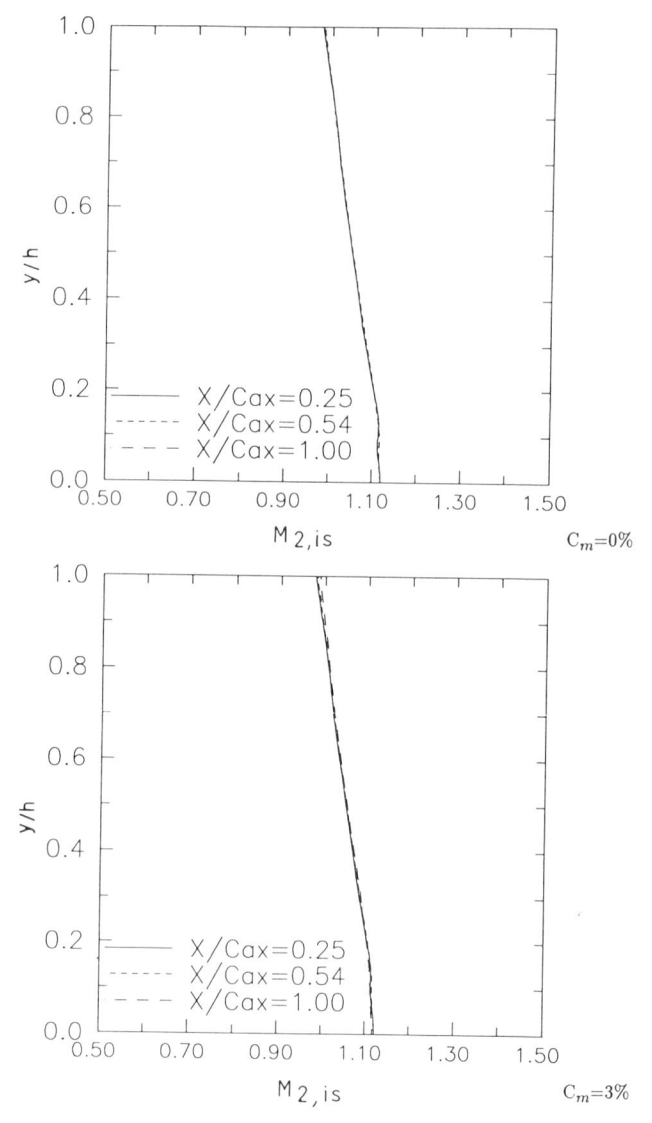

Figure 3.136 Area averaged M_{is}.

8 Comparison of experimental and numerical results

8.1 Comparison of experimental results in straight and annular cascades

8.1.1 Blade performance

The straight cascade geometry corresponds to the midspan geometry of the annular cascade and a direct comparison between the performances of both cascades is necessarily limited to the midspan flow conditions, i.e. at an isentropic outlet Mach number $M_{2,is} = 1.05$.

Comparison of experimental and numerical results

The measurement planes are, respectively, at $X/c_{ax} = 0.813$ for the straight cascade and at $X/c_{ax} = 0.25$, 0.54 and 1.0 for the annular cascade. As the flow conditions in the last plane are possibly influenced by the very thick hub end-wall boundary layer, it was preferred to use the results from the wake traverse at $X/c_{ax} = 0.54$ for comparison.

Table 3.14 presents the losses and the outlet flow angles at $M_{2,is} = 1.05$ for the nominal blowing rate $C_m = 3\%$. Remember that the coolant to main flow temperature is $T_{oc}/T_{01} = 1$ in the straight cascade tests at DLR and 0.65 in the annular cascade tests at VKI. The comparison is done on the basis of mixed-out values, which allows a fair comparison of data taken in different measurement planes. Area averaged loss and flow angles have been added for the annular cascade.

The difference in the losses is important and needs to be analysed. Besides the presence of three-dimensional flow effects in the annular cascade, there exists a non-negligible difference in the turbulence levels between both tests: $Tu = 0.5\%$ in the straight cascade and $Tu \simeq 2\%$ in the annular cascade. However, there is uncertainty that this difference in the turbulence level can explain such a difference in the loss level, because the suction-side shock boundary layer at $X/C \simeq 0.3$ (see Figure 3.46), although rather weak, will most probably induce boundary-layer transition in any event. However, one may also speculate that at low turbulence level, the transition is not completed before the flow reaccelerates for $X/C > 0.5$. If this is so, then the boundary layer may remain in a transitional state up to the trailing edge and be more affected by the trailing edge shock than a fully turbulent boundary layer in case of the higher turbulence level in the annular cascade.

The comparison at design outlet Mach number gives only a very incomplete idea of the performance difference of both cascades, because the straight cascade undergoes a dramatic increase from $\zeta_2 = 3.5\%$ at $M_{2,is} = 1.0$ to 6.7% at $M_{2,is} = 1.085$. In the annular cascade these Mach numbers occur at the span-wise positions $Y/h = 0.84$ and 0.25. Over this distance the losses remain remarkably constant. AT $Y/h = 0.25$ and $M_2 = 1.085$ the losses amount to $\zeta_2 = 3.7\%$ instead of 6.7% in the straight cascade. It is difficult to imagine that the strong Mach number effect in the straight cascade is off-set by an opposite pitch-to-chord ratio effect in the annular cascade, where g/c changes by about 8% between these two span-wise positions. In addition, the trailing-edge-thickness-to-throat ratio te/o increases with decreasing radius, and this should adversely affect the blade performance.

Finally one may not exclude three-dimensional flow effects to play an important role. It is known, and the three-dimensional viscous flow calculations in Chapter 7 have demonstrated it again, that there exists a strong

Table 3.14 Losses and outlet flow angles

Cascades	Mixed-out ζ_2 [%]	α_2	AREA averaged ζ_2 [%]	α_2
straight	4.5	73.5	—	—
annular	3.2	72.5	3.0	—

migration of low-boundary-layer momentum material from top to bottom wall in the trailing-edge base flow region. This radial migration may influence the trailing-edge base pressures and thereby the losses.

The difference in the outlet flow angle of 1° is higher than one might expect from three-dimensional flow effects. Both the three-dimensional Euler and Navier–Stokes calculations predict midspan outlet flow angles of 72.5° and 72°, respectively. The difference in the measurement planes between the two tunnels ($X/c_{ax} = 0.815$ compared to 0.54) may in this case be of importance. In fact, the outlet flow field of the straight cascade may be affected by the free shear layer developing behind the top end blade. The displacement effect of this shear layer grows with increasing downstream distance. However, it is not at all certain that this effect extends to the blade wakes traversed by the probe.

8.1.2 Wake mixing

As demonstrated in Chapters 5 and 6 the axial evolution of the wake profile can be described by Gaussian curves, in spite of the coolant flow ejection. The procedure adopted was to present the suction and pressure sides of the wake by different Gaussian curves to account for the different boundary layer development on both blade surfaces and therefore possibly for a different lateral spreading for both sides of the wake.

Remember that the study on the wake-mixing process in the straight cascade tunnel was done using CO_2 as coolant flow.

Figures 3.137 and 3.138 compare the decay of the wake maximum and the lateral spreading of the wake for the straight and annular cascade tests. The overall agreement is fairly good.

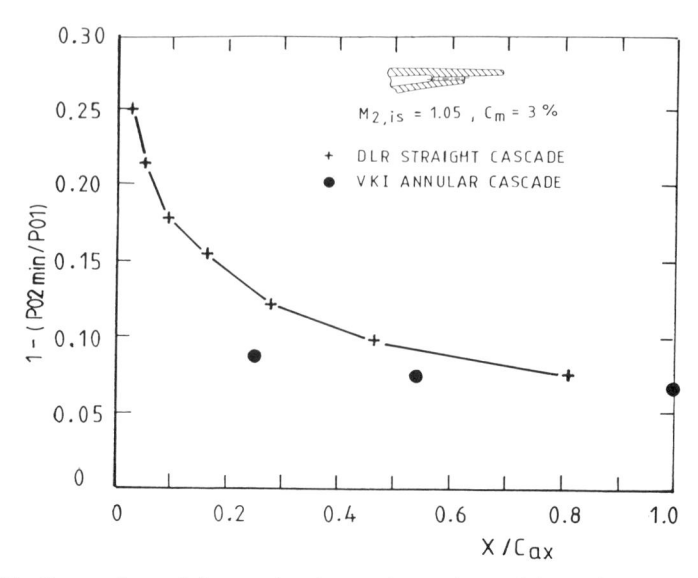

Figure 3.137 Comparison of decay of wake maximum in straight and annular cascade tests (PS-ejection blade).

Comparison of experimental and numerical results 323

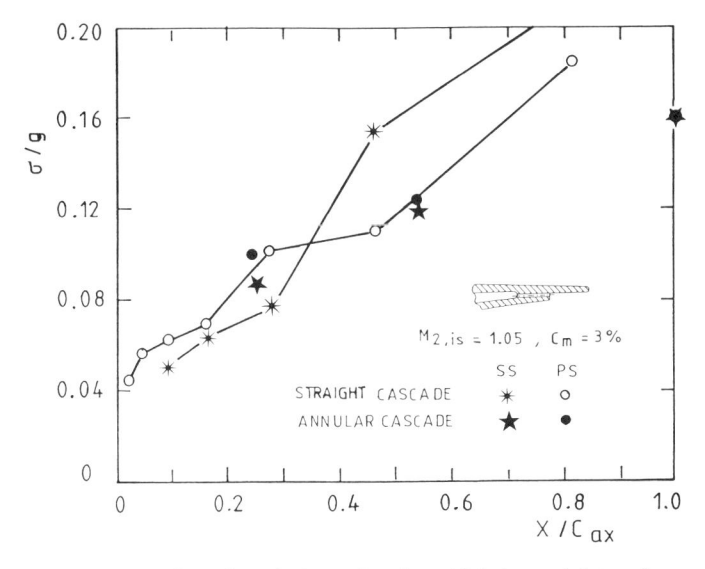

Figure 3.138 Comparison of axial evolution of wake width in straight and annular cascade (PS-ejection blade).

8.2 Comparison of numerical and experimental results

The aim of this section is to investigate the capability of the numerical method to match the experimental results and try to understand what are the aspects to be further investigated, on the numerical as well as on the experimental side.

8.2.1 Straight cascade

Blade velocity distribution Figures 3.139 and 3.140 show the blade isentropic Mach number distribution computed by the implicit solver with the BL turbulence model. As was stated in the previous chapter, no big differences have been detected between the results of this code and the explicit solver with the k–ε turbulence model. Therefore we will refer from here on to the implicit code unless otherwise stated.

The computations agree fairly well with the experiments for the subsonic and supersonic cases. At transonic outlet flow conditions, $M_{2,is} = 1.05$, the shock pattern is partially lost in both the TE and PS ejection cases. While the experiments show a flow acceleration, the computations predict a nearly constant pressure value, probably due to an excessive thickening of the boundary layer downstream of the shock. For the TE-ejection blade this behaviour is clearly due to an incorrect description of the blade boundary layer since the weak shock located at approximately 40% of the blade is well captured, while the strong shock at $x/c = 0.5$ is lost. Experiments show that the

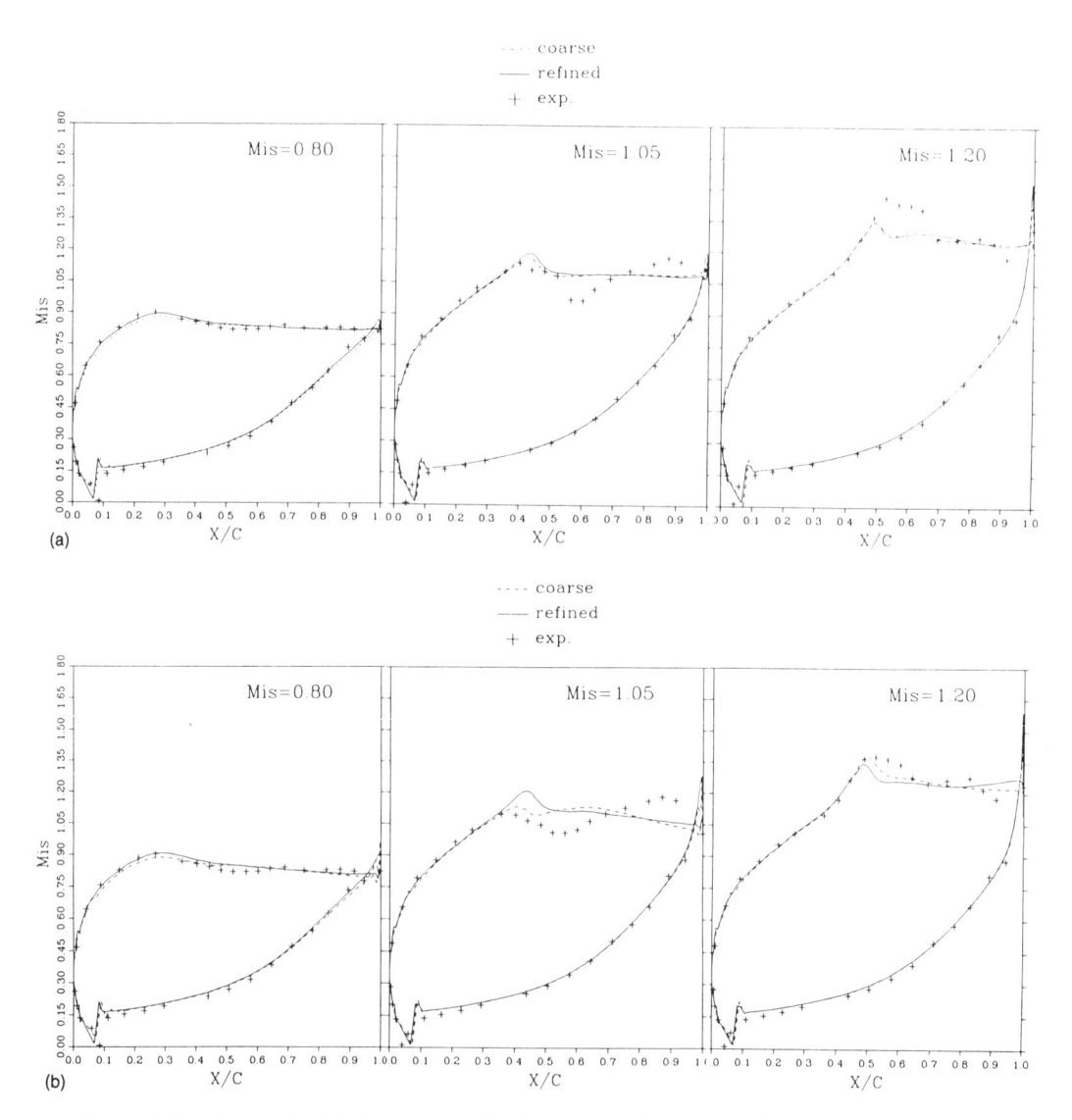

Figure 3.139 Isentropic Mach number distribution for TE ejection blade: (a) $c_m = 0\%$; (b) $c_m = 3\%$.

flow acceleration in the throat appears stronger on the PS ejection blade than on the TE ejection blade. In the transonic case the computed and measured Mach numbers differ significantly on the suction side. At zero coolant flow the experiments show a second shock, probably due to a separation bubble, which is not detected by the calculations, because of the imperfect modelling of the boundary layer obtained with the present turbulence model. When the coolant flow is switched on ($C_m = 3\%$) the second shock disappears and a large decelerated zone appears. In this region the impinging shock strength is

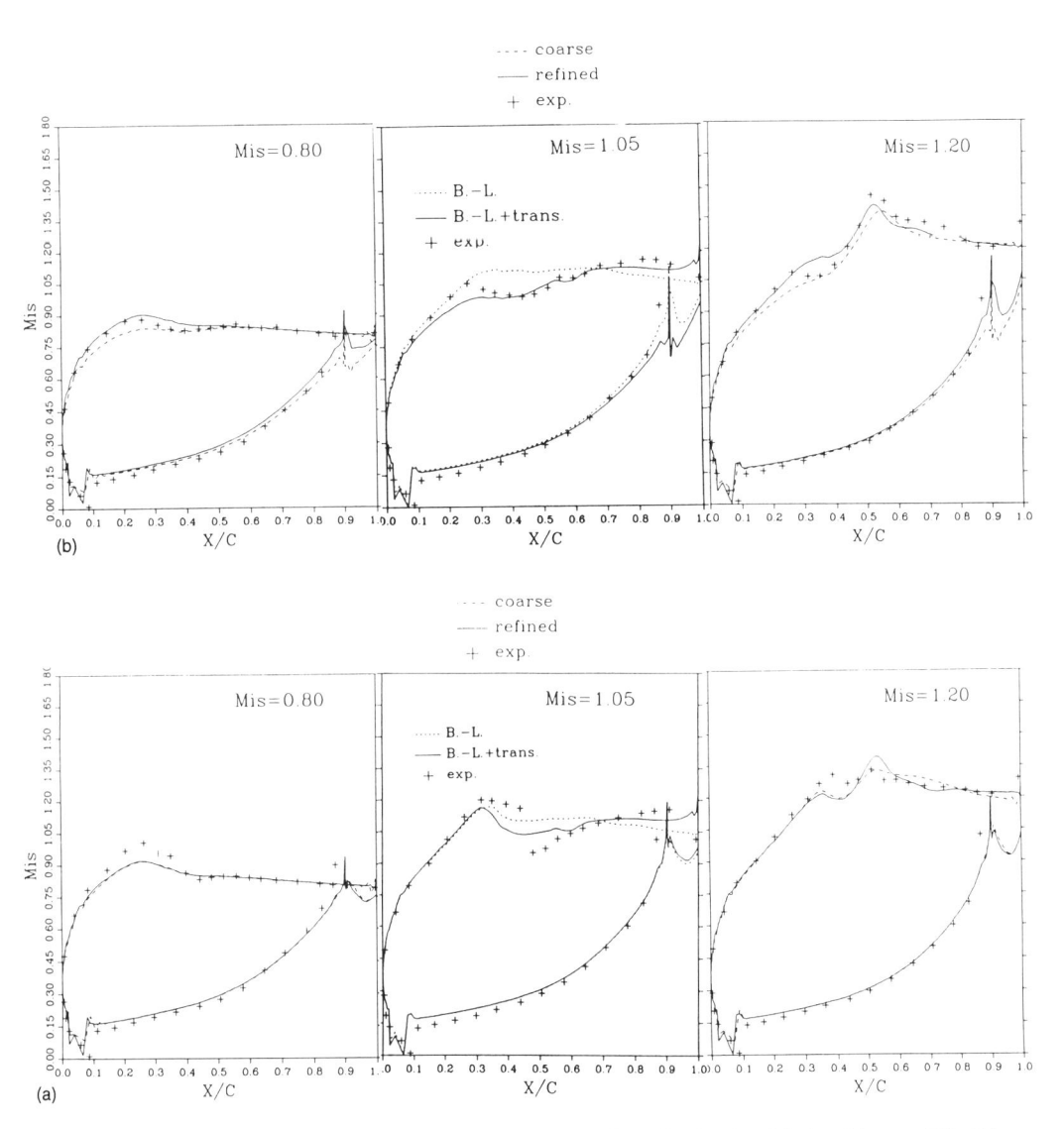

Figure 3.140 Isentropic Mach number distribution for PS ejection blade: (a) $c_m = 0\%$; (b) $c_m = 3\%$.

decreased by the coolant ejection and the calculations give a somewhat better fit with experiment, although the blade velocity peak is underestimated, and the calculated velocity is nearly constant, contradicting an experimental accelerating flow. The same behaviour was predicted by using the explicit code with a totally different turbulence model and numerical scheme. It is noteworthy that the computed result did not change switching on or off the boundary-layer transition flag in the BL model. This might indicate that the problems stem from an insufficient modelling of the shock boundary layer interaction on the suction side when this interaction extends over a long distance on the suction side. It has to be pointed out that moving from transonic flow to subsonic or supersonic regimes, the agreement with the experimental data improves significantly.

For the PS ejection blade the fit between experiments and computations generally improves. Figure 3.140 shows that in the subsonic case the flow acceleration in the cascade throat is not entirely captured when using the BL turbulence model. Analysing the $k-\varepsilon$ calculations, Figure 3.120(b), the peak, in the isentropic Mach number distribution is even lower than that obtained by the simple BL model. The up–down shape of the supersonic M_{is} is reproduced with a slight shift in the location and position of the shock. This behaviour is caused by a shock impinging on the suction-side boundary layer and can be explained only with the presence of a small separation bubble downstream of the weak shock across which the flow remains supersonic. Downstream of the reattachment the flow undergoes an expansion wave followed by another shock that can easily be seen in the $M_{is} = 1.20$ plots for all the coolant mass flow rates. The $M_{is} = 1.05$ profiles computed by the BL model with no transition flag show the same poor agreement with experiment as for the TE ejection blade. When using the transition model for the PS blade the M_{is} profile moves towards the experimental points for all the four different coolant mass flow rates under investigation. Unfortunately the same improvement was not found for the TE-ejection blade. This behaviour still lacks an explanation, but might be connected to the large differences in the trailing-edge shock pattern between the TE and PS ejection blades.

Another large difference between the two geometries lies in the pressure distribution close to the PS ejection position and can only be highlighted by means of the computer simulation because of a lack of experimental points in that region. Figure 3.140 for $M_{is} = 1.05$ and $C_m = 0\%$ shows two blade velocity peaks located at the backwards-facing step in the proximity of the coolant jet, and at the trailing edge. In case of zero C_m the blade velocity is larger than in the $C_m \neq 0$ case in the region immediately downstream of the injection. This indicates that with $C_m = 0$ the flow experiences an expansion wave at the first corner on the pressure side with a considerable acceleration, which eventually produces the departure of a strong shock from the pressure side. Conversely, the expansion wave and the subsequent shock wave are weakened, and nearly smoothed out, by the coolant ejection. This is even more evident when looking at the M_{is} profile on the suction side where the injection of the cold fluid for both the PS and TE geometries smooths, or totally removes, the shock.

Base pressure Figures 3.141 and 3.142 compare the experimental and predicted base pressure coefficients for both blade configurations. In both cases the calculations predict fairly well the general trend of the base-pressure variations with Mach number and coolant flow rate. In particular the Mach number effect for zero ejection follows closely the experimental values. The influence of the coolant flow ejection on the base pressure is well predicted at subsonic outlet Mach numbers, but is largely overestimated at transonic and

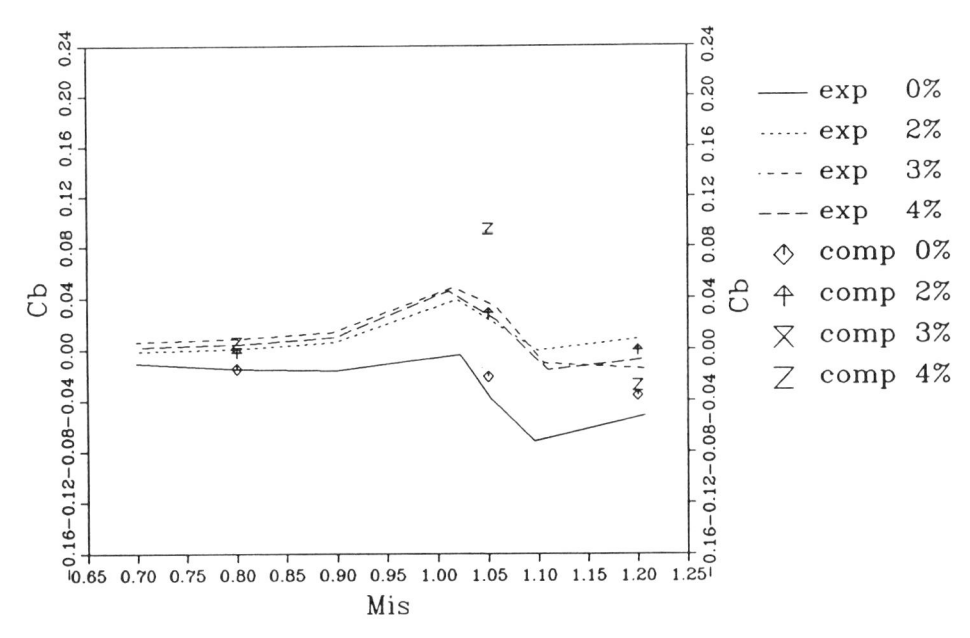

Figure 3.141 Base pressure for TE ejection blade.

supersonic Mach numbers for the pressure side lip (coefficient $C_{p,2}$ in Figure 3.142) of the PS-ejection blade.

Trailing edge flow patterns The computations already showed that the coolant ejection acts like a smoother of the step on the rear pressure side with a consequent smearing of the departing shock. This behaviour can be visualized by using the shock function S (see Chapter 7) and comparing the results of the computations with the experimental schlieren flow visualization. Figure 3.143 refers to the PS blade at $M_{is} = 1.20$ at $C_m = 0$ and 3%. The comparison is carried out for this case only since it most significantly shows the change in strength of the shocks induced by the coolant ejection. The computations compare favourably with experiments, especially for the shock smearing produced by the coolant ejection. The double expansion wave-shock wave departing from the pressure side of the PS blade exhibits the same angle with respect to the main flow in both experiments and computations. The complex pattern on the blade suction side is also well reproduced by the computations.

Evolution of wake profile Measurements with air have been performed only at the last station (81.3% of axial chord downstream of the trailing edge) so that the correctness of the wake simulation can only be guessed from the comparison between computations and experiments at this section. Figures 3.144–147 show the total pressure and outlet flow angle at the design Mach number for 0% and 3% coolant flow rate for both the TE- and PS-ejection blades. The computed total pressure profiles show a good agreement with experiments,

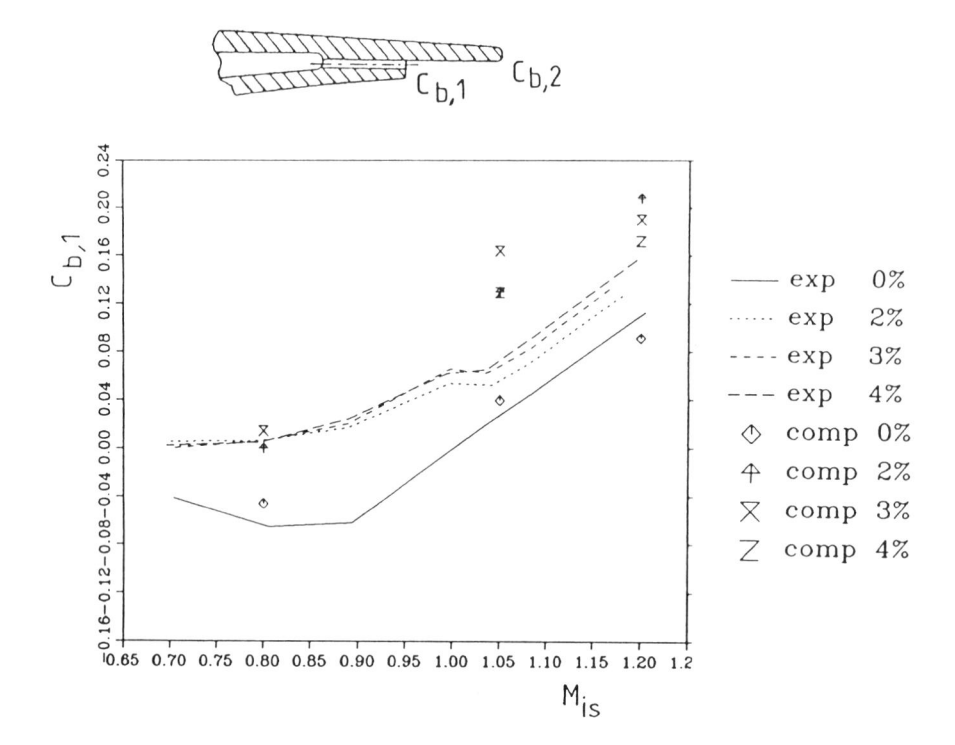

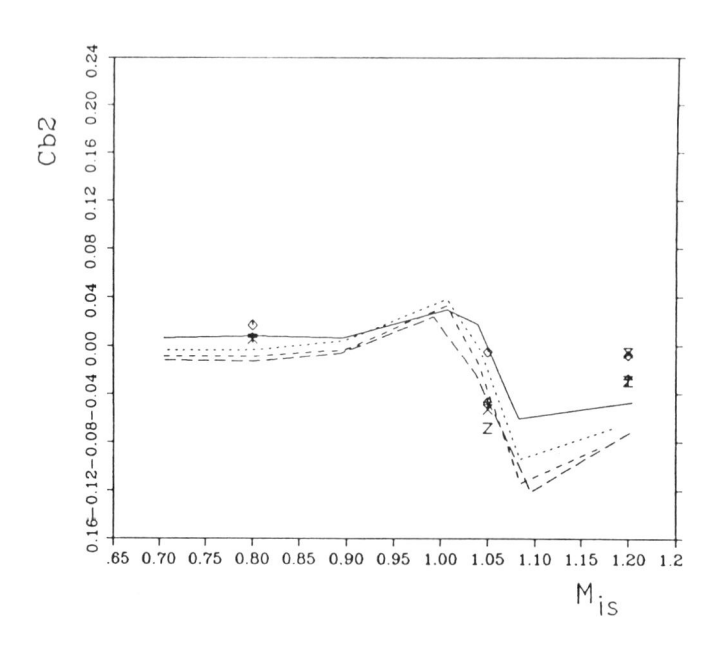

Figure 3.142 Base pressure for PS ejection blade.

Comparison of experimental and numerical results

Computations - $C_m = 0.\%$ Computations - $C_m = 3.\%$

Experiments - $C_m = 0.\%$ Experiments - $C_m = 3.\%$

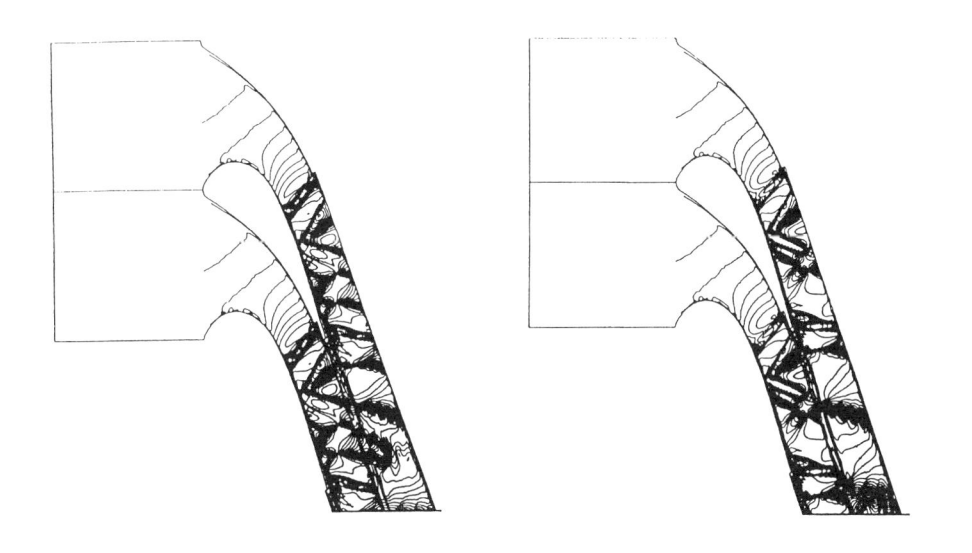

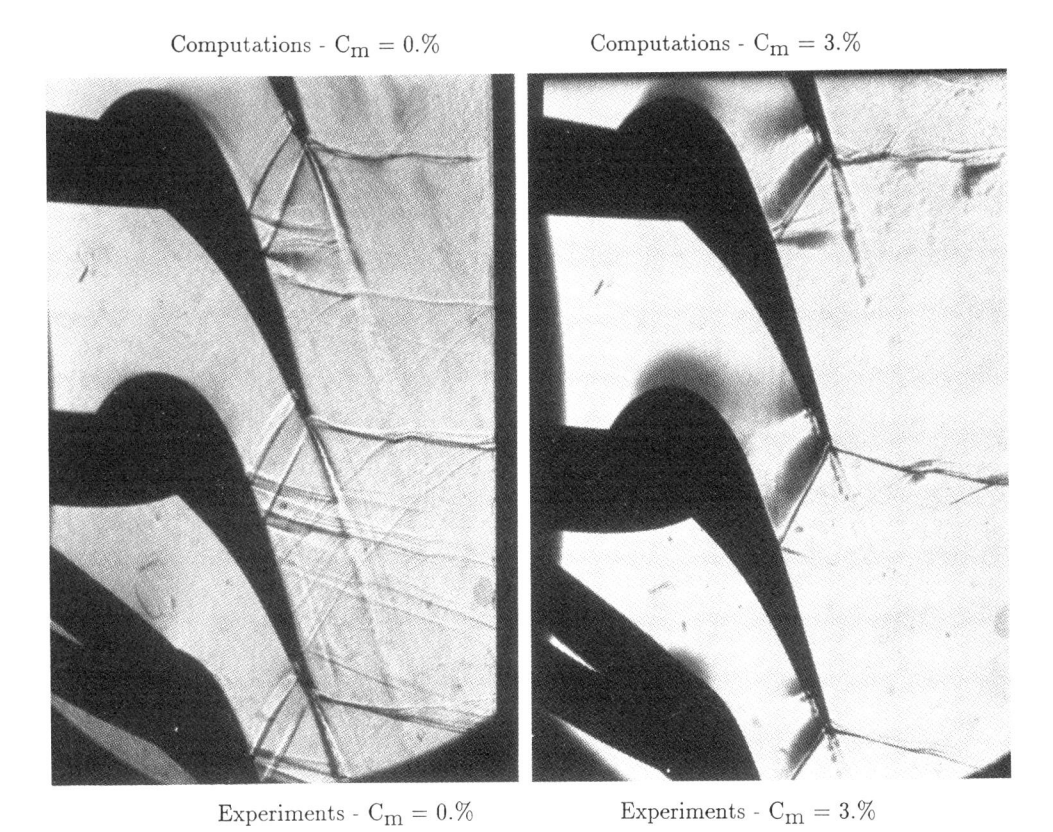

Figure 3.143 Computed and experimental shock pattern $M_{is} = 1.20$ for PS ejection blade at different coolant flow rates.

a) $C_m=0\%$

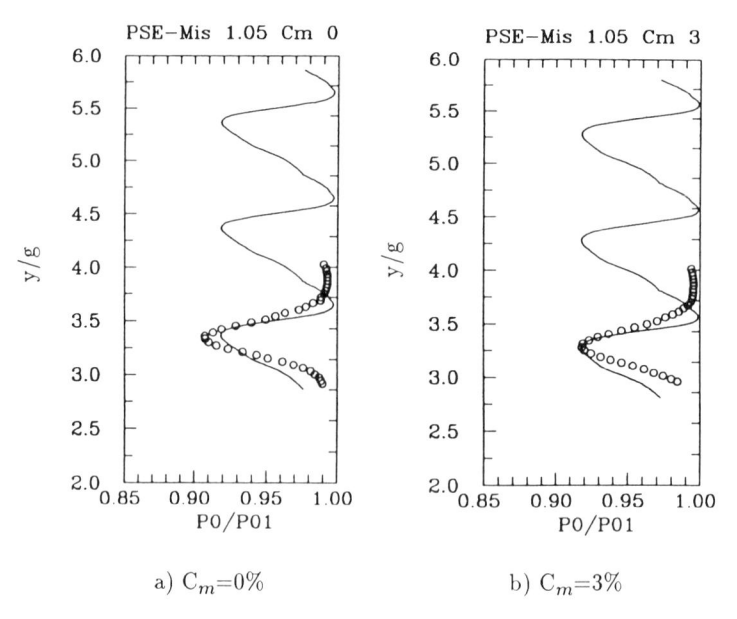

b) $C_m=3\%$

Figure 3.144 Total pressure profiles at 35 mm downstream of trailing edge for PS ejection blade: (a) $C_m = 0\%$; (b) $C_m = 3\%$.

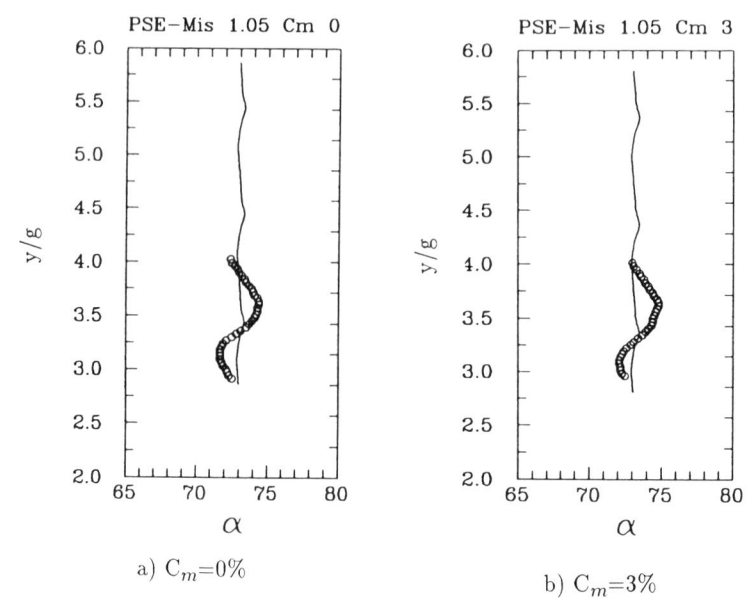

a) $C_m=0\%$

b) $C_m=3\%$

Figure 3.145 Flow angle profiles at 80% axial chord downstream of trailing edge for PS ejection blade: (a) $C_m = 0\%$; (b) $C_m = 3\%$.

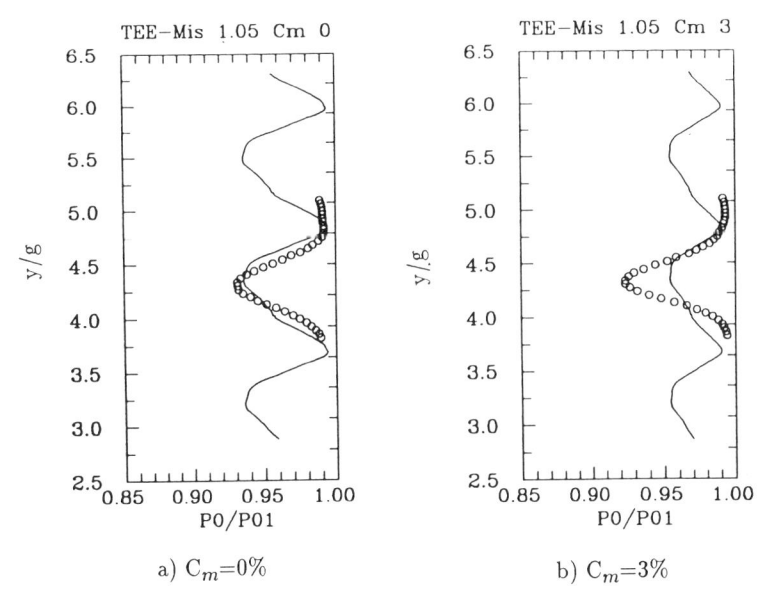

Figure 3.146 Total pressure profiles at 80% axial chord downstream of trailing edge for TE ejection blade: (a) $C_m = 0\%$; (b) $C_m = 3\%$.

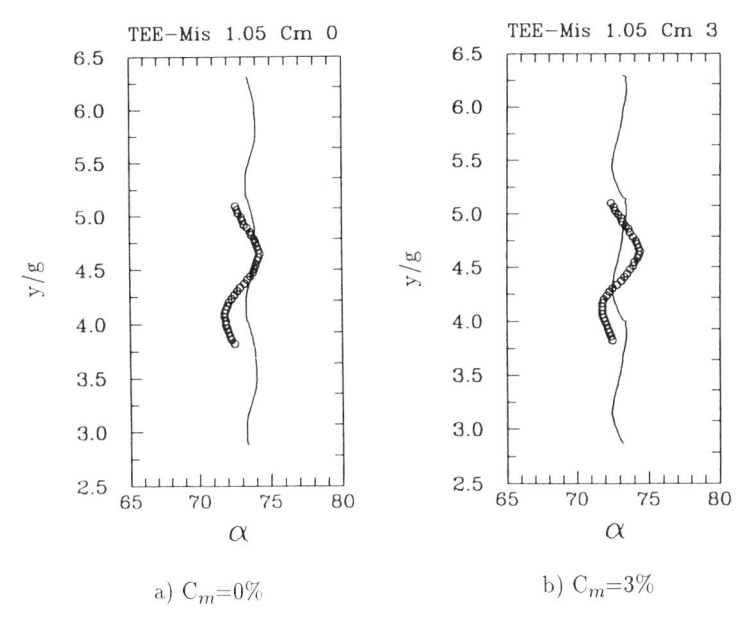

Figure 3.147 Flow angle profiles at 80% axial chord downstream of trailing edge for TE ejection blade: (a) $C_m = 0\%$; (b) $C_m = 3\%$.

although the measured wake appears somewhat more symmetric than the computed one. The wake depth is well reproduced, except for the TE-ejection blade with $C_m = 3\%$, where, surprisingly, the experiments show a deeper wake than that one measured with $C_m = 0\%$ (Figures 3.146 and 3.147). The computations show an opposite trend for the TE-ejection blade. The calculated minimum total pressure ratio in the wake centre for the TE-ejection blade is 0.95 for $C_m = 3\%$, but 0.935 for the zero mass flow rate, proving that the wake is energized by the coolant flow ejection. For the PS blade the agreement between experiments and computations is good. Since the trailing-edge recirculation is not convected away by the coolant ejection for the PS blade (see Figure 3.125) the wake is weakly energized with respect to the $C_m = 0\%$ case, as confirmed by the computed and measured total pressures shown in Figure 3.144. The exit angle profiles (Figures 3.145 and 3.147) show that the computed averaged exit flow angle matches the experiment data. However, the computations predict a nearly flat angle distribution, whereas the measurements indicate an important pitch-wise variation of the flow angle. This result might suggest that the computations predict a larger momentum diffusion than the experiments downstream of the trailing edge, which would be the main cause of the smearing of the velocity distortion produced by the blade. The combined analysis of the total pressure and flow angle indicate that the flow distortion is probably underestimated by the computations from the trailing edge onwards. In fact the computed total pressure shows a deep wake where the exit flow angle is nearly flat. This implies that the low-momentum region is convected downstream with weak mixing from the trailing edge. The agreement is not always good, but the general trend is quite well reproduced. Evidently, the total pressure profile is still impressed by the blade boundary layer, which is less than one chord length upstream of the measuring station. Neither the computations nor the experiments bear the signature of the coolant jet. The good agreement between the computations and the measurements suggest that the jet is correctly simulated, but also that not many of the changes in flow field introduced by the jet are convected downstream.

The comparison of the wake development can be performed considering the CO_2 concentration against the total temperature profile at different planes downstream of the trailing edge. In order to have a correct analysis the concentration profiles have to be transformed into total temperature profiles, taking into account that the CO_2 was selected to simulate a density ratio corresponding to a total temperature $T_{oc}/T_o = 0.638$, with the real total pressure ratio P_{oc}/P_o between the coolant jet and the main stream. In fact if we assume that the mixing process is mainly driven by the diffusion between the cold and hot mass flow, the mixing of the total temperatures of the two streams (main and coolant) is related to the mixing of air and CO_2 by the following formula, where m_c and m_g are the local mass fractions of coolant (or CO_2) and total mass mixture (air plus CO_2) flow:

Mixing of hot and cold air:

$$m_c T_{oc} + (m_g - m_c) T_{01} = m_g T_{02}$$

Comparison of experimental and numerical results

Non-dimensionalized by T_{01}:

$$m_c \frac{T_{oc}}{T_{01}} + (m_g - m_c) = m_g \frac{T_2}{T_{01}} \tag{3.44}$$

(Note: $T_{ol} \simeq$ local total temperature in wake).

Mixing of air and CO_2:

$$m_c \Theta_c + (m_g - m_c)\Theta_g = m_g \Theta_l \tag{3.45}$$

Because Θ_c, the CO_2 concentration in the jet, is 1 and Θ_g, the CO_2 concentration in the main stream at inlet, is zero, the previous formula becomes: $\Theta_l = m_c/m_g$ and therefore the conversion formulae are

$$\frac{T_{ol}}{T_o} = \Theta_l \frac{T_{oc}}{T_o} + (1 - \Theta_l) \qquad \Theta_l = \frac{T_{ol} - T_o}{T_{oc} - T_o} \tag{3.46}$$

Applying equation (3.46), the measured carbon dioxide concentration distribution was transformed into a temperature distribution and plotted against the computed total temperature profiles. In Figures 3.148 and 3.149 the wake evolution is presented for both blade geometries. In both geometries the computed thermal wake is thinner than the experimental one, which demonstrates a smaller diffusion close to the trailing edge compared with experiments. The evolution downstream maintains the same behaviour but the gap

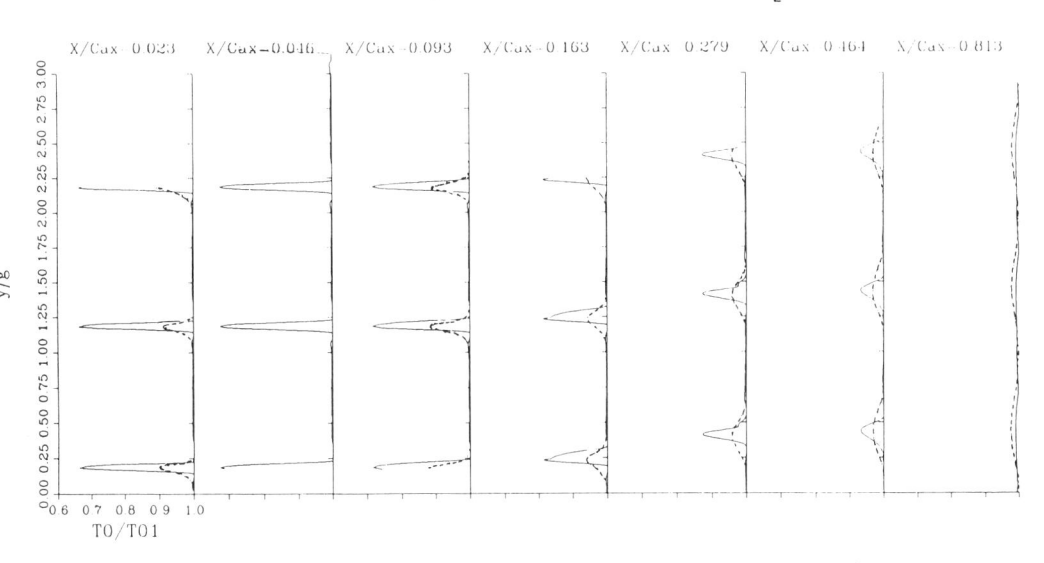

Figure 3.148 Total temperature profiles—computed and transformed measured CO_2-concentrations in several planes for TE ejection blade, $C_m = 3\%$.

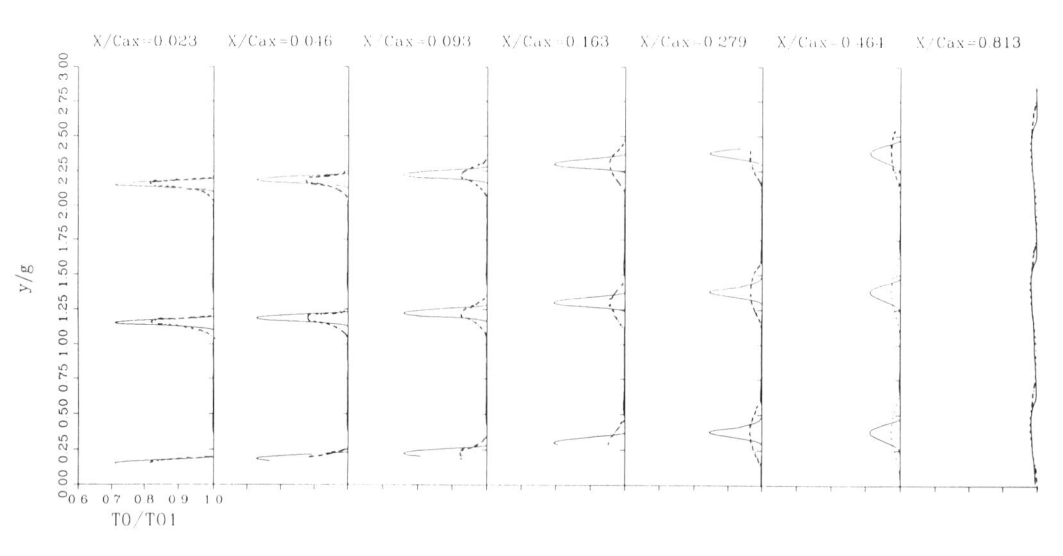

Figure 3.149 Total temperature profiles—computed and transformed measured CO_2—concentrations in several planes for PS ejection blade, $C_m = 3\%$.

between experiments and calculation decreases; at 80% axial chord downstream the profiles are very close to each other. It seems that the global mixing process is captured but the development along the wake is not the same in experiments and calculations. There are several possible reasons for this discrepancy. First, the algebraic turbulence model used does not properly simulate the turbulence decay in the wake and underestimates the viscosity value; in general, a few trailing edge diameters downstream, the eddy viscosity is quite constant, so that less diffusion is simulated by the calculations. Second, the differences may partially be ascribed to the procedure adopted to transforming the concentration into total temperatures ratios. The assumption that the CO_2 transport equation is similar to that for the total temperature with the same Prandtl number is not entirely true. Third, the resolution of the probe is not as fine as in the calculations, and the probe averages the CO_2 concentrations over the probe-head diameter. Finally, unsteady phenomena can play a rule in increasing the diffusion which can not be modelled by the present steady-state calculation. Despite these problems, the calculation seems to detect some of the features of the mixing behaviour, as can be seen by comparing the results of the two configurations. The PS-ejection blade exhibits a reduced temperature peak in the first plane when compared to the TE blade, according to what has been expected because of the longer mixing length for the coolant jet for the PS blade.

Loss calculation Compared to the BL turbulence model, the two-equation model predicts lower loss coefficients, which are in better agreement with the experimental data. The use of the conservation equations to compute the blade losses improves the agreement by reducing the influence of spurious numerical effects. The loss predictions for the TE-ejection blade (Figure 3.150(a)) show a correct trend with Mach number and the values are always within 1 point of the experimental data. At design Mach number the prediction is close to the experimental value. Also the computation of the PS-ejection blade loss (Figure 3.150(b)) shows a correct trend as Mach number increases; the sharp loss increase at supersonic flow conditions is well predicted, as well as the value at the subsonic regime, indicating that profile and shock losses are well captured by the code.

The experimental loss coefficients show that the coolant ejection from the PS blade has a beneficial effect on blade performances, though the PS ejection shows higher ζ than those of the TE-ejection blade at any coolant flow rate. Both calculations and measurements agree in not identifying any clear effect of the C_m percentage changes on the loss coefficients. In fact when C_m ranges between 2 and 4% no significant changes are found on ζ for the TE ejection

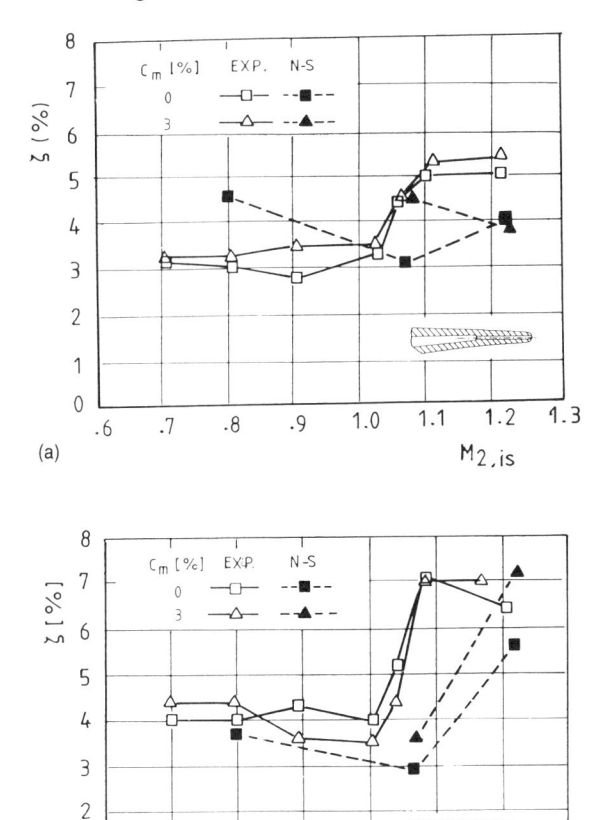

Figure 3.150 Comparison of predicted and measured blade losses for (a) TE-ejection blade; (b) PS-ejection blade.

blade and, moreover, the increase or decrease of C_m does not always have the same effect on the losses. The calculation results on the TE ejection blade predict a minor effect of coolant mass flow rate, as in the experiments, but not always in the direction indicated by the experiments. The PS-ejection blade calculations show an increase in losses for increasing coolant flow, which does not agree with the experimental data, which exhibit a changing trend as the Mach number increases. All this might prove that some parts of the mixing process are captured by the calculation, but not up to the finest structure.

Comments on straight cascade comparisons The set of calculations presented allowed the analysis of the flow characteristic for both the TE and PS blade geometries. The information gathered from the wide set of two-dimensional calculations indicate that the main weakness of the code is probably the turbulence model which, while allowing in general a good prediction of the blade pressure distribution, often prevents a good reproduction of the wake properties, although the main features of the flow are well reproduced. In fact the cited mesh refinements did not produce any significant changes in the wake development. The TE-ejection blade showed some problems in the reproduction of the pressure distribution stemming from a not entirely correct prediction of the boundary-layer development. The reader should not forget that the plots report a series of *blind* tests, in which the codes have been run without knowing the results of the experimental measurements. The area of further investigation should be, as was well known, the turbulence modelling with suitable transition model and the shock boundary-layer interaction simulation; but the assessment of the fine structure of the mixing process modelling could benefit from more detailed investigations performed experimentally. Globally the results have shown that the numerical procedure is a useful tool for the prediction of some properties of the flow field and is able to assess the main features of the blade geometries, together with the coolant mass flow ejection.

8.2.2 Annular cascade

The comparison of the measured and predicted spanwise loss and outlet flow angle distribution is presented in Figure 3.151 for the two most important axial planes at $x/c_{ax} = 0.25$ and 0.54. The angles present area averaged values. The losses are based on area-averaged total and static pressure values. The area-averaging technique was used because it appeared to be the most straightforward method to evaluate the loss prediction capability of the NS calculations.

Losses The NS calculations considerably overpredict the losses over most of the blade span. The reason for the large discrepancy may be partially attributed to the use of the BL turbulence model. In the straight cascade calculations on pp. 307 and 308, the losses were calculated with both the BL and the $K-\varepsilon$ models and the latter predicted roughly 50% lower losses.

Comparison of experimental and numerical results

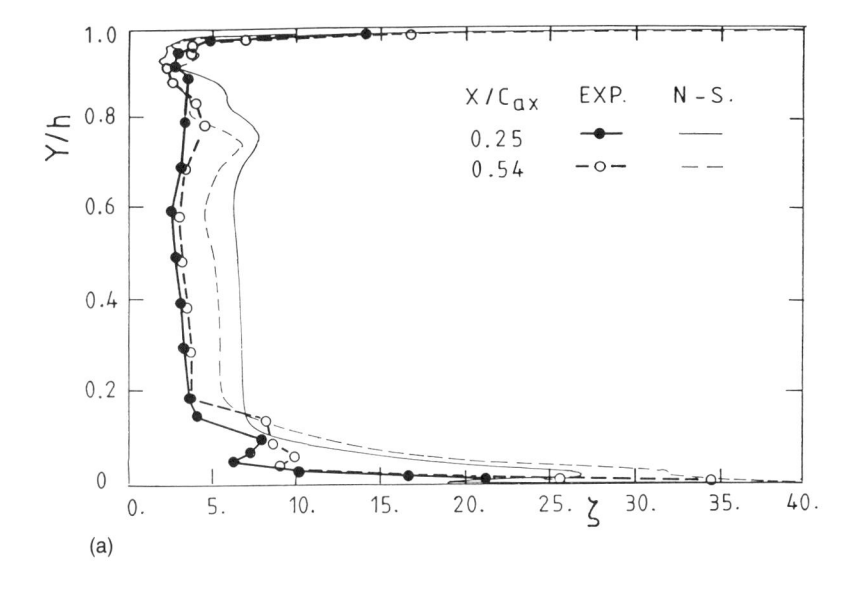

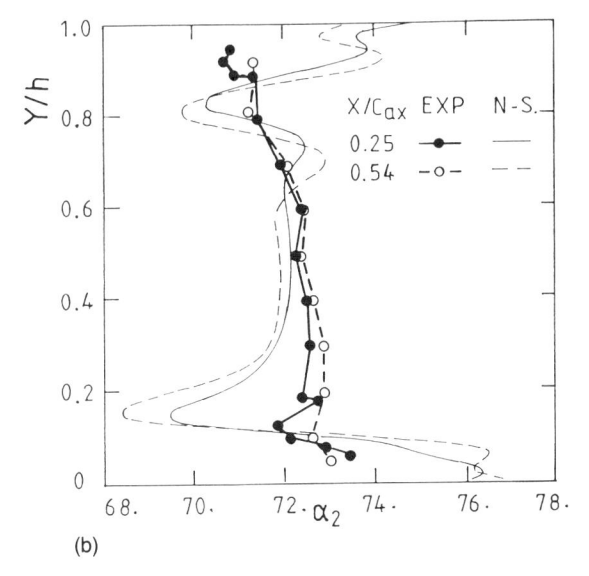

Figure 3.151 (a) Comparison of predicted and measured spanwise loss distribution in annular cascade (PS-ejection blade); (b) comparison of predicted and measured spanwise flow angle distribution in annular cascade (PS-ejection blade).

Comparing the losses in the two measurement planes shows a slight increase in the losses for the measurements, but a strong decrease for the NS calculations. This is clearly unphysical. Calculations with the $K-\varepsilon$ turbulence model would be most interesting.

The differences in the hub and tip end-wall boundary layer are qualitatively correctly predicted. Also the downwash of the upper secondary loss core through the action of the radial pressure gradient is qualitatively correct. The lower secondary loss core, which is distinctly visible in the measurements, disappears in the thick hub end-wall boundary layer.

Angles Contrary to the experimental data, the predicted flow angle distributions are characterized by strong over-/underturning regions at hub and tip. Outside these regions, i.e. between $Y/h = 0.3$ to 0.7, there is a fair agreement with the experimental results.

9 Conclusions

The blade and wake flow characteristics of two trailing-edge-cooled transonic turbine guide vanes were investigated both experimentally and numerically. The flow analysis allows a number of interesting conclusions to be drawn.

The expected advantage of blades with coolant flow ejection from the rear pressure side compared to the coolant ejection through the trailing edge, due to the apparent reduction of the trailing-edge thickness, is not confirmed by the present tests, which show lower losses for the TE-ejection blade. In fact, viscous flow calculation visualizations demonstrate clearly that the pressure-side lip of the PS-ejection blade is to be considered as a second trailing edge.

The base-pressure evolution with Mach number and coolant mass flow rate for the PS-ejection blade differs fundamentally from that for the TE-ejection blade, which has been investigated much more and is much better understood. Viscous flow calculations are particularly helpful in understanding the flow patterns in the trailing-edge region.

The total pressure wake profiles can be matched very closely by Gaussian curves. The profiles at different axial stations appear to be similar to each other. The wake decay is rather slow.

The mixing of the coolant flow with the main flow is found to be very fast in the near-wake, but very slow in the far-wake. The conclusions reached for the straight cascade tests, using CO_2 as coolant flow, are confirmed by wake temperature measurement at mid-blade height in the annular cascade.

The prediction of the wake decay and coolant flow–main flow mixing suffers from the lack of an adequate turbulence model. Overall, the evolution of the total pressure profile appears to be better predicted than the temperature profile.

The problems of the turbulence measurements could not be solved because of the difficulty in separating density and velocity fluctuations. The turbulence measurements with the Kulite pressure probe in the annular cascade are nevertheless promising but they need to be confirmed.

Because of low secondary losses and their limited spanwise extension in the axial planes $X/C_{ax} = 0.25$ and 0.54, the wake evolution at midspan can be compared to that in the straight cascade. Overall the differences are relatively small, but the wake decay is somewhat slower in the annular cascade. Beyond

$X/C_{ax} = 0.54$, the hub endwall boundary layer grows rapidly and reaches nearly mid-blade height at $X/C_{ax} = 1.0$.

Radial migration of low momentum flow in the near-wake region could play a significant role, as suggested by viscous flow calculation visualizations. Compared to the straight cascades, the annular cascade exhibits lower losses. The reasons may be differences in the inlet turbulence level (0.5% at DLR, ~ 2% at VKI) but three-dimensional effects including the radial migration along the trailing edge may be also important.

The measurement of the relative total pressure behind the annular cascade with a high-speed rotating probe revealed interesting aspects with respect to future turbine stage tests. It put into evidence the difficulty of interpreting the relative total pressure field, when the corresponding static pressure field is influenced either by the presence of the probe, as in the present case, or by the turbine rotor.

The fully integrated experiments–numerical approach throughout the project proved to be beneficial for all participants in the project.

10 References

[3.1] Heinemann, H.J., *The Test Facility for Rectilinear Cascades (EGG) of the DFVLR*, DFVLR-Report IB 222-83 A 14, 1983.

[3.2] Sieverding, C.H. and Arts, T., *The VKI Annular Compression Tube Cascade Facility CT3*, ASME 92-GT-336.

[3.3] Sieverding, C.H., Arts, T. and Pasteels, H., Transonic cascade performance measurements using a high speed probe traversing mechanism in a short duration wind tunnel, *Proceedings of 9th Symposium on Measuring Techniques for Transonic and Supersonic Flow in Cascades and Turbomachines*, Oxford, March 1988.

[3.4] Brück, S., *Kalibrierung der Zuströmung des Windkanals für Ebene Gitter*, DLR IB 222-92 A 09, 1992.

[3.5] Höhler, G. and Baumgarten, D. *Boundary Layer and Turbulence Measurements in Test-Rig-ETW*, DFVLR IB 222-81 A 20, 1981.

[3.6] Amecke, J., *Auswertung von Nachlaufmessungen an transsonischen ebenen Schaufelgittern*, AVA-Report 67 A 49, 1967.

[3.7] Amecke, J., *Anwendung der transsonischen Ähnlichkeitsregel auf die Strömung durch ebene Schaufelgitter*, VDI-Forschungsheft Nr. 540, 1970.

[3.8] Amecke, J., *Data Reduction of Wake Flow Measurements with Plane Cascades*, DLR-Report IB 222-90 A 06, 1990.

[3.9] Oldfield, M.L.G., Schultz, D.L. and Nicholson, J.H., Loss measurements using a fast traverse in an ILPT transient cascade, *Proceedings of 6th Meeting on Measuring Techniques in Transonic and Supersonic Flows in Cascades and Turbomachines*, Lyon, 15–16, October, 1981.

[3.10] Ng, W. and Epstein, A., High frequency temperature and pressure probe for unsteady compressible flow, *Review of Scientific Instruments*, **54**(12), 1678–1683, 1983.

[3.11] Sieverding, C.H., Recent progress in the understanding of basic aspects of secondary flows in turbine blade passages, *Journal of Engineering for Gas Turbines and Power*, **107**(2), 248–257, 1985.

[3.12] Dejc, M.E. and Trojanovskij, B.M., *Untersuchung und Berechnung axialer Turbinenstufen*, VEB Verlag Technik, Berlin.
[3.13] Sieverding, C.H., Van Hove, W. and Boletis, E., Experimental study of the 3D flow field in an annular turbine nozzle guide vane, *Journal of Engineering for Power*, **106**(2) 437–44, 1984.
[3.14] Pulliam, T., Efficient solution methods for the Navier–Stokes equations, Von Karman Institute Lecture Series 1986-02, *Numerical Techniques for Viscous Flows Calculations in Turbomachinery Bladings*, 20–24 January, 1986.
[3.15] Michelassi, V., Liou, M.-S. and Povinelli, L.A., *Implicit Solution of 3D Internal Turbulent Flows*, NASA TN-103099, ICOMP-90-13.
[3.16] Martelli, F. and Michelassi, V., 3-D Viscous Analysis of Turbine Bladings. Yokohama International Gas Turbine Congress, Yokohama, October, 1991.
[3.17] Michelassi, V. and Martelli, F., Efficient solution of turbulent incompressible separated flows, *8th GAMM Conference on Numerical methods in Fluid Mechanics*, Delft, September 1989.
[3.18] Martelli, F. and Boretti, A.A., Improvement in gas turbine blade loss prediction methods, *ASME Cogen-Turbo Conference and Exhibit*, Montreux, 30 Aug–1 Sept, 1988.
[3.19] Martelli, F. and Boretti, A.A., Accuracy and efficiency of the time-marching approach for combustor modeling, *AGARD CP-437 Symposium on Validation of Computational Fluid Dynamics*, Lisbon, Portugal, 1988.
[3.20] Baker, A.J., Manhardt, P.D., Orzechowski, J.A. and Yen, K.T., A 3D finite element algorithm for prediction of V/Stol jet induced flowfields, *AGARD Fluid dynamic of jet with application to V/STOL*, 1981.
[3.21] Barata, J.M.M., Durao, D.F.G. and McGuirk, J.J., *Numerical Study of Single Impinging Jets*. AIAA Paper 89-0449.
[3.22] Barata, J.M.M., Durao, D.F.G. Heitor, M.V. and McGuirk, J.J., On the validation of 3-D numerical simulations of turbulent impinging jets through a crossflow, *AGARD Validation of computational Fluid Dynamics*, 1989.
[3.23] Catalano, G.D., Chang, K.S. and Mathis, J.S., *An Analytical and Experimental Investigation of Turbulent Jet Impingement in a Confined Crossflow*, AIAA Paper 89-0665.
[3.24] Childst, R.E. and Nixon, D., *Simulation of Impinging Turbulent Jets*, AIAA Paper 85-0047.
[3.25] Chleboun, P.V., Sebbowa, F.B. and Sheppard, C.G.W., A study of transverse turbulent jets in a cross flow, *Combustion Science and Technology*, **29**, 107–111, 1982.
[3.26] Claus, R.W., Numerical calculations of subsonic jets in crossflow with reduced numerical diffusion, NASA-TM-87003, 1985.
[3.27] Demuren, A.O. Numerical calculation of steady three-dimensional turbulent jets in cross flow, *Compute Methods in Applied Mechanics and Engineering*, **37**, 309–328, 1983.
[3.28] Khan, Z.A., McGuirk, J.J. and Whitelaw, J.H., A row jets in a crossflow, *AGARD Fluid dynamic of jet with application to V/STOL*, 1981.
[3.29] Launder, B.E. and Spalding, D.B., The numerical computation of turbulent flows, *Computer Methods in Applied Mechanics and Engineering*, **1**, 131–138, 1974.
[3.30] Leschziner, M.A. and Rodi, W., Calculation of annular and twin parallel jets using various discretization schemes and turbulence—model variations. *Trans. ASME, Journal of Fluid Engineering*, **103**, 352–360, 1981.
[3.31] Onvani, A., Ollivier, C., Bario, F. and Leboeuf, F., Etude experimentale et théorique d'un jet tridimensional introduit dans l'écoulement secondaire d'une

grille distributrice de turbine, *AGARD Conference Proceedings No. 390, 65th PEP Symposium on Heat Transfer and Cooling in Gas Turbines, Bergen, Norway, 6–10 May 1985, AGARD-CP-390*, pp. 40–1, 40.18, 1985.

[3.32] Patankar, S.V., Basu, D.K. and Alpay, S.A., Prediction of the three-dimensional velocity field of a deflected turbulent jet, *Trans. ASME, Journal of Fluid Engineering*, **99**, 758–762, 1977.

[3.33] Rodi, W. and Scheuerer, G., Calculation of laminar-turbulent boundary layer transition turbine blades. *AGARD Conference Proceedings No. 390, 65th PEP Symposium on Heat Transfer and Cooling in Gas Turbines, Bergen, Norway, 6–10 May 1985, AGARD-CP-390*.

[3.34] Sharif, M.A.R. and Busnaina, A.A., Prediction of lateral jets injected into swirling cross flow, *Proceedings of 5th International Conference on Numerical Methods in Laminar and Turbulent Flows*, vol. 5, part 1, 6–10 July, pp. 374–385, 1987.

[3.35] Viegas, J.R., Rubesin, M.W. and MacCormack, R.W., On the validation of a code and turbulence model appropriate to circular airfoils, *AGARD CP-437 Symposium on Validation of Computational Fluid Dynamics*, Lisbon, Portugal, May 1988.

[3.36] White, A.J., *The Prediction of the Flow and Heat Transfer in the Vicinity of a Jet in Crossflow*, ASME Paper 80-WA/HT-26, 1980.

[3.37] Kar, S. and Poulose, C.N., Prediction of the trajectory of a turbulent jet injected in a cross flowing stream, *Proceedings Advances in Aerodynamics, Fluid Mechanics and Hydraulics*, Minneapolis, USA, June, pp. 239–244, 1986.

[3.38] Makihata, T. and Miyai, Y., Trajectories of single and double jets injected into a cross flow of arbitrary velocity distribution, *Trans. ASME, Journal of Fluid Engineering*, **101**, 217–223, 1979.

[3.39] Makihata, T. and Miyai Y., Prediction of the trajectory of triple jets in a uniform crossflow, *Trans. ASME, Journal of Fluid Engineering*, **105**, 91–97, 1983.

[3.40] Sucec, J. and Bowley, W.W. Prediction of the trajectory of a turbulent jet injected into a crossflowing stream, *Trans. ASME, Journal of Fluid Engineering*, **98**, 667–673, 1976.

[3.41] Kollen, O. and Koschel, W., Effect of film cooling on the aerodynamic performance of a turbine cascade, *AGARD Conference Proceedings No. 390, 65th PEP Symposium on Heat Transfer and Cooling in Gas Turbines, Bergen, Norway, 6–10 May*, AGARD-CP-390, pp. 39–1, 39.16, 1985.

[3.42] Baldwin, B. and Lomax, H., *Thin-layer Approximation and Algebraic Model for Separated Turbulent Flows*, AIAA-78-257, 1978.

[3.43] Jones, N.P. and Launder, B.E., The calculation of low-Reynolds number phenomena with a two-equation model of turbulence, *Int. Journal of Heat and Mass Transfer*, **16**, 1119–1130, 1973.

[3.44] Jones, W.P. and McGuirk, J.J., Mathematical modeling of gas turbine combustion chambers, *Combustor Modeling AWARD CP 275*, 1979.

[3.45] Nagano, Y. and Hishida, M., Improved form of the k-model for wall turbulent shear flows, *Trans. ASME, Journal of Fluid Engineering*, **109**, 156–160, 1987.

[3.46] Wilcox, D.C. and Rubesin, M.W., *Progress in Turbulence Modeling for Complex Flow Fields Including Effects of Compressibility*, NASA TP 1517, 1980.

Appendix 1: Coordinates, surface angle and curvature of BRITE-22N blade

Table A1 Coordinates

X_{ax} (m)	Y_{suction} (m)	Y_{pressure} (m)	X_{ax} (m)	Y_{suction} (m)	Y_{pressure} (m)
0.000167	0.000074	0.000074	0.006565	0.008493	−0.006174
0.000293	0.001022	−0.000875	0.006736	0.008594	−0.006299
0.000421	0.001405	−0.001258	0.006908	0.008691	−0.006426
0.000549	0.001693	−0.001546	0.007081	0.008785	−0.006554
0.000678	0.001931	−0.001783	0.007255	0.008875	−0.006683
0.000809	0.002158	−0.001986	0.007430	0.008961	−0.006813
0.000941	0.002384	−0.002165	0.007606	0.009044	−0.006945
0.001073	0.002608	−0.002325	0.007783	0.009124	−0.007078
0.001207	0.002830	−0.002468	0.007960	0.009200	−0.007212
0.001342	0.003048	−0.002599	0.008139	0.009272	−0.007347
0.001478	0.003262	−0.002718	0.008318	0.009341	−0.007483
0.001616	0.003473	−0.002825	0.008498	0.009406	−0.007620
0.001754	0.003680	−0.002923	0.008679	0.009467	−0.007759
0.001893	0.003885	−0.003016	0.008861	0.009526	−0.007898
0.002034	0.004086	−0.003106	0.009043	0.009580	−0.008040
0.002175	0.004284	−0.003194	0.009227	0.009631	−0.008182
0.002318	0.004479	−0.003282	0.009441	0.009679	−0.008326
0.002462	0.004671	−0.003370	0.009596	0.009723	−0.008472
0.002607	0.004860	−0.003459	0.009782	0.009764	−0.008616
0.002753	0.005046	−0.003551	0.009968	0.009801	−0.008766
0.002899	0.005228	−0.003645	0.010156	0.009834	−0.008915
0.003048	0.005407	−0.003741	0.010344	0.009864	−0.009065
0.003197	0.005583	−0.003839	0.010533	0.009891	−0.009216
0.003347	0.005756	−0.003939	0.010722	0.009913	−0.009368
0.003498	0.005925	−0.004039	0.010912	0.009932	−0.009521
0.003650	0.006091	−0.004140	0.011103	0.009947	−0.009676
0.003804	0.006255	−0.004242	0.011295	0.009958	−0.009831
0.003958	0.006414	−0.004345	0.011487	0.009965	−0.009989
0.004113	0.006570	−0.004450	0.011680	0.009969	−0.010147
0.004270	0.006723	−0.004555	0.011874	0.009969	−0.010307
0.004427	0.006871	−0.004663	0.012068	0.009965	−0.010468
0.004586	0.007017	−0.004771	0.012263	0.009958	−0.010630
0.004745	0.007159	−0.004881	0.012459	0.009946	−0.010793
0.004906	0.007297	−0.004992	0.012655	0.009931	−0.010958
0.005067	0.007432	−0.005105	0.012852	0.009912	−0.011123
0.005230	0.007564	−0.005219	0.013050	0.009889	−0.011290
0.005393	0.007692	−0.005334	0.013248	0.009861	−0.011458
0.005558	0.007817	−0.005451	0.013446	0.009830	−0.011627
0.005723	0.007939	−0.005568	0.013645	0.009794	−0.011798
0.005890	0.008057	−0.005687	0.013845	0.009755	−0.011969
0.006057	0.008171	−0.005807	0.014045	0.009711	−0.012142
0.006225	0.008282	−0.005928	0.014246	0.009663	−0.012316
0.006395	0.008389	−0.006050	0.014447	0.009610	−0.012491

Appendix 1: Coordinates, surface angle and curvature of BRITE-22N blade

Table A1 (cont.)

X_{ax} (m)	Y_{suction} (m)	Y_{pressure} (m)	X_{ax} (m)	Y_{suction} (m)	Y_{pressure} (m)
0.014649	0.009554	−0.012668	0.024428	0.000897	−0.022593
0.014851	0.009493	−0.012845	0.024636	0.000550	−0.022846
0.015054	0.009427	−0.013024	0.024845	0.000194	−0.023101
0.015257	0.009358	−0.013205	0.025054	−0.000171	−0.023358
0.015460	0.009283	−0.013386	0.025262	−0.000546	−0.023618
0.015664	0.009204	−0.013569	0.025470	−0.000930	−0.023880
0.015868	0.009121	−0.013753	0.025677	−0.001322	−0.024143
0.016073	0.009033	−0.013938	0.025885	−0.001723	−0.024409
0.016278	0.008939	−0.014125	0.026092	−0.002134	−0.024678
0.016484	0.008842	−0.014313	0.026299	−0.002554	−0.002948
0.016690	0.008739	−0.014502	0.026505	−0.002982	−0.025220
0.016896	0.008631	−0.014693	0.026712	−0.003417	−0.025495
0.017102	0.008519	−0.014885	0.026917	−0.003860	−0.025771
0.017309	0.008401	−0.015078	0.027123	−0.004310	−0.026050
0.017516	0.008279	−0.015273	0.027328	−0.004767	−0.026331
0.017724	0.008151	−0.015469	0.027533	−0.005231	−0.026615
0.017932	0.008019	−0.015667	0.027737	−0.005700	−0.026900
0.018140	0.007881	−0.015866	0.027941	−0.006175	−0.027188
0.018348	0.007738	−0.016067	0.028145	−0.006655	−0.027478
0.018556	0.007590	−0.016269	0.028348	−0.007141	−0.027771
0.018765	0.007437	−0.016472	0.028550	−0.007631	−0.028066
0.018974	0.007278	−0.016677	0.028753	−0.008124	−0.028364
0.019183	0.007114	−0.016884	0.028954	−0.008621	−0.028663
0.019392	0.006944	−0.017091	0.029156	−0.009120	−0.028965
0.019601	0.006769	−0.017301	0.029356	−0.009623	−0.029269
0.019811	0.006588	−0.017511	0.029556	−0.010127	−0.029574
0.020020	0.006402	−0.017724	0.029756	−0.010633	−0.029882
0.020230	0.006209	−0.017938	0.029955	−0.011141	−0.030193
0.020440	0.006010	−0.018154	0.030154	−0.011652	−0.030506
0.020650	0.005805	−0.018371	0.030352	−0.012164	−0.030821
0.020860	0.005593	−0.018590	0.030549	−0.012677	−0.031136
0.021070	0.005375	−0.018811	0.030746	−0.013191	−0.031458
0.021280	0.005150	−0.019033	0.030942	−0.013706	−0.031779
0.021490	0.004918	−0.019257	0.031138	−0.014222	−0.032103
0.021701	0.004679	−0.019483	0.031333	−0.014740	−0.032429
0.021911	0.004434	−0.019711	0.031527	−0.015261	−0.032756
0.022121	0.004180	−0.019940	0.031721	−0.015784	−0.033085
0.022331	0.003919	−0.020172	0.031914	−0.016313	−0.033417
0.022541	0.003652	−0.020405	0.032106	−0.016849	−0.033750
0.022751	0.003377	−0.020640	0.032298	−0.017390	−0.034086
0.022961	0.003094	−0.020877	0.032489	−0.017936	−0.034423
0.023171	0.002804	−0.021117	0.032679	−0.018486	−0.034762
0.023381	0.002506	−0.021358	0.032869	−0.019041	−0.035103
0.023590	0.002200	−0.021601	0.033058	−0.019600	−0.035446
0.023800	0.001887	−0.021845	0.033246	−0.020163	−0.035791
0.024009	0.001566	−0.022092	0.033433	−0.020727	−0.036137
0.024219	0.001235	−0.022341	0.033620	−0.021292	−0.036484

Table A1 (cont.)

X_{ax} (m)	$Y_{suction}$ (m)	$Y_{pressure}$ (m)	X_{ax} (m)	$Y_{suction}$ (m)	$Y_{pressure}$ (m)
0.033805	−0.021856	−0.036834	0.039132	−0.039370	−0.048112
0.033990	−0.022419	−0.037184	0.039288	−0.039912	−0.048479
0.034175	−0.022982	−0.037537	0.039443	−0.040451	−0.048845
0.034358	−0.023544	−0.037890	0.039598	−0.040988	−0.049209
0.034541	−0.024107	−0.038245	0.039751	−0.041521	−0.049569
0.034722	−0.024673	−0.038601	0.039903	−0.042050	−0.049926
0.034903	−0.025241	−0.038959	0.040055	−0.042576	−0.050282
0.035083	−0.025811	−0.039317	0.040205	−0.043099	−0.050640
0.035263	−0.026382	−0.039677	0.040354	−0.043617	−0.051002
0.035441	−0.026956	−0.040038	0.040502	−0.044132	−0.051371
0.035619	−0.027532	−0.040399	0.040649	−0.044644	−0.051748
0.035795	−0.028112	−0.040762	0.040795	−0.045151	−0.052126
0.035971	−0.028693	−0.041126	0.040939	−0.045655	−0.052501
0.036146	−0.029274	−0.041491	0.041083	−0.046155	−0.052867
0.036320	−0.029853	−0.041856	0.041226	−0.046651	−0.053218
0.036493	−0.030431	−0.042222	0.041367	−0.047144	−0.053550
0.036665	−0.031006	−0.042589	0.041508	−0.047632	−0.053860
0.036836	−0.031579	−0.042957	0.041647	−0.048115	−0.054142
0.037007	−0.032149	−0.043325	0.041786	−0.048594	−0.054327
0.037176	−0.032717	−0.043693	0.041923	−0.049067	−0.054436
0.037344	−0.033282	−0.044062	0.042059	−0.049535	−0.054507
0.037512	−0.033844	−0.044431	0.042194	−0.049998	−0.054550
0.037678	−0.034405	−0.044800	0.042328	−0.050457	−0.054569
0.037844	−0.034962	−0.045169	0.042461	−0.050912	−0.054568
0.038008	−0.035519	−0.045538	0.042592	−0.051364	−0.054545
0.038172	−0.036074	−0.045906	0.042723	−0.051814	−0.054501
0.038334	−0.036628	−0.046274	0.042852	−0.052261	−0.054430
0.038496	−0.037180	−0.046641	0.042981	−0.052708	−0.054326
0.038656	−0.037730	−0.047009	0.043108	−0.053155	−0.054166
0.038816	−0.038279	−0.047376	0.043234	−0.053721	−0.053720
0.038974	−0.038825	−0.047744			

Solid blade-BRITE-22N
mid section radius = 369.850 mm
chord = 72.00 mm
pitch/chord = 0.7506
Number of points = 241

Appendix 1: Coordinates, surface angle and curvature of BRITE-22N blade

Table A2 Surface angle contour (no staggered blade)

Upper Surface

$X_{pressure}$	$Y_{pressure}$	θ	$X_{pressure}$	$Y_{pressure}$	θ
0.00000	0.13294	89.65994	0.25180	0.28111	−8.37719
0.00020	0.14045	87.16968	0.26412	0.27914	−9.70901
0.00073	0.14772	84.64108	0.27688	0.27682	−10.86589
0.00154	0.15475	82.12660	0.29010	0.27415	−11.99254
0.00264	0.16157	79.56577	0.30376	0.27113	−12.90309
0.00401	0.16817	76.96269	0.31781	0.26780	−13.73838
0.00564	0.17456	74.32619	0.33225	0.26417	−14.43708
0.00753	0.18075	71.67463	0.34700	0.26029	−15.04661
0.00967	0.18675	69.07450	0.36210	0.25615	−15.58522
0.01205	0.19257	66.53001	0.37743	0.25182	−15.97399
0.01465	0.19821	64.00169	0.39299	0.24731	−16.31708
0.01746	0.20368	61.45593	0.40873	0.24266	−16.59351
0.02050	0.20897	58.87656	0.42464	0.23788	−16.86445
0.02376	0.21410	56.20975	0.44071	0.23297	−17.09205
0.02724	0.21904	53.60775	0.45692	0.22795	−17.31734
0.03093	0.22383	51.15353	0.47331	0.22280	−17.62012
0.03482	0.22846	48.72511	0.48996	0.21743	−18.16640
0.03891	0.23292	46.21607	0.50707	0.21171	−18.74314
0.04322	0.23722	43.66640	0.52454	0.20570	−19.21672
0.04774	0.24136	41.20246	0.54240	0.19940	−19.62657
0.05248	0.24532	38.77166	0.56048	0.19291	−19.81325
0.05742	0.24913	36.40490	0.57869	0.18633	−19.92778
0.06257	0.25277	34.03725	0.59703	0.17965	−20.16446
0.06794	0.25624	31.71404	0.61564	0.17275	−20.49202
0.07353	0.25954	29.37787	0.63455	0.16562	−20.84010
0.07935	0.26265	27.08042	0.65380	0.15822	−21.18888
0.08539	0.26260	24.86835	0.67332	0.15061	−21.36301
0.09166	0.26836	22.69196	0.69294	0.14293	−21.41094
0.09817	0.27094	20.62343	0.71259	0.13522	−21.45174
0.10490	0.27334	18.56497	0.73229	0.12746	−21.50187
0.11188	0.27555	16.61460	0.75207	0.11965	−21.61692
0.11910	0.27757	14.70271	0.77200	0.11173	−21.76110
0.12657	0.27940	12.83177	0.79206	0.10369	−21.88613
0.13429	0.28103	10.99273	0.81225	0.09556	−21.97951
0.14228	0.28245	9.12666	0.83250	0.08738	−22.01333
0.15055	0.28364	7.33596	0.85278	0.07918	−22.01975
0.15912	0.28461	5.49921	0.87305	0.07098	−22.02707
0.16801	0.28532	3.75273	0.89333	0.06278	−21.98969
0.17722	0.28579	2.01957	0.91360	0.05459	−22.10617
0.18677	0.28598	0.34579	0.93378	0.04646	−21.43890
0.19667	0.28591	−1.19742	0.95383	0.03844	−23.39160
0.20692	0.28555	−2.73325	0.97391	0.03039	−16.31179
0.21754	0.28491	−4.15256	0.99561	0.02107	−52.12606
0.22854	0.28397	−5.62305	1.00000	0.00996	−75.49093
0.23997	0.28270	−7.00161			

Lower Surface

$X_{pressure}$	$Y_{pressure}$	θ	$X_{pressure}$	$Y_{pressure}$	θ
0.00000	0.13294	89.65994	0.38926	0.07374	−172.87350
0.00012	0.12518	92.04026	0.39774	0.07477	−173.26514
0.00057	0.11715	94.39018	0.40630	0.07575	−173.68961
0.00138	0.10884	96.79908	0.41492	0.07667	−174.10687
0.00259	0.10022	99.12289	0.42362	0.07753	−174.55775
0.00420	0.09129	101.21559	0.43240	0.07834	−175.00203
0.00621	0.08204	103.29668	0.44125	0.07908	−175.47899
0.00866	0.07245	105.36332	0.45020	0.07974	−175.96628
0.01158	0.06249	107.21438	0.45924	0.08034	−176.45256
0.01499	0.05215	109.45486	0.46837	0.08087	−176.96913
0.01897	0.04136	110.40589	0.47760	0.08132	−177.46321
0.02347	0.03016	115.42281	0.48693	0.08169	−177.99355
0.03857	0.01066	141.51907	0.49636	0.08197	−178.51512
0.06113	0.00067	170.31775	0.50590	0.08218	−179.07306
0.07179	0.00000	−177.63766	0.51557	0.08228	−179.63924
0.08022	0.00108	−167.50812	0.52535	0.08230	179.82643
0.08736	0.00316	−161.07365	0.53525	0.08222	179.27086
0.09406	0.00559	−159.53677	0.54527	0.08205	178.74382
0.10073	0.00804	−160.45709	0.55542	0.08178	178.17535
0.10755	0.01037	−161.50313	0.56571	0.08140	177.61845
0.11443	0.01266	−161.68587	0.57613	0.08091	177.02214
0.12133	0.01493	−161.99591	0.58671	0.08030	176.42674
0.12829	0.01716	−162.48048	0.59743	0.07958	175.87772
0.13530	0.01934	−162.91449	0.60830	0.07874	175.31848
0.14236	0.02148	−163.27846	0.61932	0.07779	174.76805
0.14947	0.02360	−163.56955	0.63051	0.07671	174.17024
0.15660	0.02568	−163.88635	0.64186	0.07549	173.62044
0.16379	0.02773	−164.27321	0.65338	0.07415	173.07903
0.17102	0.02974	−164.62534	0.66506	0.07267	172.55991
0.17829	0.03172	−164.89874	0.67691	0.07107	172.02440
0.18560	0.03368	−165.14662	0.68894	0.06933	171.49571
0.19293	0.03561	−165.39682	0.70114	0.06745	170.98901
0.20031	0.03751	−165.69768	0.71352	0.06543	170.48743
0.20772	0.03938	−166.05760	0.72607	0.06327	170.01294
0.21519	0.04121	−166.36035	0.73881	0.06097	169.54022
0.22268	0.04301	−166.56659	0.75173	0.05854	169.08595
0.23021	0.04479	−166.78432	0.76483	0.05596	168.63173
0.23777	0.04655	−167.07715	0.77810	0.05324	168.20277
0.24537	0.04827	−167.37785	0.79156	0.05037	167.78287
0.25301	0.04996	−167.65166	0.80520	0.04737	167.36728
0.26069	0.05162	−167.90936	0.81903	0.04422	166.95834
0.26840	0.05326	−168.15169	0.83304	0.04092	166.57422
0.27615	0.05487	−168.39641	0.84723	0.03749	166.22545
0.28393	0.05645	−168.65833	0.86157	0.03393	165.93083
0.29175	0.05800	−168.94664	0.87606	0.03026	165.64857
0.29962	0.05951	−169.23141	0.89071	0.02647	165.25935
0.30753	0.06100	−169.52654	0.90557	0.02250	164.96400
0.31548	0.06244	−169.81458	0.92048	0.01850	165.04204
0.32347	0.06386	−170.10587	0.93544	0.01447	164.50461
0.33151	0.06524	−170.40729	0.95099	0.00997	163.25122
0.33959	0.06658	−170.73235	0.96698	0.00512	163.49210
0.34773	0.06789	−171.08345	0.98191	0.00110	166.99409
0.35592	0.06915	−171.43036	0.99230	0.00064	−162.09309
0.36417	0.07036	−171.79832	0.99746	0.00428	−127.07835
0.37248	0.07153	−172.14964	1.00000	0.00996	−107.74371
0.38084	0.07266	−172.50731			

Appendix 1: Coordinates, surface angle and curvature of BRITE-22N blade

Table A3 Suction-side curvature

S/S_{tot}	Curvature (1/mm)	S/S_{tot}	Curvature (1/mm)
0.000000	0.000000	0.317415	0.162893
0.016425	6.621252	0.326458	0.172860
0.023739	0.109586	0.335877	0.173051
0.030666	0.348521	0.345687	0.191465
0.037471	0.282594	0.355922	0.198376
0.044124	0.241167	0.366608	0.215405
0.050649	0.230997	0.377781	0.228459
0.057056	0.219014	0.389463	0.212327
0.063342	0.221541	0.401636	0.224464
0.069507	0.191885	0.414291	0.178877
0.075566	0.191335	0.427360	0.192075
0.081515	0.188101	0.440815	0.142317
0.087353	0.197822	0.454583	0.161866
0.093074	0.162476	0.468635	0.094590
0.098702	0.165406	0.482868	0.105982
0.104240	0.150712	0.497250	0.069444
0.109695	0.158328	0.511729	0.090719
0.115069	0.138339	0.526297	0.056554
0.120375	0.141968	0.540914	0.081707
0.125618	0.129128	0.555613	0.132182
0.130809	0.132680	0.570510	0.250621
0.135951	0.122402	0.585755	0.144345
0.141055	0.124601	0.601245	0.213421
0.146129	0.113243	0.616969	0.064587
0.151185	0.111199	0.632728	0.047315
0.156231	0.103756	0.648440	0.066522
0.161276	0.108500	0.664152	0.177657
0.166327	0.103870	0.679987	0.115134
0.171394	0.111245	0.695916	0.214346
0.176486	0.099839	0.711993	0.079768
0.181611	0.098788	0.728052	0.027393
0.186780	0.096649	0.743971	0.016877
0.192003	0.106213	0.759717	0.025110
0.197289	0.100232	0.775291	0.023905
0.202651	0.104566	0.790712	0.095636
0.208099	0.098101	0.806032	0.054279
0.213642	0.106462	0.821231	0.072208
0.219293	0.100997	0.836296	0.043004
0.225066	0.112893	0.851191	0.014781
0.230974	0.106468	0.865872	0.003782
0.237031	0.114699	0.880320	0.000668
0.243249	0.107967	0.894522	0.001064
0.249638	0.118102	0.908476	0.006175
0.256214	0.111736	0.922181	0.001099
0.262987	0.121453	0.935614	0.064826
0.269973	0.118192	0.948727	0.113889
0.277183	0.132673	0.961510	0.120883
0.284641	0.132717	0.974043	0.365905
0.292367	0.149883	0.986413	2.258480
0.300391	0.149228	1.000000	0.000000
0.308731	0.166147		

Appendix 2: Flow properties and thermodynamics

Table A4 Flow properties of plane cascade of TE-blades with air ejection

c_m (%)	M_{2is}	M_1	p_{01} (kPa)	p_{02} (kPa)	p_2 (kPa)	T_{01} (K)	α_2 (°)	p_{0c} (kPa)	p_c (kPa)	T_{0c} (K)	C_b
0.00	1.207	0.178	100.73	95.89	41.17	298.8	72.9	36.02	35.99	290.4	−0.0514
0.00	1.097	0.178	100.77	96.68	47.38	301.4	73.1	40.24	40.24	290.6	−0.0709
0.00	1.052	0.178	100.76	97.46	50.04	301.8	73.1	46.13	46.13	290.8	−0.0388
0.00	1.024	0.179	100.77	98.41	51.78	303.0	73.2	51.51	51.51	291.0	−0.0027
0.00	0.900	0.176	100.80	99.27	59.63	304.1	73.0	57.99	57.99	291.4	−0.0162
0.00	0.801	0.171	100.82	99.47	66.11	304.5	72.8	64.68	64.68	291.4	−0.0142
0.00	0.702	0.163	100.85	99.79	72.59	304.5	72.6	71.46	71.46	291.5	−0.0111
2.04	1.203	0.179	100.80	95.62	41.38	302.8	72.8	58.95	42.08	285.2	0.0070
2.03	1.099	0.178	100.84	96.49	47.31	303.8	73.0	62.05	47.29	284.4	−0.0002
2.04	1.057	0.177	100.89	97.48	49.78	299.9	73.2	66.68	53.44	283.6	0.0363
2.02	1.020	0.177	100.92	98.48	52.11	297.8	73.2	68.73	56.05	288.4	0.0391
2.07	0.398	0.174	100.91	99.03	59.78	301.0	72.7	71.95	60.34	283.2	0.0056
2.11	0.801	0.170	100.83	99.37	66.05	300.9	72.7	76.74	66.06	288.1	0.0001
2.23	0.701	0.161	100.86	99.69	72.62	301.4	72.7	82.07	72.47	284.2	−0.0015
3.12	1.206	0.178	100.75	95.53	41.21	300.7	72.7	82.57	43.62	280.9	−0.0154
3.12	1.104	0.177	100.76	96.40	46.97	300.7	72.9	82.54	45.95	280.2	−0.0102
3.01	1.058	0.177	100.83	97.44	49.72	295.9	73.1	83.26	53.25	285.3	0.0350
3.02	1.015	0.176	100.84	98.40	52.32	296.1	73.1	84.58	57.11	280.2	0.0475
3.04	0.899	0.175	100.81	98.92	59.67	295.7	72.6	86.30	60.97	279.5	0.0129
3.18	0.801	0.170	100.71	99.30	66.02	293.0	72.6	91.24	66.84	283.8	0.0082
3.32	0.701	0.162	100.70	99.60	72.53	293.5	72.4	94.89	73.13	282.2	0.0060
3.97	1.205	0.177	100.34	95.11	41.11	292.5	72.3	106.24	56.13	282.2	−0.0086
3.88	1.110	0.177	100.30	96.15	46.38	293.8	72.8	103.12	54.48	279.1	−0.0157
3.87	1.057	0.177	100.28	96.96	49.52	294.8	72.9	102.78	54.30	279.2	0.0238
3.88	1.012	0.176	100.27	97.94	52.25	296.6	73.0	102.79	57.04	279.2	0.0478
3.96	0.899	0.175	100.19	98.50	59.33	297.3	72.6	104.23	60.46	278.6	0.0113
4.04	0.801	0.170	100.18	99.07	65.67	297.9	72.5	105.21	65.99	279.1	0.0031
4.26	0.701	0.161	100.30	99.51	72.25	297.4	72.3	108.19	72.67	279.2	0.0041

Table A5 Flow properties of plane cascade of PS-blades with air ejection

c_m (%)	M_{2is}	M_1	p_{01} (kPa)	p_{02} (kPa)	p_2 (kPa)	T_{01} (K)	α_2 (°)	p_{0c} (kPa)	p_c (kPa)	T_{0c} (K)	C_{b1}	C_{b2}
0.00	1.207	0.178	99.27	93.08	40.54	291.1	72.1	51.59	51.57	291.2	0.1111	−0.0473
0.00	1.085	0.178	99.12	93.53	47.30	291.6	72.9	51.58	51.56	291.5	0.0430	−0.0596
0.00	1.041	0.179	99.06	95.23	49.88	291.9	73.2	51.66	51.65	291.5	0.0179	0.0180
0.00	1.007	0.177	98.98	96.25	51.85	292.2	73.4	51.67	51.66	291.6	−0.0020	0.0309
0.00	0.894	0.177	98.96	96.62	58.88	292.3	73.1	52.26	52.25	291.6	−0.0670	0.0057
0.00	0.802	0.173	98.57	96.80	64.54	293.3	72.8	58.07	58.07	283.5	−0.0657	0.0097
0.00	0.701	0.165	98.58	97.25	70.98	294.2	72.6	66.94	66.94	286.8	−0.0410	0.0082
1.97	1.184	0.175	99.47	93.33	41.90	293.9	72.1	67.60	54.40	286.4	0.1257	−0.0690
1.95	1.085	0.177	99.18	93.96	47.32	293.8	73.1	67.37	54.30	289.9	0.0704	−0.0951
1.96	1.044	0.177	99.11	95.79	49.69	294.6	73.5	67.69	54.93	286.6	0.0529	−0.0026
1.95	1.006	0.175	99.16	96.74	52.03	294.1	73.8	69.51	57.36	286.7	0.0538	0.0382
1.97	0.898	0.175	99.36	97.55	58.90	295.0	73.3	72.05	60.51	286.3	0.0162	0.0038
2.02	0.798	0.170	99.15	97.36	65.15	295.4	73.0	76.22	65.69	286.6	0.0054	−0.0038
2.11	0.702	0.162	99.41	97.96	71.52	295.5	72.9	81.58	72.00	286.2	0.0049	−0.0035
3.00	1.168	0.178	99.33	92.95	42.68	293.6	71.9	85.49	55.66	281.0	0.1306	−0.0859
3.02	1.084	0.178	99.32	93.79	47.43	294.1	72.7	86.05	55.65	281.1	0.0827	−0.1141
3.00	1.040	0.177	99.40	96.13	50.10	294.8	73.7	86.82	56.29	289.7	0.0623	−0.0147
2.98	1.004	0.176	99.44	97.00	52.30	295.3	74.0	87.02	58.90	284.9	0.0664	0.0341
3.01	0.893	0.174	99.46	97.50	59.27	295.8	73.4	88.06	61.23	284.9	0.0197	−0.0030
3.08	0.795	0.169	99.47	97.57	65.57	296.4	73.2	90.56	66.07	284.9	0.0050	−0.0098
3.27	0.698	0.161	99.51	98.02	71.90	296.8	73.0	94.72	71.97	284.9	0.0007	−0.0084
3.83	1.205	0.177	98.60	91.57	40.38	291.3	71.6	105.57	55.77	282.5	0.1555	−0.0721
3.89	1.097	0.178	98.58	92.93	46.34	292.1	72.7	106.22	56.11	278.6	0.0951	−0.1208
3.93	1.037	0.177	99.24	95.71	50.18	290.8	73.5	109.23	57.71	283.7	0.0643	−0.0225
3.83	0.993	0.177	99.28	96.67	52.88	291.7	73.7	105.74	58.98	279.2	0.0614	0.0242
3.89	0.891	0.174	99.33	97.39	59.33	292.7	73.3	105.95	61.56	279.1	0.0224	−0.0065
3.98	0.794	0.170	99.34	97.84	65.54	293.4	73.1	106.83	65.97	279.1	0.0044	−0.0119
4.19	0.696	0.161	99.36	98.29	71.88	293.9	72.9	108.78	71.71	279.1	−0.0018	−0.0120

Table A6 Thermodynamics properties of air and CO_2

	Air	CO_2
Molecular mass (kg/kmol)	28.96	44.01
Gas constant (J/kg K)	287.2	188.9
Spec. Heat at const. volume (J/kg K)	720	656
Spec. Heat at const. pressure (J/kg K)	1007	845
Isentropic exponent	1.40	1.29

Appendix 3: Description of opto-electronic transmission system

The development of the hereafter presented opto-electronic data transmission system was part of the development of a high-speed rotating probe-traversing system for the VKI Compression Tube Annular Cascade Facility. The system had to meet the following requirements:

(a) transmission of an analogue signal of 3 kHz requiring a bandwidth of about 30–36 kHz;

(b) operation in a low-density environment with pressure levels down to 0.05 to 0.1 bar (condition in test section before test run);

(c) a high signal-to-noise ratio;

(d) low costs.

None of the existing systems appeared convincingly to meet the above conditions. It was therefore decided to attempt the development of a transmission system based on the use of optical diodes. In an advanced stage of the development work, the authors became aware that similar systems had been developed by Heil et al. [1] and by Kappler et al. [2]. However, these two sets of workers applied their optical transmission techniques to the slow cooling process of respectively the rotor of an asynchronous motor [1] and the disc of a gas turbine rotor [2]. Neither of these sets of authors addresses any of the issues relevant for high-frequency aerodynamic measurements.

The use of optical diodes for signal transmission calls for digitization of the signal before its transmission. The digitization has the advantage that all noise below a certain threshold is rejected, while the direct transmission of an analogue signal via sliprings implies the integral transmission of the signal, noise included. There exist two alternative principles for the digitization of an analog signal: the analog–digital conversion and the voltage–frequency conversion. The second solution was preferred because of its relative simplicity, lower cost and smaller size of the converter. The size of the components was important because of the confined space for the electronic in the shaft of the

Appendix 3: Description of opto-electronic transmission system

rotating-probe traversing mechanism. The precision of a V/F converter is superior to that of an A/D converter but its conversion rate is much smaller.

The choice of the diodes, laser or infrared diodes, was made rapidly in favour of the latter because of the prohibitive cost of laser diodes.

The transmission system is shown in the form of a block diagram in Figure A1.

Components in rotating frame

- measurement device (e.g. kulite pressure transducer) with a voltage output range from 0–50 mV;
- an amplifier with a gain of $\simeq 100$;
- a V/F converter transforming the analogue signal into a frequency between 500 kHz and 1 MHz;
- a circuit with a second amplifier driving the infrared diodes

Components in fixed frame

- a receiver diode with a very high-impedance pre-amplifier and an operational amplifier;
- an F/V converter integrated into a PLL circuit;
- a low-pass filter used to filer the residual frequencies of the PLL.

The emitting diode is the Hamamatsu LED 1939 with a peak wavelength of 890 mm and a frequency band of 1 MHz. It is characterized by a very large emission angle. At 90° from its principal emission direction it conserves a power output of 75% of the maximum power.

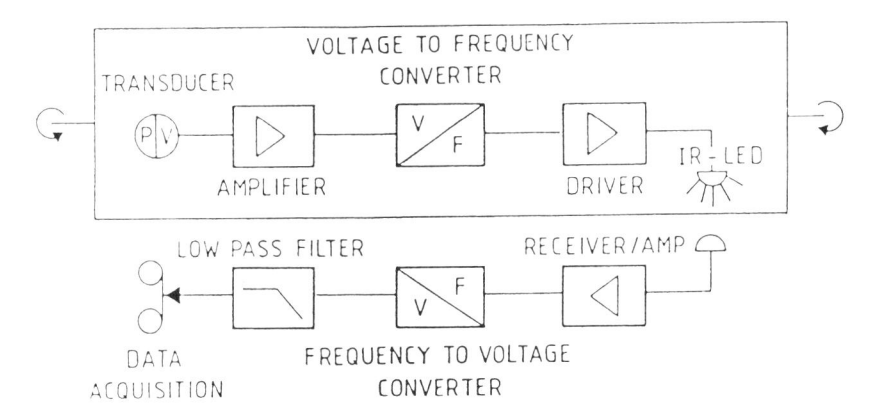

Figure A1 Block diagram for data transmission system.

The receiver diode S1223 is also from Hamamatsu. It has a peak wavelength of 920 ± 50 nm and a bandwidth of 30 MHz.

The dynamic range of the entire system depends on the frequency-to-voltage phase-locked loop, composed of a phase frequency comparator, a filter, an integrator amplifier and the F/V converter (Figure A2). The natural frequency and the damping of the PLL is controlled by the integrator amplifier. For the low-pass filter following the F/V circuit, a fifth-order Butterworth filter was chosen. Since the low-pass filter influences the PLL, the optimization asks for a careful tuning of the natural frequency and the damping of the PLL on one side and the cut-off frequency and the steepness of the attenuation of the low-pass filter on the other side.

In a first step, a one channel prototype was designed, built and tested with both the emitter and receiver optics in fixed positions facing each other. The frequency response curve of the optimized transmission system to a square-wave input signal is presented in Figure A3. The bandwidth is 100 kHz (defined by a gain of −3 dB). A maximum positive gain of +0.558 dB occurs at 52 kHz.

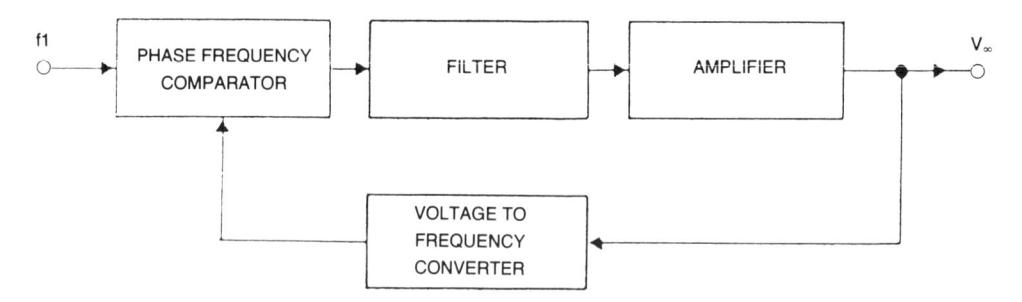

Figure A2 Phase-locked loop (PLL) of frequency-to-voltage converter.

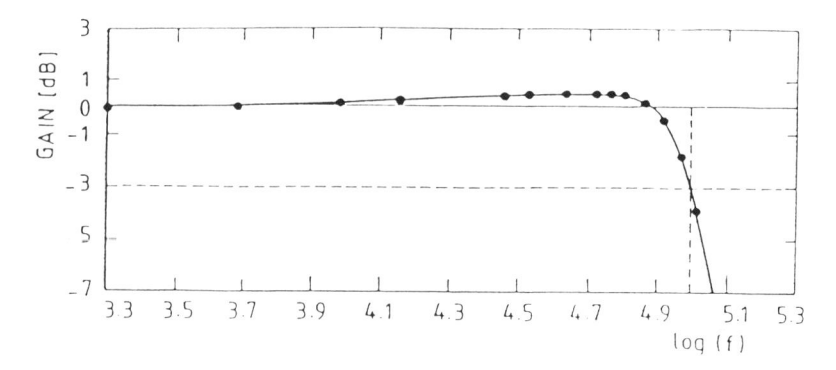

Figure A3 Frequency response curve.

Appendix 3: Description of opto-electronic transmission system

Figure A4 presents both the phase-angle shift and the corresponding time delay of the transmission system. The time delay is of the order of 8 μs in the frequency range up to 10 Hz and rises then progressively to 10.5 μs at 100 kHz. To properly appreciate the significance of this delay on the probe measurements at a blade-passing frequency of 3 kHz, let us assume that the wake covers 25% of the blade pitch. Considering the wake as a half-sine wave, the characteristic frequency for the wake is equal to twice the blade passing frequency, i.e., 6 kHz with the corresponding time period of 166 μs. A delay of 8 μs for a time period of 166 μs appears to be quite significant, but one has to keep in mind that this delay corresponds to a square-wave input signal, while the wake profile is close to a semi-sinusoid, which will considerably reduce the delay time.

Figure A5 finally shows the response of the transmission system to a d.c. input. The deviation of the output voltage from the input voltage is of the order of ± 0.7‰ over the measured voltage range. Testing of several transmission systems showed a typical error bandwidth of ± 1‰ over the same range.

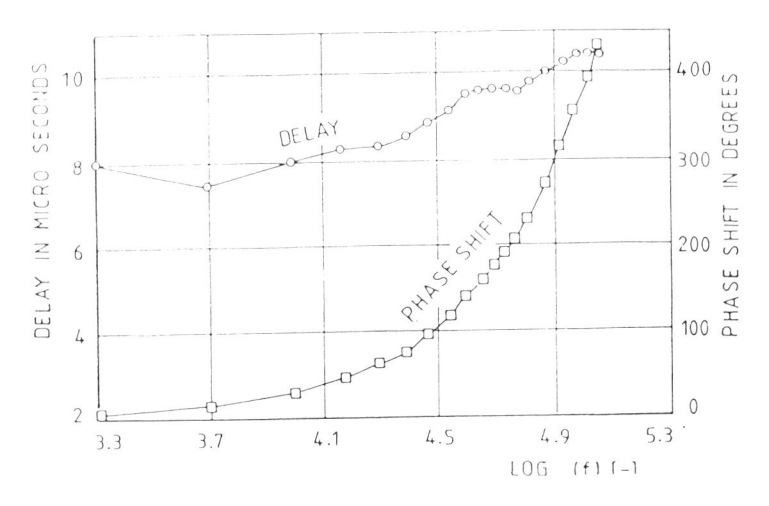

Figure A4 Delay and phase shift of transmission system.

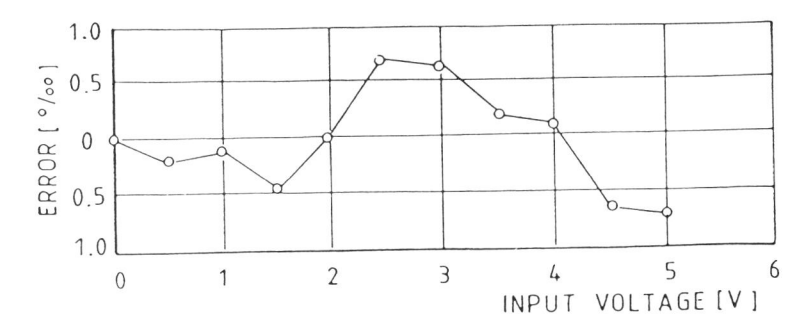

Figure A5 Error in signal transmission for DC input.

Model test rig

A small test rig was built to test the transmission system in rotation. Figure A6 shows the design of the model. A fast response total pressure probe (Kulite transducer) is rotated through air jets exiting from five equidistant closely spaced 10 mm ∅ nozzles, arranged circumferentially on a circle of 300 mm ∅ in the rear wall of an annular settling chamber. The probe is attached to a disc mounted on a shaft, which, in its central part, houses the electronics for signal amplification and transmission.

A 5-psi Kulite pressure transducer of 1.6 mm ∅, type XCS-062-5D with screen B, is mounted in the probe tip. The distance between nozzle exit and transducer is 5 mm. Because of the shear layer on the jet border and due to the finite size of the transducer, the probe does not sense a pressure discontinuity but a very strong pressure gradient.

The voltage supply for transducer and electronics is transmitted via ordinary sliprings (copper rings and carbon brushes) arranged on the right end of the shaft. A voltage supply of ± 15 V is needed. Three carbon brushes, positioned at $3 \times 120°$ around the shaft, proved to be necessary to eliminate random strong voltage peaks observed in the signal output when using single brushes only.

To the right of the sliprings the shaft carries the emitting diodes. This leaves the shaft end for the electric drive. To establish a permanent contact between the emitter and receiver unit, several emitting diodes are to be distributed around the shaft. Figure A7 shows the emission angle curves of 3, respectively 4, emitting diodes, mounted on the shaft of 37 mm ∅. The curves present the distance of equal power with respect to the power at 0° emission angle. The distance of 90 mm corresponds approximately to the maximum distance between emitting and receiving diodes for a safe transmission at 0° emission

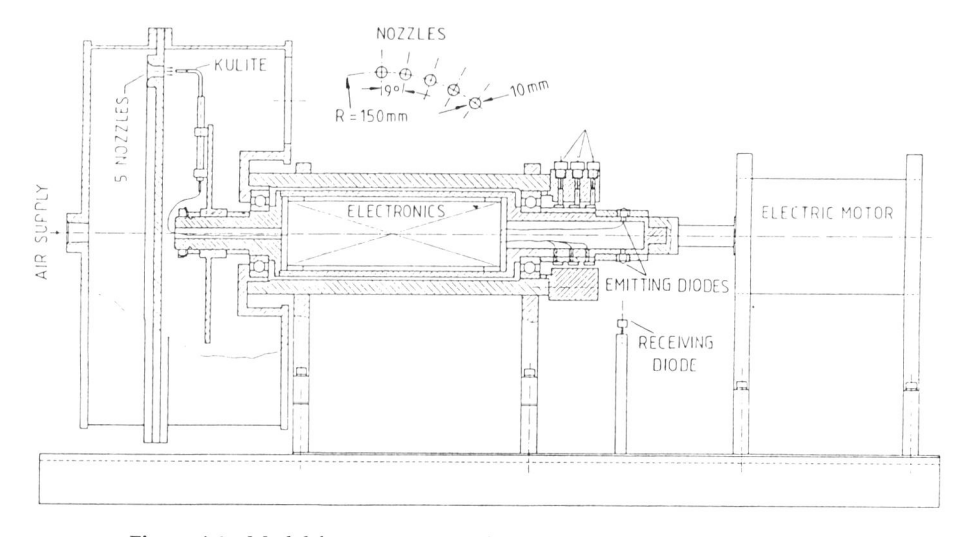

Figure A6 Model for testing opto-electronic data transmission system.

Appendix 3: Description of opto-electronic transmission system

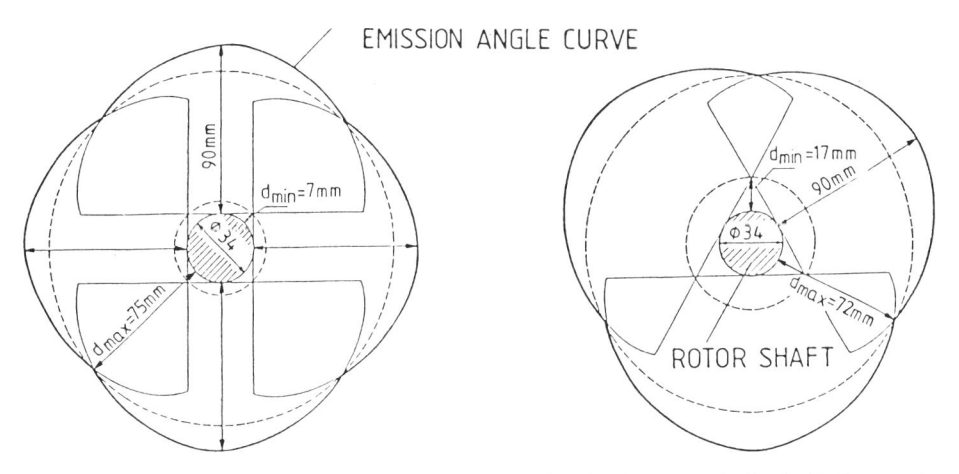

Figure A7 Area coverage of IR-diode emission angles for 3, respectively 4, diodes on the rotorshaft.

angle. Table A1 shows that a system with only one receiver diode can operate in the following limits:

Table A1 Diode operating limits

Emitting diodes LED 1939	Receiving diode S1223 Min. dist (mm)	Max. dist. (mm)
3	17	72
4	7	75

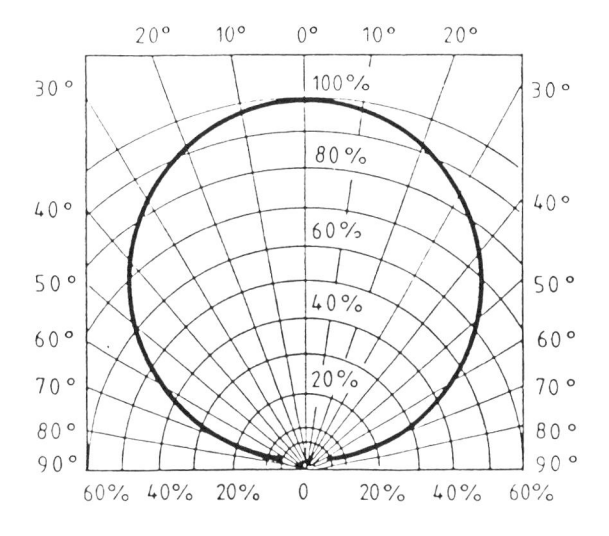

Figure A8 Relative sensitivity of receiver diode.

The wide angular sensitivity of the receiver diode S1223, Figure A8, would enable a transmission with three emitter diodes, but at the end a system with four diodes was adopted.

Model test results

Tests were run with settling chamber gauge pressures of 150–200 mbar resulting in air jet velocities of 150–170 m/s. The rotational speed of the probe was varied from 1000 to 3000 r.p.m., corresponding to peripheral speeds of 15.7–47.1 m/s. With a nozzle spacing of 9° the jet passing frequency at 3000 r.p.m. amounts to 2 kHz.

The data are acquisitioned with a BE 490 plug-in board high-speed data acquisition card with a maximum sampling frequency of 1 MHz. A maximum of 8 channels may be acquisitioned with this card which implies, of course, a corresponding decrease of the maximum sample frequency.

Figure A9 shows a typical pressure signal trace taken at 1000 r.p.m., i.e. at a jet-passing frequency of 666 Hz. The traces show very clearly the pressure pulses due to the jets. Typical pressure rise and pressure fall times are of the order of 0.06 ms. A rise time of 0.06 ms at 1000 r.p.m. corresponds to a pressure rise over approximately 1 mm distance. The reason for this gradual change can entirely be attributed to the combined effect of a shear layer thickness of ≈ 1.1 mm and a probe head diameter of 1.6 mm.

Superimposed on the jet pressure signal is a high-frequency noise. Stretching the time scale of the signal reveals a periodic signal with a frequency of

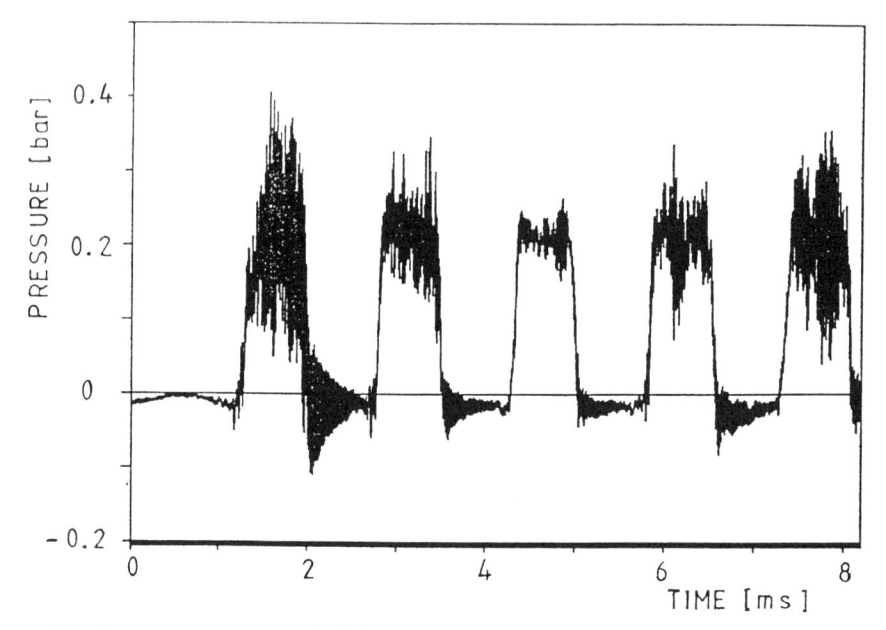

Figure A9 Pressure trace recorded by Kulite pressure probe traversing 5 air jets at 1000 r.p.m.

Appendix 3: Description of opto-electronic transmission system

94 kHz. Figure A10 shows this stretched signal for the probe displacement between jet 1 and 2. The trace is typical for a dynamic system of the second order with slight damping. The power spectrum density of the signal containing all five air jets confirms the existence of a resonance frequency at 94 kHz, Figure A11. The fast drop of the signal for higher frequencies is due to the fifth-order Butterworth filter.

To eliminate the high-frequency noise, a numerical filter with a cut-off frequency of 40 kHz was used. The result of the filtered signal is presented in Figure A12. The pressure trace between the air jets is remarkably flat, which

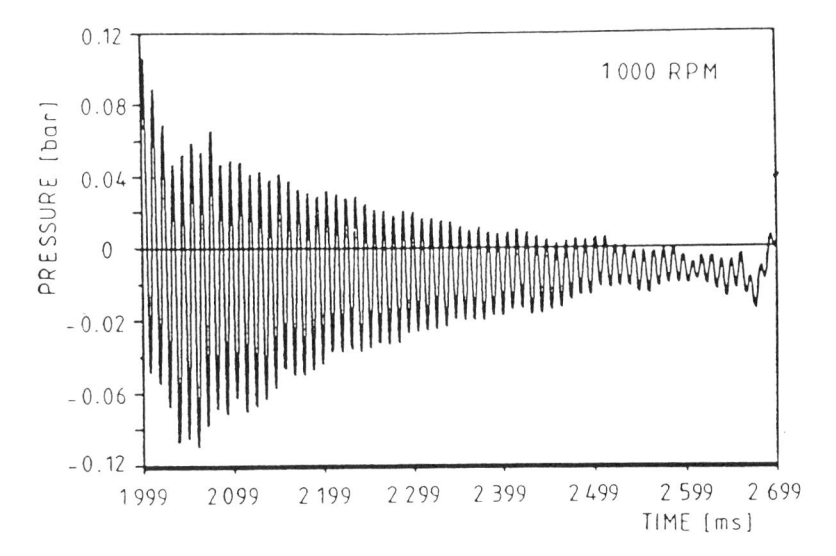

Figure A10 Pressure trace between jet 1 and 2.

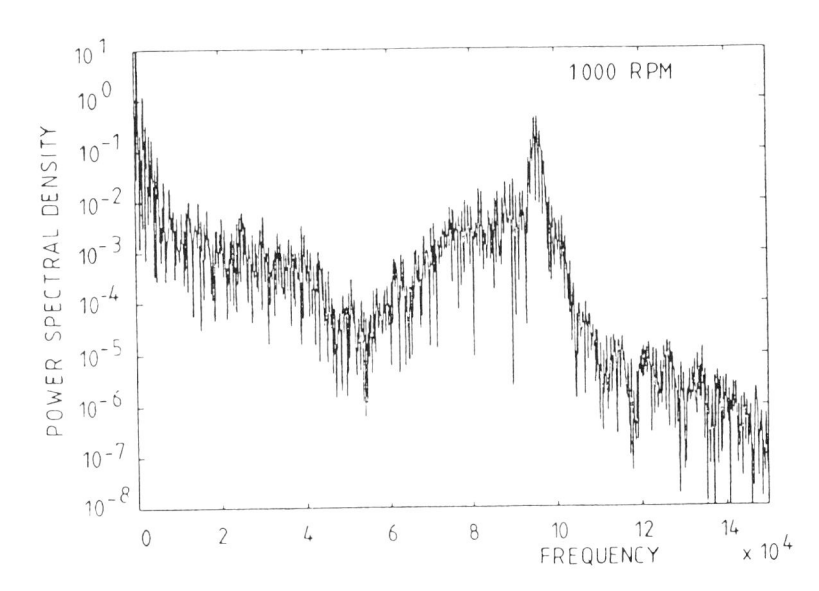

Figure A11 Power spectrum density of the signal containing the five air jets.

indicates an overall very low noise level. The low noise level between the jets leads to the conclusion that the strong fluctuations in the jets represent flow turbulence.

Figure A13 presents test results taken at 3000 r.p.m. for a settling chamber

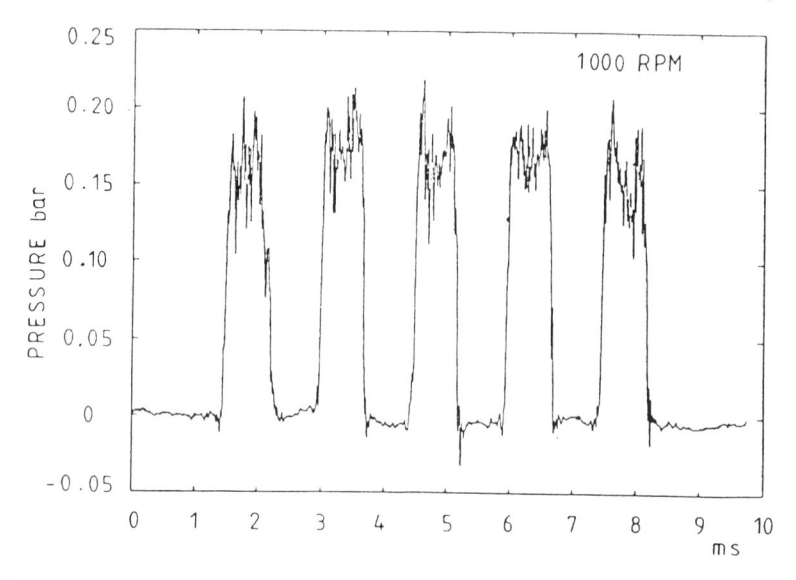

Figure A12 Filtered pressure signal of the rotating pressure transducer at 1000 r.p.m.

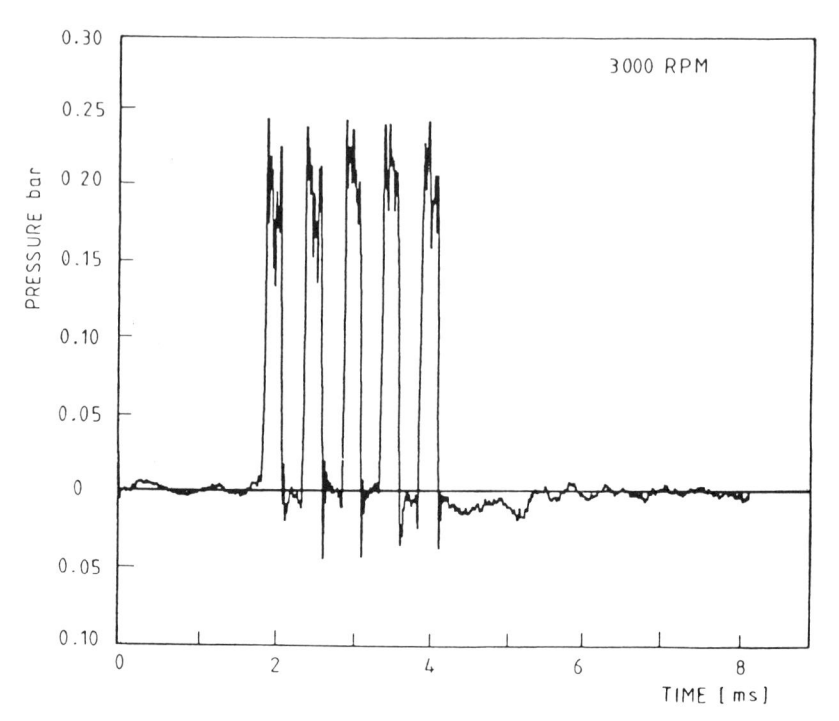

Figure A13 Pressure traces of Kulite pressure probe at 3000 r.p.m.

Appendix 3: Description of opto-electronic transmission system

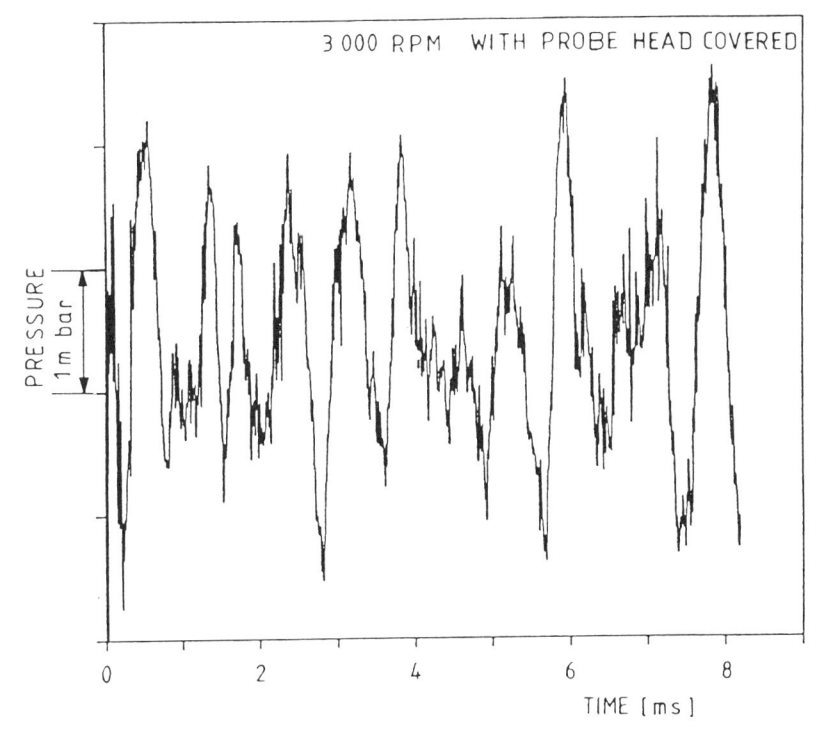

Figure A14 Noise of transmission system at 3000 r.p.m.

gauge pressure of 200 mbar. The recorded signal has already been filtered at 40 kHz. The pressure rise and fall are again very sharp. On leaving the jet the pressure trace shows a strong negative pressure peak. A similar peak occurs probably also when entering the jet, but it is less evident because of the general high level of fluctuations recorded in the jet. It is not quite clear whether these peaks are related to the dynamic load variation on the transducer or whether they have to be imputed to the transmission system.

To conclude, a test was run at 3000 r.p.m. at zero flow with a cover on the Kulite pressure transducer, Figure A14. The pressure transducer calibration was the same as for the test in Figure A13. The peak-to-peak pressure variation for the zero flow test is 4 mbar. This is representative for the overall noise of the transmission system for the considered test set-up and running conditions including electronic, optical and mechanical (bearings) effects.

References

[1] Heil, J., Kruckow, W., Wehr, A. Zeitmultiplexe berührungslose Meßwertübertragung aus dem Läufer einer elektrischen Maschine, *Technisches Messen*, **54**, Jahrgang, Heft 2, 61–65, 1987.

[2] Kappler, G., Erhard, W., Braun, H. *Rotating Optoelectronic Data Transmitter for Local Heat Transfer Measurements*. ICIASF'87 Record, pp. 77–82.

Index

Aero engines, future, combustor conditions for 106, 112, 115, 125, 143
  present, pollution levels of 112
Aircraft emissions, atmospheric impact of 105–52
Air distribution, influence on homogeneity 108, 114, 120–2, 136–7, 147–8
Air ejection, tests with 199–219
Anemometer, three-dimensional 2
Annular cascade
  design 182–4
  facility 40
  tests 235–7
ARA 159
ASCII files 7, 8
Atomizers, air blast 106–7, 117, 118, 129, 144
  atomization process of 113, 144
Axial flow compressors
  database of test cases 6–8
Axial compressor rotor of Inoue 76

Base pressure coefficients 207–11
Blade
  boundary-layer calculation 164
  clearance region 3
  geometry 166, 298–300
  instrumentation 178
  manufacture 182
  mechanical design of 175–84
  performance calculations 306–8
  pressure fields 9
  profile design 160–5
  static pressure tappings 57
  tip, square edged 35–6
  trailing edge losses 158
  velocity distribution 162–3, 167, 295–8, 312–14, 323–6

Bound vorticity 14
BRITE annular inlet guide vane,
  aerodynamic design of 169
  cascade tests 275–9
  turbine inlet guide vane, design of 159
  22N blade coordinates 342–7

Carbon dioxide
  ejection 219–35
  volume concentration, wake
    measurements of 223–34
CARS (coherent antiStokes Raman
    scattering) 107
Cartesian coordinate system, rotating 33
Cascade, annular 336–8
  facility 5–6, 40, 184–5
  flow fields 312–14
  mechanical design of 175–6
  tests, annular 235–84
    straight 195–7
  tunnel 188–90
Clearance grid 36
Coherent antiStokes Raman scattering
    (CARS) 107
Combustion, fundamental aspects 112
Combustor design technology 106
  tubular 107, 119, 120–2, 131–2, 144
Compressor 26
  civil core 2
  linear, cascade of Flot 33, 73–6
  linear, cascade of Storer 33, 70–3
  low-speed 26, 57
  rotating, of Cranfield 33, 84–8
  rotating, of Inoue 33
  test cases 41–2
Compressor rotor, tip clearance of 3
  Cranfield 84–8
Computer run-times 39

Coolant flow 241
  ejection 169, 287–8, 301–6
  supply system 186–8
Cranfield 2
  rotating compressor 33, 84–8
CT-3 facility, measurements in 276–9
Cylindrical blades 159

Database 4, 6–8
  format 7
  subdirectories 7
  test cases 41, 42, 43
Data reduction of wake flow measurements 219–23
Data transmission from rotating frame 271–2
Distribution, loss and angle 317–20
DLR research institute 159
  straight cascade facility 184–5
Droplet size, influence of 108, 114, 120, 133–6, 147

Euler calculation on TE ejection blade 170–5

Filefrmt 7
Flot 33, 73–6
Flow conditions 197–8, 237–41
  calculations, three-dimensional 309–11
  patterns, trailing edge 327
  patterns, upstream 237–9
  patterns, viscous 284–320
  properties 348–50
  visualization 315–17
Flow field 15
  calculations 129
  cascade 312–4
  outlet, axial evolution of 156–8
Fuel distribution, influence of 108, 114, 120, 133–6, 147

Gap flow model 10–14, 96
Gap losses, calculation of 11
Geometry, blade 166, 298–300
  inlet guidance 169
Guide vane 190

High-pressure civil core compressors 2
High-speed cascade rig (HSCR) 2
Homogeneity 108, 114, 120, 133, 147
HSCR (high-speed cascade rig) 2

Inlet guide vane geometry 169
Inoue, axial compressor rotor of 76

Jet velocity magnitude 17–18
  component, calculation results 46–7

Kinematic field 14
Kulite pressure probe 272–4
  transducer 272–4

L2F (laser two focus) 2, 28
  measurements 65
  three-dimensional 39–41
  see also Laser Doppler anemometer
Laser anemometry 65–7
Laser diagnostics 107, 113, 118–119, 130–1
Laser Doppler anemometer 39, 99
Laser two focus (L2F) 2, 28
Leading edge modelling 37
Leakage vortex 9
  formation 18
  radius, calculation of 17, 18
Lean blow-out tests 119, 132
  modelling 120
Lifting line theory 14
Line vortex diffusion model 23–5
Linear compressor cascade of Storer 33, 70–3
  of Flot 33, 73–6
Low emission combustor technology (LOWNOX 1) 106–149
LOWNOX 1 106–149
Low Reynolds turbulence model 38
Low-speed research compressor (LSRC) 2
LSRC construction 26
  experiments 25–6

Mass conservation analysis 37
Measurements in rotation 268–70
Meridional calculation method 9
  flow calculation 22
Mesh generation 290–1
Models
  air jet tests 275
  basic 10–21
  first level 5, 9
  gap flow 10–14
  second level 31–9
  simple vortex 14–17
  trailing edge 37
  turbulence, Reynolds 38

Index

two-dimensional 2
  validity assessment 30–31
Moment of momentum theorem 16
MTU 159

Navier–Stokes codes 156
  pressure-corrected solver 2
  three-dimensional 31–9, 98
Nitrogen oxides 106, 110, 111
Nozzle guide vane 235–7
NTUA 2
  meeting reports 8–41

Opto-electronic transmission system 274, 350–9
Outlet flow angle 259–60
  conditions 239–41
  field, axial evolution of 156

PS-ejection blade 166, 178
Parabolic profile 12
PDA 107
Performance measurements 245–62
Phase Doppler anemometry (PDA) 107
Pollutant calculation 119
  formation 108
  future emissions 132
  levels 106, 112, 115–17, 125, 143–4
Pollution reduction methods 106, 109, 112, 115, 117, 125–8, 143, 144, 149
Premix duct 109, 123–5, 148
  modelling of 139–42, 148
Pre-mixing, modelling of 115, 123–5, 138, 148
Pressure distribution 17, 18, 204–7
  loss profile 23
  taps 2
Pressure probes, design of 242–3
Pressure-side trailing edge 159
Primary zone architecture 107, 113, 119, 131–2, 144, 146
Probe blockage effect 243–4
  double hot wire aspirating 262–4
  measurements 192

Reynolds turbulence model 38
Rolls Royce research 2
Rotating-probe measurements 193–4
  in BRITE cascade 275–9
Rotor relative velocities 66–69

Schlieren pictures 196, 199–204
Secondary flow calculation code 42
Shed circulation 14, 17, 18

Smoke production 108, 114–15, 122, 137–8, 148
Snecma research 2, 159
Software 7–8
  compressor aerodynamic design 97
Soot production 108
Spray diagnostics 117
  formation prediction model 106–7
Stationary measurements 192, 242–68
Stoichiometry, influence of 107, 119, 131–2
  on pollution formation 113, 144, 146
Storer 33, 70–3
Straight cascade blades, mechanical design of 176–7
  tests 195–235
Streamline curvature analysis 64
Suction-side trailing edge 159
Symbols 154–5

TE-ejection blade 162, 177
  Euler calculation on 170–5
Temperature probes 262–5
Tests
  facilities 184–94
  with air ejection 199–219
  with carbon dioxide ejection 219–35
Thermodynamics 348–50
Three-dimensional Euler calculation 170
Throughflow approach 2
  industrial throughflow software 97
Tip clearance
  flow 21, 41, 44
  jet, flow rate of 11
  level, pressure difference at the 45
  model 9, 17–21, 42, 96–7
  vortex strength 14
Tip leakage
  jet flow 9
  vortex, rotation of 2
Tip, square-edge blade 35–6
Trailing edge flow patterns 327
  losses 158
  modelling 37
Transducer 272–4
Tubular combustor 107, 119, 120–2, 131–2, 144

Turbine blade with pressure-side ejection 157
  design 156
  testing 156
Turbine guide vane, transonic 156
TURBO3D code 39, 98
Turbomeca 2
Turbulence 260–2
  measurements 196, 214–19
  models 157, 292–3
  Reynolds model 38
Two-dimensional flow calculations 293–4

University of Florence 159

Vena contracta 9, 10, 11
Viscous flow calculations 284–320

VKI Research Institute 159
  compression tube annular cascade facility 188–90
Vortex model, diffusion 23–5
  revised 17
  simple 14–17

Wake flow measurements 219–23
Wake-mixing
  process 156
  profiles 223–34
Wake profile 252–8
  evolution of 308–9, 327
  measurements 196, 211–13
  relative total pressure 279–80
  temperature profile 265–8
Wall motion 14

Index compiled by Sheila Shephard